Fluid Mechanics and Thermodynamics of Turbomachinery

Fourth Edition, in SI/Metric units

In memory of
Avril and baby Paul

Fluid Mechanics and Thermodynamics of Turbomachinery

Fourth Edition, in SI/Metric units

S. L. Dixon, B.Eng., Ph.D
Senior Fellow at the University of Liverpool

Boston   Oxford   Johannesburg
Melbourne   New Delhi   Singapore

Butterworth-Heinemann

 A member of the Reed Elsevier Group

First published by Pergamon Press Ltd. 1966
Second edition 1975
Third edition 1978
Reprinted 1979, 1982 (twice), 1984, 1989, 1992, 1995
Foruth edition 1998

 Recognizing the importance of preserving what has been written, Butterworth-Heinemann prints its books on acid-free paper whenever possible.

 Butterworth-Heinemann supports the efforts of American Forests
and the Global ReLeaf program in its campaign for the betterment
of trees, forests, and our environment.

Library of Congress Cataloging-in-Publication Data
A catalog record for this book is available from the Library of Congress

ISBN 0-7506-7059-2

British Library Cataloguing-in-Publication Data
A catalogue record for this book is available from the British Library

The publisher offers special discounts on bulk orders of this book.
For information, please contact:

> Manager of Special Sales
> Butterworth-Heinemann
> 225 Wildwood Avenue
> Woburn, MA 01801-2041
> Tel: (781) 904-2500
> Fax: (781) 904-2620

For information on all Butterworth-Heinemann publications available, contact our
World Wide Web home page at: http://www.bh.com

10   9 8 7 6 5 4 3 2
Printed in the United States of America
Typeset by Laser Words, Madras, India

Contents

6. Three-dimensional Flows in Axial Turbomachines   169

7. Centrifugal Pumps, Fans and Compressors   199

Preface to the Fourth Edition

It is now twenty years since the third edition of this book was published and in that period many advances have been made to the art and science of turbomachinery design. Knowledge of the flow processes within turbomachines has increased dramatically resulting in the appearance of new and innovative designs. Some of the long-standing, apparently intractable, problems such as surge and rotating stall have begun to yield to new methods of control. New types of flow machine have made their appearance (e.g. the Wells turbine and the axi-fuge compressor) and some changes have been made to established design procedures. Much attention is now being given to blade and flow passage design using computational fluid dynamics (CFD) and this must eventually bring forth further design and flow efficiency improvements. However, the fundamentals do not change and this book is still concerned with the basics of the subject as well as looking at new ideas.

The book was originally perceived as a text for students taking an Honours degree in engineering which included turbomachines as well as assisting those undertaking more advanced postgraduate courses in the subject. The book was written for engineers rather than mathematicians. Much stress is laid on physical concepts rather than mathematics and the use of specialised mathematical techniques is mostly kept to a minimum. The book should continue to be of use to engineers in industry and technological establishments, especially as brief reviews are included on many important aspects of turbomachinery giving pointers to more advanced sources of information. For those looking towards the wider reaches of the subject area some interesting reading is contained in the bibliography. It might be of interest to know that the third edition was published in four languages.

A fairly large number of additions and extensions have been included in the book from the new material mentioned as well as "tidying up" various sections no longer to my liking. Additions include some details of a new method of fan blade design, the determination of the design point efficiency of a turbine stage, sections on centrifugal stresses in turbine blades and blade cooling, control of flow instabilities in axial-flow compressors, design of the Wells turbine, consideration of rothalpy conservation in impellers (and rotors), defining and calculating the optimum efficiency of inward flow turbines and comparison with the nominal design. A number of extensions of existing topics have been included such as updating and extending the treatment and application of diffuser research, effect of prerotation of the flow in centrifugal compressors and the use of backward swept vanes on their performance, also changes in the design philosophy concerning the blading of axial-flow compressors. The original chapter on radial flow turbines has been split into two chapters; one dealing with radial gas turbines with some new extensions and the other on hydraulic turbines. In a world striving for a 'greener' future it was felt that there would now be more than just a little interest in hydraulic turbines. It is a subject that is usually included in many mechanical engineering courses. This chapter includes a few new ideas which could be of some interest.

A large number of illustrative examples have been included in the text and many new problems have been added at the end of most chapters (answers are given at the end of the book)! It is planned to publish a new supplementary text called Solutions Manual, hopefully, shortly after this present text book is due to appear, giving the complete and detailed solutions of the unsolved problems.

S. Lawrence Dixon

Preface to Third Edition

Several modifications have been incorporated into the text in the light of recent advances in some aspects of the subject. Further information on the interesting phenomenon of cavitation has been included and a new section on the optimum design of a pump inlet together with a worked example have been added which take into account recently published data on cavitation limitations. The chapter on *three-dimensional flows in axial turbomachines* has been extended; in particular the section concerning the *constant specific mass flow design* of a turbine nozzle has been clarified and now includes the flow equations for a following rotor row. Some minor alterations on the definition of blade shapes were needed so I have taken the opportunity of including a simplified version of the parabolic arc camber line as used for some low camber blading.

Despite careful proof reading a number of errors still managed to elude me in the second edition. I am most grateful to those readers who have detected errors and communicated with me about them.

In order to assist the reader I have (at last) added a list of symbols used in the text.

S.L.D.

Acknowledgements

The author is indebted to a number of people and manufacturing organisations for their help and support; in particular the following are thanked:

Professor W. A. Woods, formerly of Queen Mary College, University of London and a former colleague at the University of Liverpool for his encouragement of the idea of a fourth edition of this book as well as providing papers and suggestions for some new items to be included. Professor F. A. Lyman of Syracuse University, New York and Professor J. Moore of Virginia Polytechnic Institute and State University, Virginia, for their helpful correspondence and ideas concerning the vexed question of the conservation of rothalpy in turbomachines. Dr Y. R. Mayhew is thanked for supplying me with generous amounts of material on units and dimensions and the latest state of play on SI Units.

Thanks are also given to the following organisations for providing me with illustrative material for use in the book, product information and, in one case, useful background historical information:

Sulzer Hydro of Zurich, Switzerland; Rolls-Royce of Derby, England; Voith Hydro Inc., Pennsylvania; and Kvaerner Energy, Norway.

Last, but by no means least, to my wife Rose, whose quiet patience and support enabled this new edition to be prepared.

List of Symbols

A	area
a	sonic velocity, position of maximum camber
b	passage width, maximum camber
C_f	tangential force coefficient
C_L, C_D	lift and drag coefficients
C_p	specific heat at constant pressure, pressure coefficient, pressure rise coefficient
C_{pi}	ideal pressure rise coefficient
C_v	specific heat at constant volume
C_X, C_Y	axial and tangential force coefficients
c	absolute velocity
c_0	spouting velocity
D	drag force, diameter
D_{eq}	equivalent diffusion ratio
D_h	hydraulic mean diameter
E, e	energy, specific energy
F_c	centrifugal force in blade
f	acceleration, friction factor
g	gravitational acceleration
H	head, blade height
H_E	effective head
H_f	head loss fue to friction
H_G	gross head
H_S	net positive suction head (NPSH)
h	specific enthalpy
I	rothalpy
i	incidence angle
K, k	constants
K_N	nozzle velocity coefficient
L	lift force, length of diffuser wall
l	blade chord length, pipe length
M	Mach number
m	mass, molecular 'weight'
N	rotational speed, axial length of diffuser
N_S	specific speed (rev)
N_{SP}	power specific speed (rev)
N_{SS}	suction specific speed (rev)
n	number of stages, polytropic index
p	pressure

p_a	atmospheric pressure
p_v	vapour pressure
Q	heat transfer, volume flow rate
q	dryness fraction
R	reaction, specific gas constant
Re	Reynolds number
R_H	reheat factor
R_0	universal gas constant
r	radius
S	entropy, power ratio
s	blade pitch, specific entropy
T	temperature
t	time, thickness
U	blade speed, internal energy
u	specific internal energy
V, v	volume, specific volume
W	work transfer
ΔW	specific work transfer
w	relative velocity
X	axial force
x, y, z	Cartesian coordinate directions
Y	tangential force, actual tangential blade load per unit span
Y_{id}	ideal tangential blade load per unit span
Y_k	tip clearance loss coefficient
Y_p	profile loss coefficient
Y_S	net secondary loss coefficient
Z	number of blades, Ainley blade loading parameter
α	absolute flow angle
β	relative flow angle
Γ	circulation
γ	ratio of specific heats
δ	deviation angle
ε	fluid deflection angle, cooling effectiveness
ζ	enthalpy loss coefficient, total pressure loss coefficient
η	efficiency
Θ	minimum opening at cascade exit
θ	blade camber angle, wake momentum thickness
λ	profile loss coefficient
μ	dynamic viscosity
ν	kinematic viscosity, blade stagger angle, velocity ratio
ρ	density
σ	slip factor, solidity
σ_b	blade cavitation coefficient
σ_c	Thoma's coefficient, centrifugal stress
τ	torque

ϕ	flow coefficient, velocity ratio
Ψ	stage loading factor
Ω	speed of rotation (rad/s)
Ω_S	specific speed (rad)
Ω_{SP}	power specific speed (rad)
Ω_{SS}	suction specific speed (rad)
ω	vorticity
$\bar{\omega}$	stagnation pressure loss coefficient

Subscripts

av	average
c	compressor, critical
D	diffuser
e	exit
h	hydraulic, hub
i	inlet, impeller
id	ideal
is	isentropic
m	mean, meridional, mechanical, material
N	nozzle
n	normal component
o	stagnation property, overall
p	polytropic, constant pressure
R	reversible process, rotor
r	radial
rel	relative
s	isentropic, stall condition
ss	stage isentropic
t	turbine, tip, transverse
v	velocity
x, y, z	cartesian coordinate components
θ	tangential

Superscript

	time rate of change
-	average
$'$	blade angle (as distinct from flow angle)
*	nominal condition

CHAPTER 1

Introduction: Dimensional Analysis: Similitude

If you have known one you have known all. (TERENCE, *Phormio.*)

Definition of a turbomachine

We classify as turbomachines all those devices in which energy is transferred either to, or from, a continuously flowing fluid by the *dynamic action* of one or more moving blade rows. The word *turbo* or *turbinis* is of Latin origin and implies that which spins or whirls around. Essentially, a rotating blade row, a *rotor* or an *impeller* changes the stagnation enthalpy of the fluid moving through it by either doing positive or negative work, depending upon the effect required of the machine. These enthalpy changes are intimately linked with the pressure changes occurring simultaneously in the fluid.

The definition of a turbomachine as stated above, is rather too general for the purposes of this book as it embraces *open* turbomachines such as propellers, wind turbines and unshrouded fans, all of which influence the state of a not readily quantifiable flow of a fluid. The subject *fluid mechanics, thermodynamics of turbomachinery*, therefore, is limited to machines enclosed by a closely fitting casing or shroud through which a readily measurable quantity of fluid passes in unit time. The subject of open turbomachines is covered by the classic text of Glauert (1959) or by Duncan *et al.* (1970), the elementary treatment of propellers by general fluid mechanics textbooks such as Streeter and Wylie (1979) or Massey (1979), and the important, still developing subject of wind turbines, by Freris (1990).

Two main categories of turbomachine are identified: firstly, those which *absorb* power to increase the fluid pressure or head (ducted fans, compressors and pumps); secondly, those that *produce* power by expanding fluid to a lower pressure or head (hydraulic, steam and gas turbines). Figure 1.1 shows, in a simple diagrammatic form, a selection of the many different varieties of turbomachine encountered in practice. The reason that so many different types of either pump (compressor) or turbine are in use is because of the almost infinite range of service requirements. Generally speaking, for a given set of operating requirements there is one type of pump or turbine best suited to provide optimum conditions of operation. This point is discussed more fully in the section of this chapter concerned with specific speed.

Turbomachines are further categorised according to the nature of the flow path through the passages of the rotor. When the path of the *through-flow* is wholly or mainly parallel to the axis of rotation, the device is termed an *axial flow turbomachine* (e.g.

1

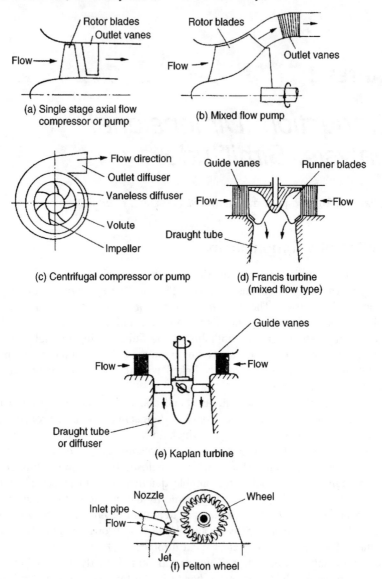

Fɪɢ. 1.1. Diagrammatic form of various types of turbomachine.

Figure 1.1(a) and (e)). When the path of the *through-flow* is wholly or mainly in a plane perpendicular to the rotation axis, the device is termed a *radial flow turbomachine* (e.g. Figure 1.1(c)). More detailed sketches of radial flow machines are given in Figures 7.1, 7.2, 8.2 and 8.3. *Mixed flow turbomachines* are widely used. The term *mixed flow* in this context refers to the direction of the through-flow at rotor outlet when both radial and axial velocity components are present in significant amounts. Figure 1.1(b) shows a mixed flow pump and Figure 1.1(d) a mixed flow hydraulic turbine.

One further category should be mentioned. All turbomachines can be classified as either *impulse* or *reaction* machines according to whether pressure changes are absent or present respectively in the flow through the rotor. In an impulse machine all the pressure change takes place in one or more nozzles, the fluid being directed onto the rotor. The Pelton wheel, Figure 1.1(f), is an example of an impulse turbine.

The main purpose of this book is to examine, through the laws of fluid mechanics and thermodynamics, the means by which the energy transfer is achieved in the chief types of turbomachine, together with the differing behaviour of individual types in operation. Methods of analysing the flow processes differ depending upon the geometrical configuration of the machine, on whether the fluid can be regarded as incompressible or not, and whether the machine absorbs or produces work. As far as possible, a unified treatment is adopted so that machines having similar configurations and function are considered together.

Units and dimensions

The International System of Units, SI (le Système International d'Unités) is a unified self-consistent system of measurement units based on the MKS (metre–kilogram–second) system. It is a simple, logical system based upon decimal relationships between units making it easy to use. The most recent detailed description of SI has been published in 1986 by HMSO. For an explanation of the relationship between, and use of, physical quantities, units and numerical values see *Quantities, Units and Symbols*, published by The Royal Society (1975) or refer to ISO 31/0-1981.

Great Britain was the first of the English-speaking countries to begin, in the 1960s, the long process of abandoning the old Imperial System of Units in favour of the International System of Units, and was soon followed by Canada, Australia, New Zealand and South Africa. In the USA a ten year voluntary plan of conversion to SI units was commenced in 1971. In 1975 US President Ford signed the Metric Conversion Act which coordinated the metrication of units, but did so without specifying a schedule of conversion. Industries heavily involved in international trade (cars, aircraft, food and drink) have, however, been quick to change to SI for obvious economic reasons, but others have been reluctant to change.

SI has now become established as the only system of units used for teaching engineering in colleges, schools and universities in most industrialised countries throughout the world. The Imperial System was derived arbitrarily and has no consistent numerical base, making it confusing and difficult to learn. In this book all numerical problems involving units are performed in metric units as this is more convenient than attempting to use a mixture of the two systems. However, it is recognised that some problems exist as a result of the conversion to SI units. One of these is that many valuable papers and texts written prior to 1969 contain data in the old system of units and would need converting to SI units. A brief summary of the conversion factors between the more frequently used Imperial units and SI units is given in Appendix 1 of this book.

Some SI units

The SI basic units used in fluid mechanics and thermodynamics are the *metre* (m), *kilogram* (kg), *second* (s) and *thermodynamic temperature* (K). All the other units used in this book are derived from these basic units. The *unit of force* is the *newton* (N), defined as that force which, when applied to a mass of 1 kilogram, gives an acceleration to the mass of $1 \, \text{m/s}^2$. The recommended *unit of pressure* is the *pascal* (Pa) which is the pressure produced by a force of 1 newton uniformly distributed over an area of 1 square metre. Several other units of pressure are in wide-spread use, however, foremost of these being the *bar*. Much basic data concerning properties of substances (steam and gas tables, charts, etc.) have been prepared in SI units with pressure given in bars and it is acknowledged that this alternative unit of pressure will continue to be used for some time as a matter of expediency. It is noted that 1 bar equals $10^5 \, \text{Pa}$ (i.e. $10^5 \, \text{N/m}^2$), roughly the pressure of the atmosphere at sea level, and is perhaps an inconveniently large unit for pressure in the field of turbomachinery anyway! In this book the convenient size of the *kilopascal* (kPa) is found to be the most useful multiple of the recommended unit and is extensively used in most calculations and examples.

In SI the units of all forms of energy are the same as for work. The *unit of energy* is the *joule* (J) which is the work done when a force of 1 newton is displaced through a distance of 1 metre in the direction of the force, e.g. kinetic energy ($\frac{1}{2}mc^2$) has the dimensions $\text{kg} \times \text{m}^2/\text{s}^2$; however, $1 \, \text{kg} = 1 \, \text{N s}^2/\text{m}$ from the definition of the newton given above. Hence, the units of kinetic energy must be $\text{Nm} = \text{J}$ upon substituting dimensions.

The *watt* (W) is the *unit of power*; when 1 watt is applied for 1 second to a system the input of energy to that system is 1 joule (i.e. 1 J).

The *hertz* (Hz) is the number of repetitions of a regular occurrence in 1 second. Instead of writing c/s for cycles/sec, Hz is used instead.

The unit of thermodynamic temperature is the *kelvin* (K), written without the ° sign, and is the fraction 1/273.16 of the thermodynamic temperature of the triple point of water. The degree celsius (°C) is equal to the unit kelvin. Zero on the celsius scale is the temperature of the ice point (273.15 K). Specific heat capacity, or simply specific heat, is expressed as J/kg K or as J/kg°C.

Dynamic viscosity, dimensions $ML^{-1}T^{-1}$, has the SI units of pascal seconds, i.e.

$$\frac{M}{LT} \equiv \frac{kg}{m.s} = \frac{N.s^2}{m.^2 s} = Pa \, s.$$

Hydraulic engineers find it convenient to express pressure in terms of *head* of a liquid. The static pressure at any point in a liquid at rest is, relative to the pressure acting on the free surface, proportional to the vertical distance of the free surface above that point. The head H is simply the height of a column of the liquid which can be supported by this pressure. If ρ is the mass density (kg/m^3) and g the local gravitational acceleration (m/s^2), then the static pressure p (relative to atmospheric pressure) is $p = \rho g H$, where H is in metres and p is in pascals (or N/m^2). This is left for the student to verify as a simple exercise.

Dimensional analysis and performance laws

The widest comprehension of the general behaviour of all turbomachines is, without doubt, obtained from *dimensional analysis*. This is the formal procedure whereby the group of variables representing some physical situation is reduced into a smaller number of dimensionless groups. When the number of independent variables is not too great, dimensional analysis enables experimental relations between variables to be found with the greatest economy of effort. Dimensional analysis applied to turbomachines has two further important uses: (a) prediction of a prototype's performance from tests conducted on a scale model (similitude); (b) determination of the most suitable type of machine, on the basis of maximum efficiency, for a specified range of head, speed and flow rate. Several methods of constructing non-dimensional groups have been described by Douglas *et al.* (1995) and by Shames (1992) among other authors. The subject of dimensional analysis was made simple and much more interesting by Edward Taylor (1974) in his comprehensive account of the subject. It is assumed here that the basic techniques of forming non-dimensional groups have already been acquired by the student.

Adopting the simple approach of elementary thermodynamics, an imaginary envelope (called a *control surface*) of fixed shape, position and orientation is drawn around the turbomachine (Figure 1.2). Across this boundary, fluid flows steadily, entering at station 1 and leaving at station 2. As well as the flow of fluid there is a flow of work across the control surface, transmitted by the shaft either to, or from, the machine. For the present all details of the flow within the machine can be ignored and only externally observed features such as shaft speed, flow rate, torque and change in fluid properties across the machine need be considered. To be specific, let the turbomachine be a *pump* (although the analysis could apply to other classes of turbomachine) driven by an electric motor. The speed of rotation N, can be adjusted by altering the current to the motor; the volume flow rate Q, can be *independently* adjusted by means of a throttle valve. For fixed values of the set Q and N, all other variables such as torque τ, head H, are thereby established. The choice of Q and N as *control variables* is clearly arbitrary and any other pair of independent variables such as τ and H could equally well have been chosen. The important point to recognise is, that there are for this pump, *two* control variables.

If the fluid flowing is changed for another of different density ρ, and viscosity μ, the performance of the machine will be affected. Note, also, that for a turbomachine handling compressible fluids, other *fluid properties* are important and are discussed later.

So far we have considered only one particular turbomachine, namely a pump of a given size. To extend the range of this discussion, the effect of the *geometric*

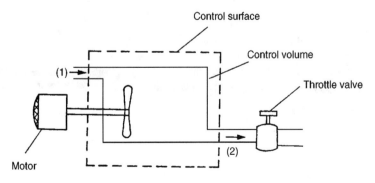

FIG. 1.2. Turbomachine considered as a control volume.

variables on the performance must now be included. The size of machine is characterised by the impeller diameter D, and the shape can be expressed by a number of length ratios, l_1/D, l_2/D, etc.

Incompressible fluid analysis

The performance of a turbomachine can now be expressed in terms of the control variables, geometric variables and fluid properties. For the hydraulic pump it is convenient to regard the net energy transfer gH, the efficiency η, and power supplied P, as dependent variables and to write the three functional relationships as

$$gH = f_1 \left(Q, N, D, \rho, \mu, \frac{l_1}{D}, \frac{l_2}{D}, \ldots \right), \tag{1.1a}$$

$$\eta = f_2 \left(Q, N, D, \rho, \mu, \frac{l_1}{D}, \frac{l_2}{D}, \ldots \right), \tag{1.1b}$$

$$P = f_3 \left(Q, N, D, \rho, \mu, \frac{l_1}{D}, \frac{l_2}{D}, \ldots \right), \tag{1.1c}$$

By the procedure of dimensional analysis using the three primary dimensions, mass, length and time, or alternatively, using three of the independent variables we can form the dimensionless groups. The latter, more direct procedure, requires that the variables selected, ρ, N, D, do not of themselves form a dimensionless group. The selection of ρ, N, D as common factors avoids the appearance of special fluid terms (e.g. μ, Q) in more than one group and allows gH, η and P to be made explicit. Hence the three relationships reduce to the following easily verified forms.

Energy transfer coefficient, sometimes called head coefficient

$$\psi = \frac{gH}{(ND)^2} = f_4 \left(\frac{Q}{ND^3}, \frac{\rho ND^2}{\mu}, \frac{l_1}{D}, \frac{l_2}{D}, \ldots \right), \tag{1.2a}$$

$$\eta = f_5 \left(\frac{Q}{ND^3}, \frac{\rho ND^2}{\mu}, \frac{l_1}{D}, \frac{l_2}{D}, \ldots \right). \tag{1.2b}$$

Power coefficient

$$\hat{P} = \frac{P}{\rho N^3 D^5} = f_6 \left(\frac{Q}{ND^3}, \frac{\rho ND^2}{\mu}, \frac{l_1}{D}, \frac{l_2}{D}, \ldots \right). \tag{1.2c}$$

The non-dimensional group $Q/(ND^3)$ is a volumetric flow coefficient and $\rho ND^2/\mu$ is a form of Reynolds number, Re. In axial flow turbomachines, an alternative to $Q/(ND^3)$ which is frequently used is the velocity (or flow) coefficient $\phi = c_x/U$ where U is blade tip speed and c_x the average axial velocity. Since

$$Q = c_x \times \text{ flow area} \propto c_x D^2$$

and $U \propto ND$.

then

$$\frac{Q}{ND^3} \propto \frac{c_x}{U}.$$

Because of the large number of independent groups of variables on the right-hand side of eqns. (1.2), those relationships are virtually worthless unless certain terms can be discarded. In a family of *geometrically similar* machines l_1/D, l_2/D are constant and may be eliminated forthwith. The kinematic viscosity, $\nu = \mu/\rho$ is very small in turbomachines handling water and, although speed, expressed by ND, is low the Reynolds number is correspondingly high. Experiments confirm that effects of Reynolds number on the performance are small and may be ignored in a first approximation. The functional relationships for geometrically similar hydraulic turbomachines are then,

$$\psi = f_4[Q/(ND^3)] \tag{1.3a}$$

$$\eta = f_5[Q/(ND^3)] \tag{1.3b}$$

$$\hat{P} = f_6[Q/(ND^3)]. \tag{1.3c}$$

This is as far as the reasoning of dimensional analysis alone can be taken; the actual *form* of the functions f_4, f_5 and f_6 must be ascertained by experiment.

One relation between ψ, ϕ, η and $\hat{P}$ may be immediately stated. For a pump the *net hydraulic power*, P_N equals ρQgH which is the minimum shaft power required in the absence of all losses. No real process of power conversion is free of losses and the actual shaft power P must be larger than P_N. We define pump efficiency (more precise definitions of efficiency are stated in Chapter 2) $\eta = P_N/P = \rho QgH/P$. Therefore

$$P = \frac{1}{\eta}\left(\frac{Q}{ND^3}\right)\frac{gH}{(ND)^2}\rho N^3 D^5. \tag{1.4}$$

Thus f_6 may be derived from f_4 and f_5 since $\hat{P} = \phi\psi/\eta$. For a turbine the net hydraulic power P_N supplied is greater than the actual shaft power delivered by the machine and the efficiency $\eta = P/P_N$. This can be rewritten as $\hat{P} = \eta\phi\psi$ by reasoning similar to the above considerations.

Performance characteristics

The operating condition of a turbomachine will be *dynamically similar* at two different rotational speeds if all fluid velocities at *corresponding points* within the machine are in the same direction and proportional to the blade speed. If two points, one on each of two different head–flow characteristics, represent dynamically similar operation of the machine, then the non-dimensional groups of the variables involved, ignoring Reynolds number effects, may be expected to have the same numerical value for both points. On this basis, non-dimensional presentation of performance data has the important practical advantage of collapsing into virtually a single curve, results that would otherwise require a multiplicity of curves if plotted dimensionally.

Evidence in support of the foregoing assertion is provided in Figure 1.3 which shows experimental results obtained by the author (at the University of Liverpool) on a simple centrifugal laboratory pump. Within the normal operating range of this pump, $0.03 < Q/(ND^3) < 0.06$, very little systematic scatter is apparent which might be associated with a Reynolds number effect, for the range of speeds $2500 \leq N \leq 5000$ rev/min. For smaller flows, $Q/(ND^3) < 0.025$, the flow became unsteady and the manometer readings of uncertain accuracy but, nevertheless, dynamically similar conditions still appear to hold true. Examining the results at high flow rates one is struck by a marked systematic deviation away from the "single-curve" law at increasing speed. This effect is due to *cavitation*, a high speed phenomenon of hydraulic machines caused by the release of vapour bubbles at low pressures, which is discussed later in this chapter. It will be clear at this stage that under cavitating flow conditions, dynamical similarity is not possible.

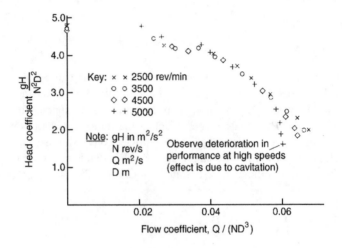

FIG. 1.3. Dimensionless head-volume characteristic of a centrifugal pump.

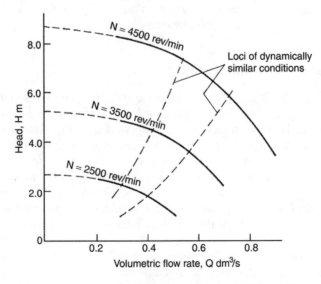

FIG. 1.4. Extrapolation of characteristic curves for dynamically similar conditions at $N = 3500$ rev/min.

The non-dimensional results shown in Figure 1.3 have, of course, been obtained for a particular pump. They would also be approximately valid for a range of *different* pump sizes so long as all these pumps are geometrically similar and cavitation is absent. Thus, neglecting any change in performance due to change in Reynolds number, the dynamically similar results in Figure 1.3 can be applied to predicting the dimensional performance of a given pump for a series of required speeds. Figure 1.4 shows such a dimensional presentation. It will be clear from the above discussion that the locus of dynamically similar points in the $H-Q$ field lies on a parabola since H varies as N^2 and Q varies as N.

Variable geometry turbomachines

The efficiency of a fixed geometry machine, ignoring Reynolds number effects, is a unique function of flow coefficient. Such a dependence is shown by line (b) in Figure 1.5. Clearly, off-design operation of such a machine is grossly inefficient and designers sometimes resort to a *variable geometry* machine in order to obtain a better match with changing flow conditions. Figure 1.6 shows a sectional sketch of a mixed-flow pump in which the impeller vane angles may be varied *during* pump operation. (A similar arrangement is used in Kaplan turbines, Figure 1.1.) Movement of the vanes is implemented by cams driven from a servomotor. In some very large installations involving many thousands of kilowatts and where operating

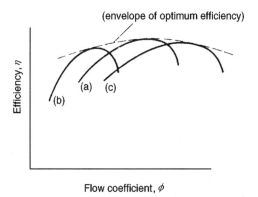

FIG. 1.5. Different efficiency curves for a given machine obtained with various blade settings.

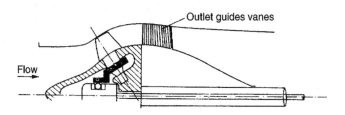

FIG. 1.6. Mixed-flow pump incorporating mechanism for adjusting blade setting.

conditions fluctuate, sophisticated systems of control may incorporate an electronic computer.

The lines (a) and (c) in Figure 1.5 show the efficiency curves at other blade settings. Each of these curves represents, in a sense, a different constant geometry machine. For such a variable geometry pump the desired operating line intersects the points of maximum efficiency of each of these curves.

Introducing the additional variable β into eqn. (1.3) to represent the setting of the vanes, we can write

$$\psi = f_1(\phi, \beta); \eta = f_2(\phi, \beta). \tag{1.5}$$

Alternatively, with $\beta = f_3(\phi, \eta) = f_4(\phi, \psi)$, β can be eliminated to give a new functional dependence

$$\eta = f_5(\phi, \psi) = f_5 \left(\frac{Q}{ND^3}, \frac{gH}{N^2D^2} \right) \tag{1.6}$$

Thus, efficiency in a variable geometry pump is a function of both flow coefficient and energy transfer coefficient.

Specific speed

The pump or hydraulic turbine designer is often faced with the basic problem of deciding what type of turbomachine will be the best choice for a given duty. Usually the designer will be provided with some preliminary design data such as the head H, the volume flow rate Q and the rotational speed N when a pump design is under consideration. When a turbine preliminary design is being considered the parameters normally specified are the shaft power P, the head at turbine entry H and the rotational speed N. A non-dimensional parameter called the *specific speed*, N_s, referred to and conceptualised as the *shape number*, is often used to facilitate the choice of the most appropriate machine. This new parameter is derived from the non-dimensional groups defined in eqn. (1.3) in such a way that the characteristic diameter D of the turbomachine is eliminated. The value of N_s gives the designer a guide to the type of machine that will provide the normal requirement of high efficiency at the design condition.

For any one hydraulic turbomachine *with fixed geometry* there is a unique relationship between efficiency and flow coefficient if Reynolds number effects are negligible and cavitation absent. As is suggested by any one of the curves in Figure 1.5, the efficiency rises to a maximum value as the flow coefficient is increased and then gradually falls with further increase in ϕ. This optimum efficiency $\eta = \eta_{\max}$, is used to identify a unique value $\phi = \phi_1$ and corresponding unique values of $\psi = \psi_1$ and $\hat{P} = \hat{P}_1$. Thus,

$$\frac{Q}{ND^3} = \phi_1 = \text{constant}, \tag{1.7a}$$

$$\frac{gH}{N^2D^2} = \psi_1 = \text{constant}, \tag{1.7b}$$

$$\frac{P}{\rho N^3 D^5} = \hat{P}_1 = \text{constant}. \tag{1.7c}$$

It is a simple matter to combine any pair of these expressions in such a way as to eliminate the diameter. For a pump the customary way of eliminating D is to divide $\phi_1^{1/2}$ by $\psi_1^{3/4}$. Thus

$$N_s = \frac{\phi_1^{1/2}}{\psi_1^{3/4}} = \frac{NQ^{1/2}}{(gH)^{3/4}}, \tag{1.8}$$

where N_s is called the *specific speed*. The term specific speed is justified to the extent that N_s is directly proportional to N. In the case of a turbine the *power specific speed* N_{sp}, is more useful and is defined by,

$$N_{sp} = \frac{\hat{P}_1^{1/2}}{\psi_1^{5/4}} = \frac{N(P/\rho)^{1/2}}{(gH)^{5/4}} \tag{1.9}$$

Both eqns. (1.8) and (1.9) are *dimensionless*. It is always safer and less confusing to calculate specific speed in one or other of these forms rather than dropping the factors g and ρ which would make the equations *dimensional* and any values of specific speed obtained using them would then depend upon the choice of the units employed. The dimensionless form of N_s (and N_{sp}) is the only one used in this book. Another point arises from the fact that the rotational speed, N, is expressed in the units of revolutions per unit of time so that although N_s is dimensionless, numerical values of specific speed need to be thought of as revs. Alternative versions of eqns. (1.8) and (1.9) in radians are also in common use and are written

$$\Omega_s = \frac{\Omega Q^{1/2}}{(gH)^{3/4}}, \tag{1.8a}$$

$$\Omega_{sp} = \frac{\Omega\sqrt{P/\rho}}{(gH)^{5/4}}. \tag{1.9a}$$

There is a simple connection between N_s and N_{sp} (and between Ω_s and Ω_{sP}). By dividing eqn. (1.9) by eqn. (1.8) we obtain

$$\frac{N_{sp}}{N_s} = \frac{N(P/\rho)^{1/2}}{(gH)^{5/4}}\frac{(gH)^{3/4}}{NQ^{1/2}} = \left(\frac{P}{\rho g Q H}\right)^{1/2}.$$

From the definition of hydraulic efficiency, for a pump we obtain:

$$\frac{N_{sp}}{N_s} = \frac{\Omega_{sp}}{\Omega_s} = \frac{1}{\sqrt{\eta}}, \tag{1.9b}$$

and, for a turbine we obtain:

$$\frac{N_{sp}}{N_s} = \frac{\Omega_{sp}}{\Omega_s} = \sqrt{\eta}. \tag{1.9c}$$

Remembering that specific speed, as defined above, is at the point of maximum efficiency of a turbomachine, it becomes a parameter of great importance in selecting the type of machine required for a given duty. The maximum efficiency condition *replaces* the condition of geometric similarity, so that any alteration in specific

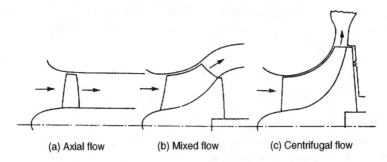

(a) Axial flow   (b) Mixed flow   (c) Centrifugal flow

FIG. 1.7. Range of pump impellers of equal inlet area.

speed implies that the machine design changes. Broadly speaking, each different class of machine has its optimum efficiency within its own fairly narrow range of specific speed.

For a pump, eqn. (1.8) indicates, for constant speed N, that N_s is increased by an increase in Q and decreased by an increase in H. From eqn. (1.7b) it is observed that H, at a constant speed N, increased with impeller diameter D. Consequently, to increase N_s the entry area must be made large and/or the maximum impeller diameter small. Figure 1.7 shows a range of pump impellers varying from the axial-flow type, through mixed flow to a centrifugal- or radial-flow type. The size of each inlet is such that they all handle the same volume flow Q. Likewise, the head developed by each impeller (of different diameter D) is made equal by adjusting the speed of rotation N. Since Q and H are constant, then N_s varies with N alone. The most noticeable feature of this comparison is the large change in size with specific speed. Since a higher specific speed implies a smaller machine, for reasons of economy, it is desirable to select the *highest possible specific speed* consistent with good efficiency.

Cavitation

In selecting a hydraulic turbomachine for a given head H and capacity Q, it is clear from the definition of specific speed, eqn. (1.8), that the highest possible value of N_s should be chosen because of the resulting reduction in size, weight and cost. On this basis a turbomachine could be made extremely small were it not for the corresponding increase in the fluid velocities. For machines handling liquids the lower limit of size is dictated by the phenomenon of *cavitation*.

Cavitation is the boiling of a liquid at normal temperature when the static pressure is made sufficiently low. It may occur at the entry to pumps or at the exit from hydraulic turbines in the vicinity of the moving blades. The dynamic action of the blades causes the static pressure to reduce locally in a region which is already normally below atmospheric pressure and cavitation can commence. The phenomenon is accentuated by the presence of dissolved gases which are released with a reduction in pressure.

For the purpose of illustration consider a centrifugal pump operating at constant speed and capacity. By steadily reducing the inlet pressure head a point is reached when streams of small vapour bubbles appear within the liquid and close to solid surfaces. This is called *cavitation inception* and commences in the regions of lowest pressure. These bubbles are swept into regions of higher pressure where they collapse. This condensation occurs suddenly, the liquid surrounding the bubbles either hitting the walls or adjacent liquid. The pressure wave produced by bubble collapse (with a magnitude of the order 400 MPa) momentarily raises the pressure level in the vicinity and the action ceases. The cycle then repeats itself and the frequency may be as high as 25 kHz (Shepherd 1956). The repeated action of bubbles collapsing near solid surfaces leads to the well-known cavitation erosion.

The collapse of vapour cavities generates noise over a wide range of frequencies – up to 1 MHz has been measured (Pearsall 1972) i.e. so-called "white noise". Apparently it is the collapsing smaller bubbles which cause the higher frequency noise and the larger cavities the lower frequency noise. Noise measurement can be used as a means of detecting cavitation (Pearsall 1966/7). Pearsall and McNulty (1968) have shown experimentally that there is a relationship between cavitation noise levels and erosion damage on cylinders and concludes that a technique could be developed for predicting the occurrence of erosion.

Up to this point no detectable deterioration in performance has occurred. However, with further reduction in inlet pressure, the bubbles increase both in size and number, coalescing into pockets of vapour which affects the whole field of flow. This growth of vapour cavities is usually accompanied by a sharp drop in pump performance as shown conclusively in Figure 1.3 (for the 5000 rev/min test data). It may seem surprising to learn that with this large change in bubble size, the solid surfaces are much less likely to be damaged than at inception of cavitation. The avoidance of cavitation inception in conventionally designed machines can be regarded as one of the essential tasks of both pump and turbine designers. However, in certain recent specialised applications pumps have been designed to operate under *super-cavitating* conditions. Under these conditions large size vapour bubbles are formed but, bubble collapse takes place *downstream* of the impeller blades. An example of the specialised application of a supercavitating pump is the fuel pumps of rocket engines for space vehicles where size and mass must be kept low at all costs. Pearsall (1966) has shown that the supercavitating principle is most suitable for axial flow pumps of high specific speed and has suggested a design technique using methods similar to those employed for conventional pumps.

Pearsall (1966) was one of the first to show that operating in the supercavitating regime was practicable for axial flow pumps and he proposed a design technique to enable this mode of operation to be used. A detailed description was later published (Pearsall 1973), and the cavitation performance was claimed to be much better than that of conventional pumps. Some further details are given in Chapter 7 of this book.

Cavitation limits

In theory cavitation commences in a liquid when the static pressure is reduced to the vapour pressure corresponding to the liquid's temperature. However, in practice, the physical state of the liquid will determine the pressure at which cavitation starts (Pearsall 1972). Dissolved gases come out of solution as the pressure is reduced forming gas cavities at pressures in excess of the vapour pressure. Vapour cavitation requires the presence of nuclei – submicroscopic gas bubbles or solid non-wetted particles – in sufficient numbers. It is an interesting fact that in the absence of such nuclei a liquid can withstand negative pressures (i.e. *tensile stresses*)! Perhaps the earliest demonstration of this phenomenon was that performed by Osborne Reynolds (1882) before a learned society. He showed how a column of mercury more than twice the height of the barometer could be (and was) supported by the internal cohesion (stress) of the liquid. More recently Ryley (1980) devised a simple centrifugal apparatus for students to test the tensile strength of both plain, untreated tap water in comparison with water that had been filtered and then de-aerated by boiling. Young (1989) gives an extensive literature list covering many aspects of cavitation including the tensile strength of liquids. At room temperature the theoretical tensile strength of water is quoted as being as high as 1000 atm (100 MPa)! Special pretreatment (i.e. rigorous filtration and pre-pressurization) of the liquid is required to obtain this state. In general the liquids flowing through turbomachines will contain some dust and dissolved gases and under these conditions negative pressure do not arise.

A useful parameter is the available suction head at entry to a pump or at exit from a turbine. This is usually referred to as the *net positive suction head*, NPSH, defined as

$$H_s = (p_o - p_v)/(\rho g) \tag{1.10}$$

where p_o and p_v are the absolute stagnation and vapour pressures, respectively, at pump inlet or at turbine outlet.

To take into account the effects of cavitation, the performance laws of a hydraulic turbomachine should include the additional independent variable H_s. Ignoring the effects of Reynolds number, the performance laws of a constant geometry hydraulic turbomachine are then dependent on two groups of variable. Thus, the efficiency,

$$\eta = f(\phi, N_{ss}) \tag{1.11}$$

where the *suction specific speed* $N_{ss} = NQ^{1/2}/(gH_s)^{3/4}$, determines the effect of cavitation, and $\phi = Q/(ND^3)$, as before.

It is known from experiment that cavitation inception occurs for an almost constant value of N_{ss} for all pumps (and, separately, for all turbines) designed to resist cavitation. This is because the blade sections at the inlet to these pumps are broadly similar (likewise, the exit blade sections of turbines are similar) and it is the *shape* of the low pressure passages which influences the onset of cavitation.

Using the alternative definition of suction specific speed $\Omega_{ss} = \Omega Q^{1/2}/(gH_s)^{1/2}$, where Ω is the rotational speed in rad/s, Q is the volume flow in m^3/s and gH_s, is in m^2/s^2, it has been shown empirically (Wislicehus 1947) that

$$\Omega_{ss} \simeq 3.0 \text{ (rad)} \tag{1.12a}$$

for pumps, and

$$\Omega_{ss} \simeq 4.0 \text{ (rad)} \tag{1.12b}$$

for turbines.

Pearsall (1973) described a supercavitating pump with a cavitation performance much better that of conventional pumps. For this pump suction specific speeds, Ω_{ss}

up to 9.0 were readily obtained and, it was claimed, even better values might be possible, but at the cost of reduced head and efficiency. It is likely that supercavitating pumps will be increasingly used in the search for higher speeds, smaller sizes and lower costs.

Compressible gas flow relations

Stagnation properties

In turbomachines handling compressible fluids, large changes in flow velocity occur across the stages as a result of pressure changes caused by the expansion or compression processes. For any point in the flow it is convenient to combine the energy terms together. The enthalpy, h, and the kinetic energy, $\frac{1}{2}c^2$ are combined and the result is called the *stagnation enthalpy*,

$$h_0 = h + \tfrac{1}{2}c^2.$$

The stagnation enthalpy is constant in a flow process that does not involve a work transfer or a heat transfer even though irreversible processes may be present. In Figure 1.8, point 1 represents the actual or static state of a fluid in an enthalpy–entropy diagram with enthalpy, h_1 at pressure p_1 and entropy s_1. The fluid velocity is c_1. The stagnation state is represented by the point 01 brought about by an irreversible deceleration. For a reversible deceleration the stagnation point would be at point 01s and the state change would be called *isentropic*.

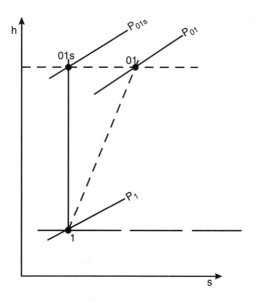

FIG. 1.8. The static state (point 1), the stagnation (point 01) and the isentropic stagnation (point 01s) of a fluid.

Stagnation temperature and pressure

If the fluid is a perfect gas, then $h = C_p T$, where $C_p = \gamma R/(\gamma - 1)$, so that the stagnation temperature can be defined as

$$T_0 = T + \tfrac{1}{2}c^2/C_p,$$

(handwritten: $h_o = h + \frac{c^2}{2}$; $C_p T_o = C_p T + \frac{c^2}{2}$ $T_o = T + \frac{c^2}{2C_p}$)

$$\frac{T_0}{T} = 1 + \tfrac{1}{2}(\gamma - 1)\frac{c^2}{\gamma RT} = 1 + \tfrac{1}{2}(\gamma - 1)M^2, \tag{1.13a}$$

where the *Mach number*, $M = c/a = c/\sqrt{\gamma RT}$.

The Gibb's relation, derived from the second law of thermodynamics (see Chapter 2), is

$$T\,ds = dh - \frac{1}{\rho}dp.$$

If the flow is brought to rest both adiabatically and isentropically (i.e. $ds = 0$), then, using the above Gibb's relation,

(handwritten: $dh = \frac{1}{\rho}dp$)

$$dh = C_p dT = \frac{dp}{p}RT$$

(handwritten: $h = C_p T$ $dh = C_p dT$)

so that

$$\frac{dp}{p} = \frac{C_p}{R}\frac{dT}{T} = \frac{\gamma}{\gamma - 1}\frac{dT}{T}.$$

(handwritten: $PV = mRT$ $P\nu = RT$ $\nu = \frac{RT}{P} = \frac{1}{\rho}$ $\frac{1}{\rho}dP = \frac{dp}{p}RT$)

Integrating, we obtain

$$\ln p = \ln \text{constant} + \frac{\gamma}{\gamma - 1}\ln T,$$

and so,

$$\frac{p_0}{p} = \left(\frac{T_0}{T}\right)^{\gamma/(\gamma-1)} = \left(1 + \frac{\gamma - 1}{2}M^2\right)^{\gamma/\gamma-1} \tag{1.13b}$$

From the gas law density, $\rho = p/(RT)$, we obtain $\rho_0/\rho = (p_0/p)(T/T_0)$ and hence,

$$\frac{\rho_0}{\rho} = \left(\frac{T_0}{T}\right)^{1/(\gamma-1)} = \left(1 + \frac{\gamma - 1}{2}M^2\right)^{1/(\gamma-1)}. \tag{1.13c}$$

Compressible fluid analysis

The application of dimensional analysis to compressible fluids increases, not unexpectedly, the complexity of the functional relationships obtained in comparison with those already found for incompressible fluids. Even if the fluid is regarded as a perfect gas, in addition to the previously used fluid properties, two further characteristics are required; these are a_{01}, the stagnation speed of sound at entry to the machine and γ, the ratio of specific heats C_p/C_v. In the following analysis the compressible fluids under discussion are either perfect gases, or else, dry vapours approximating in behaviour to a perfect gas.

Another choice of variables is usually preferred when appreciable density changes occur across the machine. Instead of volume flow rate Q, the mass flow rate $\dot{m}$ is used; likewise for the head change H, the isentropic *stagnation enthalpy* change Δh_{os} is employed.

The choice of this last variable is a significant one for, in an ideal and adiabatic process, Δh_{os} is equal to the work done by unit mass of fluid. This will be discussed still further in Chapter 2. Since heat transfer from the casings of turbomachines is, in general, of negligible magnitude compared with the flux of energy through the machine, temperature on its own may be safely excluded as a fluid variable. However, temperature is an easily observable characteristic and, for a perfect gas, can be easily introduced at the last by means of the equation of state, $p/\rho = RT$, where $R = R_0/m = C_p - C_v$, m being the molecular weight of the gas and $R_0 = 8.314\,\text{kJ/(kg mol K)}$ is the *Universal gas constant*.

The performance parameters Δh_{os}, η and P for a turbomachine handling a compressible flow, are expressed functionally as:

$$\Delta h_{os}, \eta, P = f(\mu, N, D, \dot{m}, \rho_{01}, a_{01}, \gamma). \tag{1.14a}$$

Because ρ_0 and a_0 change through a turbomachine, values of these fluid variables are selected at inlet, denoted by subscript 1. Equation (1.14a) express *three* separate functional relationships, each of which consists of eight variables. Again, selecting ρ_{01}, N, D as common factors each of these three relationships may be reduced to five dimensionless groups,

$$\frac{\Delta h_{0s}}{N^2 D^2}, \eta, \frac{P}{\rho_{01} N^3 D^5} = f \left\{ \frac{\dot{m}}{\rho_{01} N D^3}, \frac{\rho_{01} N D^2}{\mu}, \frac{ND}{a_{01}}, \gamma \right\}. \tag{1.14b}$$

Alternatively, the flow coefficient $\phi = \dot{m}/(\rho_{01} N D^3)$ can be written as $\phi = \dot{m}/(\rho_{01} a_{01} D^2)$. As ND is proportional to blade speed, the group ND/a_{01} is regarded as a *blade Mach number*.

For a machine handling a perfect gas a different set of functional relationships is often more useful. These may be found either by selecting the appropriate variables for a perfect gas and working through again from first principles or, by means of some rather straightforward transformations, rewriting eqn. (1.14b) to give more suitable groups. The latter procedure is preferred here as it provides a useful exercise.

As a concrete example consider an adiabatic compressor handling a perfect gas. The isentropic stagnation enthalpy rise can now be written $C_p(T_{02s} - T_{01})$ for the perfect gas. This compression process is illustrated in Figure 1.9a where the stagnation state point changes at constant entropy between the stagnation pressures p_{01} and p_{02}. The equivalent process for a turbine is shown in Figure 1.9b. Using the adiabatic isentropic relationship $p/\rho^\gamma = \text{constant}$, together with $p/\rho = RT$, the expression

$$\frac{T_{02s}}{T_{01}} = \left(\frac{p_{02}}{p_{01}} \right)^{(\gamma-1)/\gamma}$$

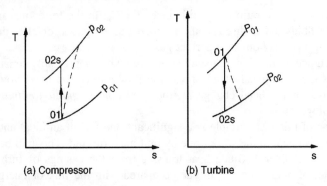

(a) Compressor   (b) Turbine

FIG. 1.9. The ideal adiabatic change in stagnation conditions across a turbomachine.

is obtained. Hence $\Delta h_{0s} = C_p T_{01}[(p_{02}/p_{01})^{(\gamma-1)/\gamma} - 1]$. Since $C_p = \gamma R/(\gamma - 1)$ and $a_{01}^2 = \gamma R T_{01}$, then

$$\Delta h_{0s}/a_{01}^2 \propto f(p_{02}/p_{01}).$$

The flow coefficient can now be more conveniently expressed as

$$\frac{\dot{m}}{\rho_{01} a_{01} D^2} = \frac{\dot{m} R T_{01}}{p_{01}\sqrt{(\gamma R T_{01})}D^2} = \frac{\dot{m}\sqrt{(R T_{01})}}{D^2 p_{01}\sqrt{\gamma}}.$$

As $\dot{m} \equiv \rho_{01} D^2(ND)$, the power coefficient may be written

$$\hat{P} = \frac{P}{\rho_{01} N^3 D^5} = \frac{\dot{m} C_p \Delta T_0}{\{\rho_{01} D^2(ND)\}(ND)^2} = \frac{C_p \Delta T_0}{(ND)^2} \equiv \frac{\Delta T_0}{T_{01}}.$$

Collecting together all these newly formed non-dimensional groups and inserting them in eqn. (1.14b) gives

$$\frac{p_{02}}{p_{01}}, \eta, \frac{\Delta T_0}{T_{01}} = f\left\{\frac{\dot{m}\sqrt{(R T_{01})}}{D^2 p_{01}}, \frac{ND}{\sqrt{(R T_{01})}}, Re, \gamma\right\}. \tag{1.15}$$

The justification for dropping γ from a number of these groups is simply that it already appears separately as an independent variable.

For a machine of a specific size and handling a single gas it has become customary, in industry at least, to delete γ, R, and D from eqn. (1.15) and similar expressions. If, in addition, the machine operates at high Reynolds numbers (or over a small speed range), Re can also be dropped. Under these conditions eqn. (1.15) becomes

$$\frac{p_{02}}{p_{01}}\eta, \frac{\Delta T_0}{T_{01}} = f\left\{\frac{\dot{m}\sqrt{T_{01}}}{p_{01}}, \frac{N}{\sqrt{T_{01}}}\right\}. \tag{1.16}$$

Note that by omitting the diameter D and gas constant R, the independent variables in eqn. (1.16) are no longer dimensionless.

Figures 1.10 and 1.11 represent typical performance maps obtained from compressor and turbine test results. In both figures the pressure ratio across the whole

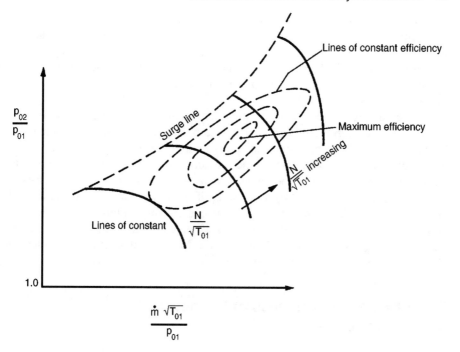

FIG. 1.10. Overall characteristic of a compressor.

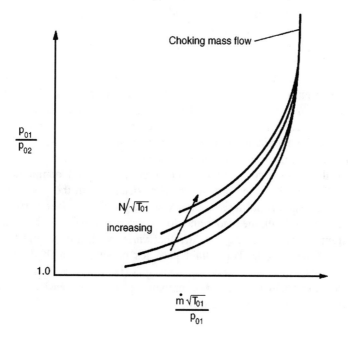

FIG. 1.11. Overall characteristic of a turbine.

machine is plotted as a function of $\dot{m}(\sqrt{T_{01}})/p_{01}$ for fixed values of $N/(\sqrt{T_{01}})$, this being a customary method of presentation. Notice that for both machines subscript 1 is used to denote conditions as inlet. One of the most striking features of these performance characteristics is the rather weak dependence of the turbine performance upon $N/\sqrt{T_{01}}$ contrasting with the strong dependence shown by the compressor on this parameter.

For the compressor, efficient operation at constant $N/\sqrt{T_{01}}$ lies to the right of the line marked "*surge*". A discussion of the phenomenon of surge is included in Chapter 5; in brief, for multistage compressors it commences approximately at the point (for constant $N/\sqrt{T_{01}}$) where the pressure ratio flattens out to its maximum value. The surge line denotes the limit of *stable operation* of a compressor, unstable operation being characterised by a severe oscillation of the mass flow rate through the machine. The choked regions of both the compressor and turbine characteristics may be recognised by the vertical portions of the constant speed lines. No further increase in $\dot{m}(\sqrt{T_{01}})/p_{01}$ is possible since the Mach number across some section of the machine has reached unity and the flow is said to be *choked*.

The inherent unsteadiness of the flow within turbomachines

A fact often ignored by turbomachinery designers, or even unknown to students, is that turbomachines can only work the way they do because of unsteady flow effects taking place within them. The fluid dynamic phenomena that are associated with the unsteady flow in turbomachines has been examined by Greitzer (1986) in a discourse which was intended to be an introduction to the subject but actually extended far beyond the technical level of this book! Basically Greitzer, and others before him, in considering the fluid mechanical process taking place on a fluid particle in an isentropic flow, deduced that *stagnation enthalpy of the particle can change only if the flow is unsteady*. Dean (1959) appears to have been the first to record that without an *unsteady flow* inside a turbomachine, no work transfer can take place. Paradoxically, both at the inlet to and outlet from the machine the conditions are such that the flow can be considered as steady.

A physical situation considered by Greitzer is the axial compressor rotor as depicted in Figure 1.12a. The pressure field associated with the blades is such that the pressure increases from the suction surface (S) to the pressure surface (P). This pressure field moves with the blades and, to an observer situated at the point * (in the absolute frame of reference), a pressure that varies with time would be recorded, as shown in Figure 1.12b. Thus, fluid particles passing through the rotor would experience a positive pressure increase with time (i.e. $\partial p/\partial t > 0$). From this fact it can then be shown that the stagnation enthalpy of the fluid particle also increases because of the unsteadiness of the flow, i.e.

$$\frac{Dh_0}{Dt} = \frac{1}{\rho}\frac{\partial p}{\partial t},$$

where D/Dt is the rate of change following the fluid particle.

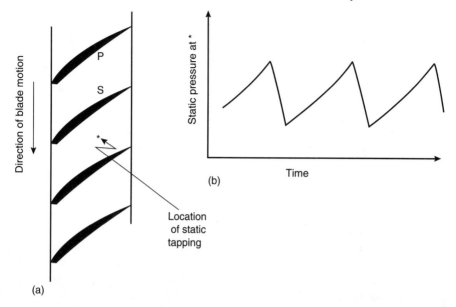

FIG. 1.12. Measuring unsteady pressure field of an axial compressor rotor. (a) Pressure is measured at point ∗ on the casing. (b) Fluctuating pressure measured at point ∗.

References

Dean, R. C. (1959). On the necessity of unsteady flow in fluid machines. *J. Basic Eng., Trans. Am. Soc. Mech. Engrs.*, **81**, pp. 24–8.

Douglas, J. F., Gasiorek, J. M. and Swafffield, J. A. (1995). *Fluid Mechanics*. Longman.

Duncan, W. J., Thom, A. S. and Young, A. D. (1970). *Mechanics of Fluids*. Edward Arnold.

Freris, L. L. (1990). *Wind Energy Conversion Systems*. Prentice Hall.

Glauert, H. (1959). *The Elements of Aerofoil and Airscrew Theory*. Cambridge University Press.

Greitzer, E. M. (1986). An introduction to unsteady flow in turbomachines. In *Advanced Topics in Turbomachinery, Principal Lecture Series No. 2*. (D. Japikse, ed.) pp. 2.1–2.29, Concepts ETI.

ISO 31/0 (1981). *General Principles Concerning Quantities, Units and Symbols*, International Standards Organisation, Paris. (Also published as BS 5775: Part 0: 1982, Specifications for Quantities, Units and Symbols, London, 1982).

Massey, B. S. (1979). *Mechanics of Fluids* (4th edn.). Van Nostrand.

Pearsall, I. S. (1966). The design and performance of supercavitating pumps. *Proc. of Symposium on Pump Design, Testing and Operation*, N.E.L., Glasgow.

Pearsall, I. S. (1967). Acoustic detection of cavitation. Symposium on Vibrations in Hydraulic Pumps and Turbines. *Proc. Instn. Mech. Engrs., London*, **181**, Pt. 3A.

Pearsall, I. S. and McNulty, P. J. (1968). Comparison of cavitation noise with erosion. *Cavitation Forum*, 6–7, Am. Soc. Mech. Engrs.

Pearsall, I. S. (1972). *Cavitation*. M & B Monograph ME/10. Mills & Boon.

Quantities, Units and Symbols (1975). A report by the Symbols Committee of the Royal Society, London.

Reynolds, O. (1882). On the internal cohesion of fluids. *Mem. Proc. Manchester Lit. Soc.*, 3rd Series, **7**, 1–19.

Ryley, D. J. (1980). Hydrostatic stress in water. *Int. J. Mech. Eng. Educ.*, **8** (2).

Shames, I. H. (1992). *Mechanics of Fluids*. McGraw-Hill.

Shepherd, D. G. (1956). *Principles of Turbomachinery*. Macmillan.

Streeter, V. L. and Wylie, E. B. (1979). *Fluid Mechanics* (7th edn). McGraw-Hill, Kogakusha.

Taylor, E. S. (1974). *Dimensional Analysis for Engineers*. Clarendon.

The International System of Units (1986). HMSO, London.

Wislicenus, G. F. (1947). *Fluid Mechanics of Turbomachinery*, McGraw-Hill.

Problems

1. A fan operating at 1750 rev/min at a volume flow rate of 4.25 m³/s develops a head of 153 mm measured on a water-filled U-tube manometer. It is required to build a larger, geometrically similar fan which will deliver the same head at the same efficiency as the existing fan, but at a speed of 1440 rev/min. Calculate the volume flow rate of the larger fan.

2. An axial flow fan 1.83 m diameter is designed to run at a speed of 1400 rev/min with an average axial air velocity of 12.2 m/s. A quarter scale model has been built to obtain a check on the design and the rotational speed of the model fan is 4200 rev/min. Determine the axial air velocity of the model so that dynamical similarity with the full-scale fan is preserved. The effects of Reynolds number change may be neglected.

A sufficiently large pressure vessel becomes available in which the complete model can be placed and tested under conditions of complete similarity. The viscosity of the air is independent of pressure and the temperature is maintained constant. At what pressure must the model be tested?

3. A water turbine is to be designed to produce 27 MW when running at 93.7 rev/min under a head of 16.5 m. A model turbine with an output of 37.5 kW is to be tested under dynamically similar conditions with a head of 4.9 m. Calculate the model speed and scale ratio. Assuming a model efficiency of 88%, estimate the volume flow rate through the model.

It is estimated that the force on the thrust bearing of the full-size machine will be 7.0 GN. For what thrust must the model bearing be designed?

4. Derive the non-dimensional groups that are normally used in the testing of gas turbines and compressors.

A compressor has been designed for normal atmospheric conditions (101.3 kPa and 15°C). In order to economise on the power required it is being tested with a throttle in the entry duct to reduce the entry pressure. The characteristic curve for its normal design speed of 4000 rev/min is being obtained on a day when the ambient temperature is 20°C. At what speed should the compressor be run? At the point on the characteristic curve at which the mass flow would normally be 58 kg/s the entry pressure is 55 kPa. Calculate the actual rate of mass flow during the test.

Describe, with the aid of sketches, the relationship between geometry and specific speed for pumps.

CHAPTER 2

Basic Thermodynamics, Fluid Mechanics: Definitions of Efficiency

Take your choice of those that can best aid your action. (SHAKESPEARE, *Coriolanus.*)

Introduction

THIS chapter summarises the basic physical laws of fluid mechanics and thermodynamics, developing them into a form suitable for the study of turbomachines. Following this, some of the more important and commonly used expressions for the efficiency of compression and expansion flow processes are given.

The laws discussed are:

(1) the *continuity of flow equation*;
(2) the *first law of thermodynamics* and the *steady flow energy equation*;
(3) the *momentum equation*;
(4) the *second law of thermodynamics*.

All of these laws are usually covered in first-year university engineering and technology courses, so only the briefest discussion and analysis is give here. Some fairly recent textbooks dealing comprehensively with these laws are those written by Cengel and Boles (1994), Douglas, Gasiorek and Swaffield (1995), Rogers and Mayhew (1992) and Reynolds and Perkins (1977). It is worth remembering that these laws are completely general; they are independent of the nature of the fluid or whether the fluid is compressible or incompressible.

Ref – ?

The equation of continuity

Consider the flow of a fluid with density ρ, through the element of area dA, during the time interval dt. Referring to Figure 2.1, if c is the stream velocity the elementary mass is $dm = \rho c dt dA \cos\theta$, where θ is the angle subtended by the normal of the area element to the stream direction. The velocity component perpendicular to the area dA is $c_n = c \cos\theta$ and so $dm = \rho c_n dA dt$. The elementary rate of mass flow is therefore

$$d\dot{m} = \frac{dm}{dt} = \rho c_n dA. \tag{2.1}$$

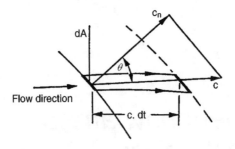

FIG. 2.1. Flow across an element of area.

Most analyses in this book are limited to one-dimensional steady flows where the velocity and density are regarded as constant across each section of a duct or passage. If A_1 and A_2 are the flow areas at stations 1 and 2 along a passage respectively, then

$$\dot{m} = \rho_1 c_{n1} A_1 = \rho_2 c_{n2} A_2 = \rho c_n A, \tag{2.2}$$

since there is no accumulation of fluid within the control volume.

The first law of thermodynamics – internal energy

The *first law of thermodynamics* states that if a system is taken through a complete cycle during which heat is supplied and work is done, then

$$\oint (dQ - dW) = 0, \tag{2.3}$$

where $\oint dQ$ represents the heat supplied to the system during the cycle and $\oint dW$ the work done by the system during the cycle. The units of heat and work in eqn. (2.3) are taken to be the same.

During a change of state from 1 to 2, there is a change in the property internal energy,

$$E_2 - E_1 = \int_1^2 (dQ - dW). \tag{2.4}$$

For an infinitesimal change of state

$$dE = dQ - dW. \tag{2.4a}$$

The steady flow energy equation

Many textbooks, e.g. Çengel and Boles (1994), demonstrate how the first law of thermodynamics is applied to the steady flow of fluid through a control volume so that the steady flow energy equation is obtained. It is unprofitable to reproduce this proof here and only the final result is quoted. Figure 2.2 shows a control volume representing a turbomachine, through which fluid passes at a steady rate of mass

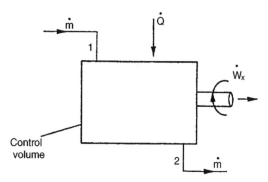

FIG. 2.2. Control volume showing sign convention for heat and work transfers.

flow $\dot{m}$, entering at position 1 and leaving at position 2. Energy is transferred from the fluid to the blades of the turbomachine, positive work being done (via the shaft) at the rate $\dot{W}_x$. In the general case positive heat transfer takes place at the rate $\dot{Q}$, *from* the surroundings *to* the control volume. Thus, with this sign convention the steady flow energy equation is

$$\dot{Q} - \dot{W}_x = \dot{m}[(h_2 - h_1) + \tfrac{1}{2}(c_2^2 - c_1^2) + g(z_2 - z_1)], \tag{2.5}$$

where h is the specific enthalpy, $\tfrac{1}{2}c^2$ the kinetic energy per unit mass and gz the potential energy per unit mass.

Apart from hydraulic machines, the contribution of the last term in eqn. (2.5) is small and usually ignored. Defining stagnation enthalpy by $h_0 = h + \tfrac{1}{2}c^2$ and assuming $g(z_2 - z_1)$ is negligible, eqn. (2.5) becomes

$$\dot{Q} - \dot{W}_x = \dot{m}(h_{02} - h_{01}). \tag{2.6}$$

Most turbomachinery flow processes are adiabatic (or very nearly so) and it is permissible to write $\dot{Q} = 0$. For work producing machines (turbines) $\dot{W}_x > 0$, so that

$$\dot{W}_x = \dot{W}_t = \dot{m}(h_{01} - h_{02}). \tag{2.7}$$

For work absorbing machines (compressors) $\dot{W}_x < 0$, so that it is more convenient to write

$$\dot{W}_c = -\dot{W}_x = \dot{m}(h_{02} - h_{01}). \tag{2.8}$$

The momentum equation – Newton's second law of motion

One of the most fundamental and valuable principles in mechanics is *Newton's second law of motion*. The momentum equation relates the sum of the external forces acting on a fluid element to its acceleration, or to the rate of change of momentum in the direction of the resultant external force. In the study of turbomachines many applications of the momentum equation can be found, e.g. the force exerted upon a blade in a compressor or turbine cascade caused by the deflection or acceleration of fluid passing the blades.

Considering a system of mass m, the sum of all the body and surface forces acting on m along some arbitrary direction x is equal to *the time rate of change of the total x-momentum of the system*, i.e.

$$\Sigma F_x = \frac{\mathrm{d}}{\mathrm{d}t}(mc_x). \tag{2.9}$$

For a control volume where fluid enters steadily at a uniform velocity c_{x1} and leaves steadily with a uniform velocity c_{x2}, then

$$\Sigma F_x = \dot{m}(c_{x2} - c_{x1}) \tag{2.9a}$$

Equation (2.9a) is the one-dimensional form of the steady flow momentum equation.

Euler's equation of motion

It can be shown for the steady flow of fluid through an elementary control volume that, in the absence of all shear forces, the relation

$$\frac{1}{\rho}\mathrm{d}p + c\mathrm{d}c + g\mathrm{d}z = 0 \tag{2.10}$$

is obtained. This is Euler's equation of motion for one-dimensional flow and is derived from Newton's second law. By shear forces being absent we mean there is neither friction nor shaft work. However, it is not necessary that heat transfer should also be absent.

Bernoulli's equation

The one-dimensional form of Euler's equation applies to a control volume whose thickness is infinitesimal in the stream direction (Figure 2.3). Integrating this equation in the stream direction we obtain

$$\int_1^2 \frac{1}{\rho}\mathrm{d}p + \frac{1}{2}(c_2^2 - c_1^2) + g(z_2 - z_1) = 0 \tag{2.10a}$$

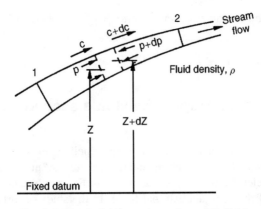

FIG. 2.3. Control volume in a streaming fluid.

which is Bernoulli's equation. For an incompressible fluid, ρ is constant and eqn. (2.10a) becomes

$$\frac{1}{\rho}(p_{02} - p_{01}) + g(z_2 - z_1) = 0, \tag{2.10b}$$

where stagnation pressure is $p_0 = p + \frac{1}{2}\rho c^2$.

When dealing with hydraulic turbomachines, the term *head H* occurs frequently and describes the quantity $z + p_0/(\rho g)$. Thus eqn. (2.10b) becomes

$$H_2 - H_1 = 0. \tag{2.10c}$$

If the fluid is a gas or vapour, the change in gravitational potential is generally negligible and eqn. (2.10a) is then

$$\int_1^2 \frac{1}{\rho} \mathrm{d}p + \frac{1}{2}(c_2^2 - c_1^2) = 0. \tag{2.10d}$$

Now, if the gas or vapour is subject to only a small pressure change the fluid density is sensibly constant and

$$p_{02} = p_{01} = p_0, \tag{2.10e}$$

i.e. the stagnation pressure is constant (this is also true for a *compressible isentropic process*).

Moment of momentum

In dynamics much useful information is obtained by employing Newton's second law in the form where it applies to the moments of forces. This form is of central importance in the analysis of the energy transfer process in turbomachines.

For a system of mass m, the vector sum of the moments of all external forces acting on the system about some arbitrary axis $A-A$ fixed in space is equal to the time rate of change of angular momentum of the system about that axis, i.e.

$$\tau_A = m\frac{\mathrm{d}}{\mathrm{d}t}(rc_\theta), \tag{2.11}$$

where r is distance of the mass centre from the axis of rotation measured along the normal to the axis and c_θ the velocity component mutually perpendicular to both the axis and radius vector r.

For a control volume the *law of moment of momentum* can be obtained. Figure 2.4 shows the control volume enclosing the rotor of a generalised turbomachine. Swirling fluid enters the control volume at radius r_1 with tangential velocity $c_{\theta 1}$ and leaves at radius r_2 with tangential velocity $c_{\theta 2}$. For one-dimensional steady flow

$$\tau_A = \dot{m}(r_2 c_{\theta 2} - r_1 c_{\theta 1}) \tag{2.11a}$$

which states that, the sum of the moments of the external forces acting on fluid temporarily occupying the control volume is equal to the net time rate of efflux of angular momentum from the control volume.

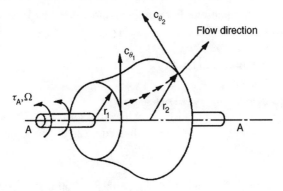

FIG. 2.4. Control volume for a generalised turbomachine.

Euler's pump and turbine equations

For a pump or compressor rotor running at angular velocity Ω, the rate at which the rotor does work on the fluid is

$$\tau_A \Omega = \dot{m}(U_2 c_{\theta 2} - U_1 c_{\theta 1}), \tag{2.12}$$

where the blade speed $U = \Omega r$.

Thus the work done on the fluid per unit mass or specific work, is

$$\Delta W_c = \frac{\dot{W}_c}{\dot{m}} = \frac{\tau_A \Omega}{\dot{m}} = U_2 c_{\theta 2} - U_1 c_{\theta 1} > 0. \tag{2.12a}$$

$U = $ blade speed
$c_\theta = $ vel. compt. $\perp r$ to r and axis

This equation is referred to as *Euler's pump equation*.

For a turbine the fluid does work *on* the rotor and the sign for work is then reversed. Thus, the specific work is

$$\Delta W_t = \frac{\dot{W}_t}{\dot{m}} = U_1 c_{\theta 1} - U_2 c_{\theta 2} > 0. \tag{2.12b}$$

Equation (2.12b) will be referred to as *Euler's turbine equation*.

Defining rothalpy

In a compressor or pump the specific work done on the fluid equals the rise in stagnation enthalpy. Thus, combining eqns. (2.8) and (2.12a),

$$\Delta W_c = \dot{W}_c / \dot{m} = U_2 C_{\theta 2} - U_1 c_{\theta 1} = h_{02} - h_{01}. \tag{2.12c}$$

This relationship is true for steady, adiabatic and irreversible flow in compressor or in pump impellers. After some rearranging of eqn. (2.12c) and writing $h_0 = h + \frac{1}{2}c^2$, then

$$h_1 + \tfrac{1}{2}c_1^2 - U_1 c_{\theta 1} = h_2 + \tfrac{1}{2}c_2^2 - U_2 c_{\theta 2} = I. \tag{2.12d}$$

According to the above reasoning a new function I has been defined having the same value at exit from the impeller as at entry. The function I has acquired the widely used name *rothalpy*, a contraction of rotational stagnation enthalpy, and is a fluid mechanical property of some importance in the study of relative flows in rotating systems. As the value of rothalpy is apparently* unchanged between entry and exit of the impeller it is deduced that it must be constant along the flow lines between these two stations. Thus, the rothalpy can be written generally as

$$I = h + \tfrac{1}{2}c^2 - Uc_\theta. \tag{2.12e}$$

The same reasoning can be applied to the thermomechanical flow through a turbine with the same result.

The second law of thermodynamics – entropy

The *second law of thermodynamics*, developed rigorously in many modern thermodynamic textbooks, e.g. Çengel and Boles (1994), Reynolds and Perkins (1977), Rogers and Mayhew (1992), enables the concept of entropy to be introduced and ideal thermodynamic processes to be defined.

An important and useful corollary of the second law of thermodynamics, known as the *Inequality of Clausius*, states that for a system passing through a cycle involving heat exchanges,

$$\oint \frac{\mathrm{d}Q}{T} \leqq 0, \tag{2.13}$$

where $\mathrm{d}Q$ is an element of heat transferred to the system at an absolute temperature T. If all the processes in the cycle are reversible then $\mathrm{d}Q = \mathrm{d}Q_R$ and the equality in eqn. (2.13) holds true, i.e.

$$\oint \frac{\mathrm{d}Q_R}{T} = 0. \tag{2.13a}$$

The property called entropy, for a finite change of state, is then defined as

$$S_2 - S_1 = \int_1^2 \frac{\mathrm{d}Q_R}{T}. \tag{2.14}$$

For an incremental change of state

$$\mathrm{d}S = m\mathrm{d}s = \frac{\mathrm{d}Q_R}{T}, \tag{2.14a}$$

where m is the mass of the system.

With steady one-dimensional flow through a control volume in which the fluid experiences a change of state from condition 1 at entry to 2 at exit,

$$\int_1^2 \frac{\mathrm{d}\dot{Q}}{T} \leqq \dot{m}(s_2 - s_1). \tag{2.15}$$

---

* A discussion of recent investigations into the conditions required for the conservation of rothalpy is deferred until Chapter 7.

If the process is adiabatic, $d\dot{Q} = 0$, then

$$s_2 \geqq s_1. \tag{2.16}$$

If the process is *reversible* as well, then

$$s_2 = s_1. \tag{2.16a}$$

Thus, for a flow which is adiabatic, the ideal process will be one in which the entropy remains unchanged during the process (the condition of *isentropy*).

Several important expressions can be obtained using the above definition of entropy. For a system of mass m undergoing a reversible process $dQ = dQ_R = mTds$ and $dW = dW_R = mpdv$. In the absence of motion, gravity and other effects the first law of thermodynamics, eqn. (2.4a) becomes

$$Tds = du + pdv. \tag{2.17}$$

With $h = u + pv$ then $dh = du + pdv + vdp$ and eqn. (2.17) then gives

$$Tds = dh - vdp. \tag{2.18}$$

Definitions of efficiency

A large number of efficiency definitions are included in the literature of turbomachines and most workers in this field would agree there are too many. In this book only those considered to be important and useful are included.

Efficiency of turbines

Turbines are designed to convert the available energy in a flowing fluid into useful mechanical work delivered at the coupling of the output shaft. The efficiency of this process, the *overall efficiency* η_0, is a performance factor of considerable interest to both designer and user of the turbine. Thus,

$$\eta_0 = \frac{\text{mechanical energy available at coupling of output shaft in unit time}}{\text{maximum energy difference possible for the fluid in unit time}}.$$

Mechanical energy losses occur between the turbine rotor and the output shaft coupling as a result of the work done against friction at the bearings, glands, etc. The magnitude of this loss as a fraction of the total energy transferred to the rotor is difficult to estimate as it varies with the size and individual design of turbomachine. For small machines (several kilowatts) it may amount to 5% or more, but for medium and large machines this loss ratio may become as little as 1%. A detailed consideration of the mechanical losses in turbomachines is beyond the scope of this book and is not pursued further.

The *isentropic efficiency* η_t or *hydraulic efficiency* η_h for a turbine is, in broad terms,

$$\eta_t(\text{or } \eta_h) = \frac{\text{mechanical energy supplied to the rotor in unit time}}{\text{maximum energy difference possible for the fluid in unit time}}.$$

Comparing the above definitions it is easily deduced that the *mechanical efficiency* η_m, which is simply the ratio of shaft power to rotor power, is

$$\eta_m = \eta_0/\eta_t \ (\text{or } \eta_0/\eta_h).$$

In the following paragraphs the various definitions of hydraulic and adiabatic efficiency are discussed in more detail.

For an incremental change of state through a turbomachine the steady flow energy equation, eqn. (2.5), can be written

$$d\dot{Q} - d\dot{W}_x = \dot{m}[dh + \tfrac{1}{2}d(c^2) + gdz].$$

From the second law of thermodynamics

$$d\dot{Q} \leqslant \dot{m}Tds = \dot{m}\left(dh - \frac{1}{\rho}dp\right).$$

Eliminating $d\dot{Q}$ between these two equations and rearranging

$$d\dot{W}_x \leqslant -\dot{m}\left[\frac{1}{\rho}dp + \frac{1}{2}d(c^2) + gdz\right]. \tag{2.19}$$

For a turbine expansion, noting $\dot{W}_x = \dot{W}_t > 0$, integrate eqn. (2.19) from the initial state 1 to the final state 2,

$$\dot{W}_x \leqslant \dot{m}\left[\int_2^1 \frac{1}{\rho}dp + \frac{1}{2}(c_1^2 - c_2^2) + g(z_1 - z_2)\right]. \tag{2.20}$$

For a reversible adiabatic process, $Tds = 0 = dh - dp/\rho$. The incremental maximum work output is then

$$d\dot{W}_{x_{max}} = -\dot{m}[dh + \tfrac{1}{2}d(c^2) + gdz]$$

Hence, the overall maximum work output between initial state 1 and final state 2 is

$$\dot{W}_{x_{max}} = \dot{m}\int_2^1 \left[dh + \frac{1}{2}d(c^2) + gdz\right]$$
$$= \dot{m}[(h_{01} - h_{02s}) + g(z_1 - z_2)] \tag{2.20a}$$

where the subscript s in eqn. (2.20a) denotes that the change of state between 1 and 2 is isentropic.

For an incompressible fluid, in the absence of friction, the maximum work output from the turbine (ignoring frictional losses) is

$$\dot{W}_{x_{max}} = \dot{m}g[H_1 - H_2], \tag{2.20b}$$

where $gH = p/\rho + \tfrac{1}{2}c^2 + gz$.

Steam and gas turbines

Figure 2.5a shows a Mollier diagram representing the expansion process through an adiabatic turbine. Line 1–2 represents the actual expansion and line 1–2s the ideal or reversible expansion. The fluid velocities at entry to and at exit from a turbine

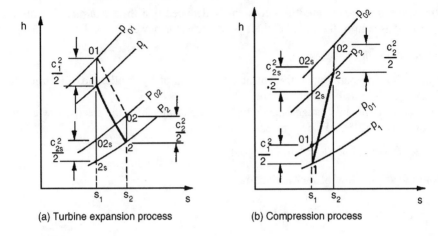

FIG. 2.5. Enthalpy-entropy diagrams for turbines and compressors.

may be quite high and the corresponding kinetic energies may be significant. On the other hand, for a compressible fluid the potential energy terms are usually negligible. Hence the *actual* turbine rotor *specific work*

$$\Delta W_x = \dot{W}_x/\dot{m} = h_{01} - h_{02} = (h_1 - h_2) + \tfrac{1}{2}(c_1^2 - c_2^2)$$

Similarly, the *ideal* turbine rotor specific work between the same two pressures is

$$\Delta W_{\max} = \dot{W}_{x_{\max}}/\dot{m} = h_{01} - h_{02s} = (h_1 - h_{2s}) + \tfrac{1}{2}(c_1^2 - c_{2s}^2).$$

In Figure 2.5a the actual turbine work/unit mass of fluid is the stagnation enthalpy change between state points 01 and 02 which lie on the stagnation pressure lines p_{01} and p_{02} respectively. The ideal turbine work per unit mass of fluid is the stagnation enthalpy change during the *isentropic process* between the same two pressures. The kinetic energy of the fluid at the end of the ideal process $\tfrac{1}{2}c_{2s}^2$ is not, however, the same as that at the end of the actual process $\tfrac{1}{2}c_2^2$. This may be adduced as follows. Taking for simplicity a perfect gas, then $h = C_p T$ and $p/\rho = RT$. Consider the constant pressure line p_2 (Figure 2.5a); as $T_2 > T_{2s}$ then $\rho_{2s} > \rho_2$. From continuity $\dot{m}/A = \rho c$ and since we are dealing with the same area, $c_2 > c_{2s}$, and the kinetic energy terms are not equal. The difference in practice is usually negligible and often ignored.

There are several ways of expressing efficiency, the choice of definition depending largely upon whether the *exit kinetic energy* is usefully employed or is wasted. An example where the exhaust kinetic energy is not wasted is from the last stage of an aircraft gas turbine where it contributes to the jet propulsive thrust. Likewise, the exit kinetic energy from one stage of a multistage turbine where it is used in the next stage, provides another example. For these two cases the turbine and stage adiabatic efficiency η, is the *total-to-total efficiency* and is defined as

$$\eta_{tt} = \Delta W_x/\Delta W_{x_{\max}} = (h_{01} - h_{02})/(h_{01} - h_{02s}). \tag{2.21}$$

If the difference between the inlet and outlet kinetic energies is small, i.e. $\frac{1}{2}c_1^2 \doteq \frac{1}{2}c_2^2$, then

$$\eta_{tt} = (h_1 - h_2)/(h_1 - h_{2s}) \tag{2.21a}$$

When the exhaust kinetic energy is not usefully employed and entirely wasted, the relevant adiabatic efficiency is the *total-to-static efficiency* η_{ts}. In this case the ideal turbine work is that obtained between state points 01 and 2s. Thus

$$\eta_{ts} = (h_{01} - h_{02})/(h_{01} - h_{02s} + \tfrac{1}{2}c_{2s}^2)$$
$$= (h_{01} - h_{02})/(h_{01} - h_{2s}). \tag{2.22}$$

If the difference between inlet and outlet kinetic energies is small, eqn. (2.22) becomes

$$\eta_{ts} = (h_1 - h_2)/(h_1 - h_{2s} + \tfrac{1}{2}c_1^2). \tag{2.22a}$$

A situation where the outlet kinetic energy is wasted is a turbine exhausting directly to the surroundings rather than through a diffuser. For example, auxiliary turbines used in rockets often do not have exhaust diffusers because the disadvantages of increased mass and space utilisation are greater than the extra propellant required as a result of reduced turbine efficiency.

Hydraulic turbines

When the working fluid is a liquid, the turbine hydraulic efficiency η_h, is defined as the work supplied by the rotor in unit time divided by the hydrodynamic energy difference of the fluid per unit time, i.e.

$$\eta_h = \frac{\Delta W_x}{\Delta W_{x_{\max}}} = \frac{\Delta W_x}{g(H_1 - H_2)}. \tag{2.23}$$

Efficiency of compressors and pumps

The isentropic *efficiency* η_c of a compressor or the *hydraulic efficiency* of a pump η_h is broadly defined as,

$$\eta_c(\text{or } \eta_h) = \frac{\text{useful (hydrodynamic) energy input to fluid in unit time}}{\text{power input to rotor}}.$$

The power input to the rotor (or impeller) is always less than the power supplied at the coupling because of external energy losses in the bearings and glands, etc. Thus, the overall efficiency of the compressor or pump is

$$\eta_o = \frac{\text{useful (hydrodynamic) energy input to fluid in unit time}}{\text{power input to coupling of shaft}}$$

Hence the mechanical efficiency is

$$\eta_m = \eta_o/\eta_c(\text{or } \eta_o/\eta_h).$$

In eqn. (2.19), for a compressor or pump process, replace $-\mathrm{d}\dot{W}_x$ with $\mathrm{d}\dot{W}_c$ and rearrange the inequality to give the incremental work input

$$\mathrm{d}\dot{W}_c \geq \dot{m}\left[\frac{1}{\rho}\mathrm{d}p + \frac{1}{2}\mathrm{d}(c^2) + g\mathrm{d}z\right]. \tag{2.24}$$

The student should carefully check the fact that the rhs of this inequality is *positive*, working from eqn. (2.19)

For a complete adiabatic compression process going from state 1 to state 2, the overall work input rate is

$$\dot{W}_c \geq \dot{m}\left[\int_1^2 \frac{\mathrm{d}p}{\rho} + \frac{1}{2}(c_2^2 - c_1^2) + g(z_2 - z_1)\right]. \tag{2.25}$$

For the corresponding *reversible* adiabatic compression process, noting that $T\mathrm{d}s = 0 = \mathrm{d}h - \mathrm{d}p/\rho$, the minimum work input rate is

$$\dot{W}_{c\,min} = \dot{m}\int_1^{2s}\left[\mathrm{d}h + \frac{1}{2}\mathrm{d}c^2 + g\mathrm{d}z\right] = \dot{m}[(h_{02s} - h_{01}) + g(z_2 - z_1)]. \tag{2.26}$$

From the steady flow energy equation, for an adiabatic process in a compressor

$$\dot{W}_c = \dot{m}(h_{02} - h_{01}). \tag{2.27}$$

Figure 2.5b shows a Mollier diagram on which the actual compression process is represented by the state change $1-2$ and the corresponding ideal process by $1-2s$. For an adiabatic compressor the only meaningful efficiency is the total-to-total efficiency which is

$$\eta_c = \frac{\text{minimum adiabatic work input per unit time}}{\text{actual adiabatic work input to rotor per unit time}}$$

$$= \frac{h_{02s} - h_{01}}{h_{02} - h_{01}}. \tag{2.28}$$

If the difference between inlet and outlet kinetic energies is small, $\frac{1}{2}c_1^2 \doteq \frac{1}{2}c_2^2$ and

$$\eta_c = \frac{h_{2s} - h_1}{h_2 - h_1}. \tag{2.28a}$$

For *incompressible* flow, eqn. (2.25) gives

$$\Delta W_p = \dot{W}_p/\dot{m} \geq [(p_2 - p_1)/\rho + \frac{1}{2}(c_2^2 - c_1^2) + g(z_2 - z_1)] \geq g[H_2 - H_1].$$

For the ideal case with no fluid friction

$$\Delta W_{p\,min} = g[H_2 - H_1]. \tag{2.29}$$

For a pump the hydraulic efficiency is defined as

$$\eta_h = \frac{\Delta W_{p\,min}}{\Delta W_p} = \frac{g[H_2 - H_1]}{\Delta W_p}. \tag{2.30}$$

Small stage or polytropic efficiency

The isentropic *efficiency* described in the preceding section, although fundamentally valid, can be misleading if used for comparing the efficiencies of turbomachines of differing pressure ratios. Now any turbomachine may be regarded as being composed of a large number of very small stages irrespective of the actual number of stages in the machine. If each small stage has the same efficiency, then the isentropic efficiency of the whole machine will be different from the small stage efficiency, the difference depending upon the pressure ratio of the machine. This perhaps rather surprising result is a manifestation of a simple thermodynamic effect concealed in the expression for isentropic efficiency and is made apparent in the following argument.

Compression process

Figure 2.6 shows an enthalpy–entropy diagram on which adiabatic compression between pressures p_1 and p_2 is represented by the change of state between points 1 and 2. The corresponding reversible process is represented by the isentropic line 1 to $2s$. It is assumed that the compression process may be divided up into a large number of small stages of equal efficiency η_p. For each small stage the actual work input is δW and the corresponding ideal work in the isentropic process is δW_{min}. With the notation of Figure 2.6,

$$\eta_p = \frac{\delta W_{min}}{\delta W} = \frac{h_{xs} - h_1}{h_x - h_1} = \frac{h_{ys} - h_x}{h_y - h_x} = \cdots$$

Since each small stage has the same efficiency, then $\eta_p = (\Sigma \delta W_{min} / \Sigma \delta W)$ is also true.

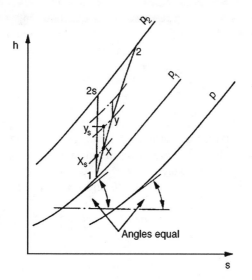

FIG. 2.6. Compression process by small stages.

From the relation $T\mathrm{d}s = \mathrm{d}h - v\mathrm{d}p$, for a constant pressure process $(\partial h / \partial s)_{p_1} = T$. This means that the higher the fluid temperature the *greater* is the slope of the constant pressure lines on the Mollier diagram. For a gas where h is a function of T, constant pressure lines diverge and the slope of the line p_2 is greater than the slope of line p_1 at the same value of entropy. At equal values of T, constant pressure lines are of equal slope as indicated in Figure 2.6. For the special case of a *perfect gas* (where C_p is constant), $C_p(\mathrm{d}T/\mathrm{d}s) = T$ for a constant pressure process. Integrating this expression results in the equation for a constant pressure line, $s = C_p \log T + \text{constant}$.

Returning now to the more general case, since

$$\Sigma \delta W = \{(h_x - h_1) + (h_y - h_x) + \cdots\} = (h_2 - h_1),$$

then

$$\eta_p = [(h_{xs} - h_1) + (h_{ys} - h_x) + \cdots]/(h_2 - h_1).$$

The adiabatic efficiency of the *whole* compression process is

$$\eta_c = (h_{2s} - h_1)/(h_2 - h_1).$$

Because of the divergence of the constant pressure lines

$$\{(h_{xs} - h_1) + (h_{ys} - h_x) + \cdots\} > (h_{2s} - h_1),$$

i.e.

$$\Sigma \delta W_{\min} > W_{\min}.$$

Therefore,

$$\eta_p > \eta_c.$$

Thus, for a compression process the isentropic efficiency of the machine is *less* than the small stage efficiency, the difference being dependent upon the divergence of the constant pressure lines. Although the foregoing discussion has been in terms of static states it can be regarded as applying to stagnation states if the inlet and outlet kinetic energies from each stage are equal.

Small stage efficiency for a perfect gas

An explicit relation can be readily derived for a perfect gas (C_p is constant) between small stage efficiency, the overall isentropic efficiency and pressure ratio. The analysis is for the limiting case of an infinitesimal compressor stage in which the incremental change in pressure is $\mathrm{d}p$ as indicated in Figure 2.7. For the actual process the incremental enthalpy rise is $\mathrm{d}h$ and the corresponding ideal enthalpy rise is $\mathrm{d}h_{is}$.

The polytropic efficiency for the small stage is

$$\eta_p = \frac{\mathrm{d}h_{is}}{\mathrm{d}h} = \frac{v\mathrm{d}p}{C_p\mathrm{d}T}, \qquad (2.31)$$

since for an isentropic process $T\mathrm{d}s = 0 = \mathrm{d}h_{is} - v\mathrm{d}p$.

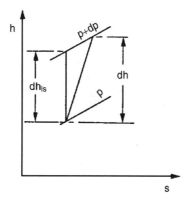

FIG. 2.7. Incremental change of state in a compression process.

Substituting $v = RT/p$ in eqn. (2.31), then

$$\eta_p = \frac{R}{C_p} \frac{T}{p} \frac{dp}{dT}$$

and hence

$$\frac{dT}{T} = \frac{(\gamma - 1)}{\gamma \eta_p} \frac{dp}{p} \tag{2.32}$$

as $\quad C_p = \gamma R/(\gamma - 1)$.

Integrating eqn. (2.32) across the whole compressor and taking equal efficiency for each infinitesimal stage gives,

$$\frac{T_2}{T_1} = \left(\frac{p_2}{p_1}\right)^{(\gamma - 1)/\eta_p \gamma} \tag{2.33}$$

$$\ln\left(\frac{T_2}{T_1}\right) = \frac{(\gamma-1)}{\eta_p \gamma} \ln\left(\frac{p_2}{p_1}\right)$$

$$\eta_p = \left(\frac{\gamma-1}{\gamma}\right) \cdot \frac{\ln\left(p_2/p_1\right)}{\ln\left(T_2/T_1\right)}$$

Now the isentropic efficiency for the whole compression process is

$$\eta_c = (T_{2s} - T_1)/(T_2 - T_1) \quad \leftarrow ? \tag{2.34}$$

if it is assumed that the velocities at inlet and outlet are equal.

For the *ideal* compression process put $\eta_p = 1$ in eqn. (2.32) and so obtain

$$\frac{T_{2s}}{T_1} = \left(\frac{p_2}{p_1}\right)^{(\gamma - 1)/\gamma} \tag{2.35}$$

which is also obtainable from $pv^\gamma = $ constant and $pv = RT$. Substituting eqns. (2.33) and (2.35) into eqn. (2.34) results in the expression

$$\eta_c = \left[\left(\frac{p_2}{p_1}\right)^{(\gamma - 1)/\gamma} - 1\right] \Big/ \left[\left(\frac{p_2}{p_1}\right)^{(\gamma - 1)/\eta_p \gamma} - 1\right]. \tag{2.36}$$

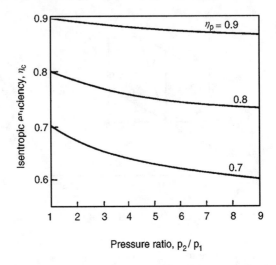

FIG. 2.8. Relationship between isentropic (overall) efficiency, pressure ratio and small stage (polytropic) efficiency for a compressor ($\gamma = 1.4$).

Values of "overall" isentropic efficiency have been calculated using eqn. (2.36) for a range of pressure ratio and different values of η_p, and are plotted in Figure 2.8. This figure amplifies the observation made earlier that the isentropic efficiency of a finite compression process is *less* than the efficiency of the small stages. Comparison of the isentropic efficiency of two machines of different pressure ratios is not a valid procedure since, for equal polytropic efficiency, the compressor with the highest pressure ratio is penalised by the *hidden* thermodynamic effect.

The term *polytropic* used above arises in the context of a reversible compressor compressing a gas from the same initial state to the same final state as the irreversible adiabatic compressor but obeying the relation pv^n = constant. The index n is called the *polytropic index*. Since an increase in entropy occurs for the change of state in both compressors, for the reversible compressor this is only possible if there is a reversible heat transfer $dQ_R = T ds$. Proceeding farther, it follows that the value of the index n must always exceed that of γ. This is clear from the following argument. For the polytropic process,

$$dQ_R = du + p dv.$$

$$= \frac{C_v}{R} d(pv) + p dv.$$

Using pv^n = constant and $C_v = R/(\gamma - 1)$, after some manipulation the expression $dQ_R = (\gamma - n)/(\gamma - 1) p dv$ is derived. For a compression process $dv < 0$ and $dQ_R > 0$ then $n > \gamma$. For an expansion process $dv > 0$, $dQ_R < 0$ and again $n > \gamma$.

EXAMPLE 2.1. An axial flow air compressor is designed to provide an overall total-to-total pressure ratio of 8 to 1. At inlet and outlet the stagnation temperatures are 300 K and 586.4 K, respectively.

Determine the overall total-to-total efficiency and the polytropic efficiency for the compressor. Assume that γ for air is 1.4.

Solution. From eqn. (2.28), substituting $h = C_p T$, the efficiency can be written as,

$$\eta_C = \frac{T_{02s} - T_{01}}{T_{02} - T_{01}} = \frac{\left(\dfrac{p_{02}}{p_{01}}\right)^{(\gamma-1)/\gamma} - 1}{T_{02}/T_{01} - 1} = \frac{8^{1/3.5} - 1}{586 \cdot 4/300 - 1} = 0.85.$$

From eqn. (2.33), taking logs of both sides and re-arranging, we get,

$$\eta_p = \frac{\gamma - 1}{\gamma} \frac{\ln(p_{02}/p_{01})}{\ln(T_{02}/T_{01})} = \frac{1}{3.5} \times \frac{\ln 8}{\ln 1.9547} = 0.8865$$

Turbine polytropic efficiency

A similar analysis to the compression process can be applied to a perfect gas expanding through an adiabatic turbine. For the turbine the appropriate expressions for an expansion, from a state 1 to a state 2, are

$$\frac{T_2}{T_1} = \left(\frac{p_2}{p_1}\right)^{\eta_p(\gamma-1)/\gamma} \tag{2.37}$$

$$\eta_t = \left[1 - \left(\frac{p_2}{p_1}\right)^{\eta_p(\gamma-1)/\gamma}\right] \Big/ \left[1 - \left(\frac{p_2}{p_1}\right)^{(\gamma-1)/\gamma}\right]. \tag{2.38}$$

The derivation of these expressions is left as an exercise for the student. "Overall" isentropic efficiencies have been calculated for a range of pressure ratio and different polytropic efficiencies and are shown in Figure 2.9. The most notable feature of these results is that, in contrast with a compression process, for an expansion, isentropic efficiency *exceeds* small stage efficiency.

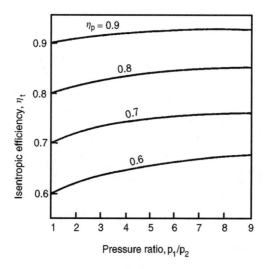

FIG. 2.9. Turbine isentropic efficiency against pressure ratio for various polytropic efficiencies ($\gamma = 1.4$).

Reheat factor

The foregoing relations obviously cannot be applied to steam turbines as vapours do not in general obey the gas laws. It is customary in steam turbine practice to use a *reheat factor* R_H as a measure of the inefficiency of the complete expansion. Referring to Figure 2.10, the expansion process through an adiabatic turbine from state 1 to state 2 is shown on a Mollier diagram, split into a number of small stages. The reheat factor is defined as

$$R_H = [(h_1 - h_{xs}) + (h_x - h_{ys}) + \cdots]/(h_1 - h_{2s}) = (\Sigma \Delta h_{is})/(h_1 - h_{2s}).$$

Due to the gradual divergence of the constant pressure lines on a Mollier chart, R_H is always greater than unity. The actual value of R_H for a large number of stages will depend upon the position of the expansion line on the Mollier chart and the overall pressure ratio of the expansion. In normal steam turbine practice the value of R_H is usually between 1.03 and 1.08. For an isentropic expansion in the superheated region with pv^n = constant, the tables of Rogers and Mayhew (1995) give a value for $n = 1.3$. Assuming this value for n is valid, the relationship between reheat factor and pressure ratio for various fixed values of the polytropic efficiency has been calculated and is shown in Figure 2.11.

Now since the isentropic efficiency of the turbine is

$$\eta_t = \frac{h_1 - h_2}{h_1 - h_{2s}} = \frac{h_1 - h_2}{\Sigma \Delta h_{is}} \cdot \frac{\Sigma \Delta h_{is}}{h_1 - h_{2s}}$$

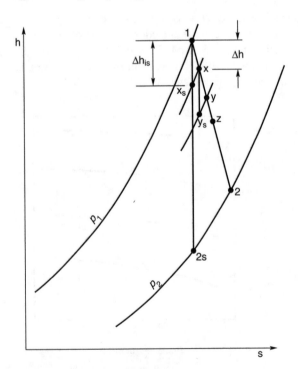

FIG. 2.10. Mollier diagram showing expansion process through a turbine split up into a number of small stages.

then

$$\eta_t = \eta_p R_H \tag{2.39}$$

which establishes the connection between polytropic efficiency, reheat factor and turbine isentropic efficiency.

Nozzle efficiency

In a large number of turbomachinery components the flow process can be regarded as a purely nozzle flow in which the fluid receives an acceleration as a result of a drop in pressure. Such a nozzle flow occurs at entry to all turbomachines and in the stationary blade rows in turbines. In axial machines the expansion at entry is assisted by a row of stationary blades (called *guide vanes* in compressors and *nozzles* in turbines) which direct the fluid on to the rotor with a large swirl angle. Centrifugal compressors and pumps, on the other hand, often have no such provision for flow guidance but there is still a velocity increase obtained from a contraction in entry flow area.

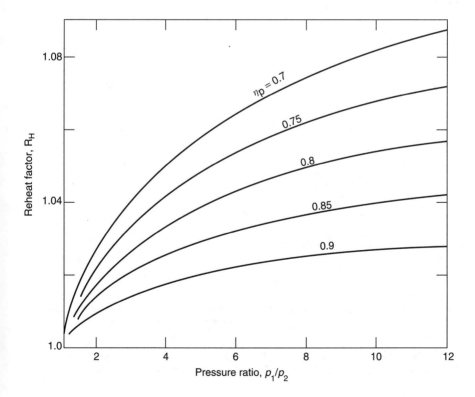

FIG. 2.11. Relationship between reheat factor, pressure ratio and polytropic efficiency ($n = 1.3$).

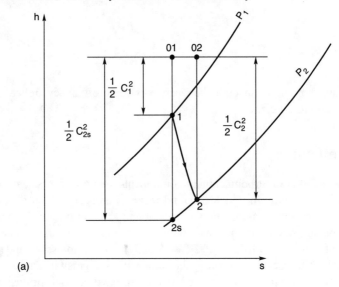

(a)

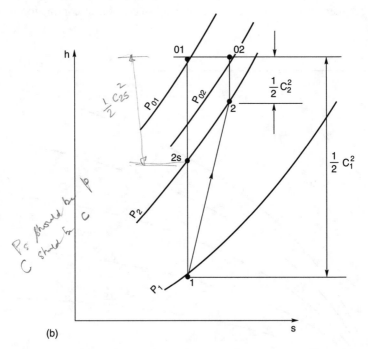

(b)

FIG. 2.12. Mollier diagrams for the flow processes through a nozzle and a diffuser: (a) nozzle; (b) diffuser.

Figure 2.12a shows the process on a Mollier diagram, the expansion proceeding from state 1 to state 2. It is assumed that the process is steady and adiabatic such that $h_{01} = h_{02}$.

According to Horlock (1966), the most frequently used definition of nozzle efficiency, η_N is, the ratio of the final kinetic energy per unit mass to the maximum theoretical kinetic energy per unit mass obtained by an isentropic expansion to the same back pressure, i.e.

$$\eta_N = (\tfrac{1}{2}c_2^2)/(\tfrac{1}{2}c_{2s}^2) = (h_{01} - h_2)/(h_{01} - h_{2s}). \tag{2.40}$$

Nozzle efficiency is sometimes expressed in terms of various loss or other coefficients. An enthalpy loss coefficient for the nozzle can be defined as

$$\zeta_N = (h_2 - h_{2s})/(\tfrac{1}{2}c_2^2), \tag{2.41}$$

and, also, a velocity coefficient for the nozzle,

$$K_N = c_2/c_{2s}. \tag{2.42}$$

It is easy to show that these definitions are related to one another by

$$\eta_N = 1/(1 + \zeta_N) = K_N^2. \tag{2.43}$$

EXAMPLE 2.2. Gas enters the nozzles of a turbine stage at a stagnation pressure and temperature of 4.0 bar and 1200 K and leaves with a velocity of 572 m/s and at a static pressure of 2.36 bar. Determine the nozzle efficiency assuming the gas has the average properties over the temperature range of the expansion of $C_p = 1.160$ kJ/kg K and $\gamma = 1.33$.

Solution. From eqns. (2.40) and (2.35) the nozzle efficiency becomes

$$\eta_N = \frac{1 - T_2/T_{01}}{1 - T_{2s}/T_{01}} = \frac{1 - T_2/T_{01}}{1 - (p_2/p_{01})^{(\gamma-1)/\gamma}}.$$

Assuming adiabatic flow $(T_{02} = T_{01})$:

$$T_2 = T_{02} - \tfrac{1}{2}c_2^2/C_p = 1200 - \tfrac{1}{2} \times 572^2/1160 = 1059 \text{ K},$$

and thus

$$\eta_N = \frac{1 - 1059/1200}{1 - (2.36/4)^{0.33/1.33}} = \frac{0.1175}{0.12271} = \underline{0.9576}.$$

Diffusers

A diffuser is a component of a fluid flow system designed to reduce the flow velocity and thereby increase the fluid pressure. All turbomachines and many other flow systems incorporate a diffuser (e.g. closed circuit wind tunnels, the duct between the compressor and burner of a gas turbine engine, the duct at exit from a gas turbine connected to the jet pipe, the duct following the impeller of a centrifugal compressor, etc.). Turbomachinery flows are, in general, subsonic ($M < 1$) and the diffuser can be represented as a channel *diverging* in the direction of flow (see Figure 2.13).

The basic diffuser is a geometrically simple device with a rather long history of investigation by many researchers. The long timespan of the research is an indicator that the fluid mechanical processes within it are complex, the research rather more

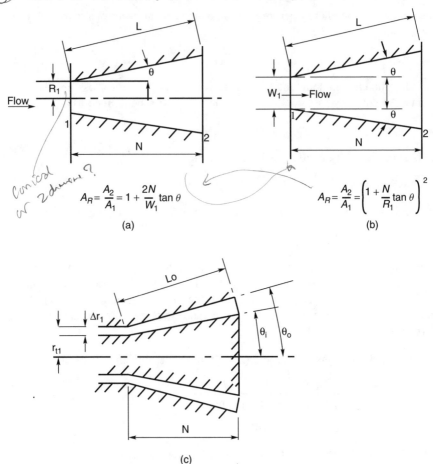

$$A_R = \frac{A_2}{A_1} = 1 + \frac{2N}{W_1}\tan\theta$$

(a)

$$A_R = \frac{A_2}{A_1} = \left(1 + \frac{N}{R_1}\tan\theta\right)^2$$

(b)

Conical or 2 dimensional ?

(c)

FIG. 2.13. Some subsonic diffuser geometries and their parameters: (a) two-dimensional; (b) conical; (c) annular.

difficult than might be anticipated, and some aspects of the flow processes are still not fully understood. There is now a vast literature about the flow in diffusers and their performance. Only a few of the more prominent investigations are referenced here. A noteworthy and recommended reference, however, which reviews many diverse and recondite aspects of diffuser design and flow phenomena is that of Kline and Johnson (1986).

The primary fluid mechanical problem of the diffusion process is caused by the tendency of the boundary layers to separate from the diffuser walls if the rate of diffusion is too rapid. The result of too rapid diffusion is always large losses in stagnation pressure. On the other hand, if the rate of diffusion is too low, the fluid is exposed to an excessive length of wall and fluid friction losses become predominant. Clearly, there must be an *optimum rate of diffusion* between these two extremes for which the losses are minimised. Test results from many sources indicate that an included angle of about $2\theta = 7$ degrees gives the optimum recovery for both two-dimensional and conical diffusers.

Diffuser performance parameters

The diffusion process can be represented on a Mollier diagram, Figure 2.12b, by the change of state from point 1 to point 2, and the corresponding changes in pressure and velocity from p_1 and c_1 to p_2 and c_2. The actual performance of a diffuser can be expressed in several different ways:

(1) as the ratio of the actual enthalpy change to the isentropic enthalpy change;
(2) as the ratio of an actual pressure rise coefficient to an ideal pressure rise co-efficient.

For steady and adiabatic flow in stationary passages, $h_{01} = h_{02}$, so that

$$h_2 - h_1 = \tfrac{1}{2}(c_1^2 - c_2^2). \qquad \text{isentropic ?} \qquad (2.44a)$$

For the equivalent reversible adiabatic process from state point 1 to state point 2s,

$$(h_{2s} - h_1) = \tfrac{1}{2}(c_1^2 - c_{2s}^2). \qquad (2.44b)$$

A *diffuser efficiency*, η_D, also called the *diffuser effectiveness*, can be defined as

$$\eta_D = (h_{2s} - h_1)/(h_2 - h_1) = (c_1^2 - c_{2s}^2)/(c_1^2 - c_2^2). \qquad (2.45a)$$

In a low speed flow or a flow in which the density ρ can be considered nearly constant,

$$h_{2s} - h_1 = (p_2 - p_1)/\rho \quad \leftarrow ? \qquad dh = \upsilon\, dp = \tfrac{1}{\rho}\, dp \; - \; \text{see } p36$$

so that the diffuser efficiency can be written

$$\eta_D = 2(p_2 - p_1)/\{\rho(c_1^2 - c_2^2)\}. \qquad (2.45b)$$

Equation (2.45a) can be expressed entirely in terms of pressure differences, by writing

$$h_2 - h_{2s} = (h_2 - h_1) - (h_{2s} - h_1)$$
$$= \tfrac{1}{2}(c_1^2 - c_2^2) - (p_2 - p_1)/\rho = (p_{01} - p_{02})/\rho,$$

then, with eqn. (2.45a),

$$\eta_D = \frac{(h_{2s} - h_1)}{(h_{2s} - h_1) - (h_{2s} - h_2)} = \frac{1}{1 - (h_{2s} - h_2)/(h_{2s} - h_1)}$$
$$= \frac{1}{1 + (p_{01} - p_{02})/(p_2 - p_1)}. \qquad (2.46)$$

Alternative expressions for diffuser performance

(1) A *pressure rise coefficient* C_p can be defined:

$$C_p = (p_2 - p_1)/q_1, \qquad (2.47a)$$

where $q_1 = \tfrac{1}{2}\rho c_1^2$.

For an *incompressible* flow through the diffuser the energy equation can be written as

$$p_1/\rho + \tfrac{1}{2}c_1^2 = p_2/\rho + \tfrac{1}{2}c_2^2 + \Delta p_0/\rho, \qquad (2.48)$$

where the loss in total pressure, $\Delta p_0 = p_{01} - p_{02}$. Also, using the continuity equation across the diffuser, $c_1 A_1 = c_2 A_2$, we obtain

$$c_1/c_2 = A_2/A_1 = A_R, \tag{2.49}$$

where A_R is the area ratio of the diffuser.

From eqn. (2.48), by setting Δp_0 to zero and with eqn. (2.49), it is easy to show that the *ideal pressure rise coefficient* is

$$C_{pi} = 1 - (c_2/c_1)^2 = 1 - \left[\frac{1}{A_R^2}\right] \tag{2.47b}$$

Thus, eqn. (2.48) can be rewritten as

$$C_p = C_{pi} - \Delta p_0/q_1. \tag{2.50}$$

Using the definition given in eqn. (2.46), then the diffuser efficiency (referred to as the *diffuser effectiveness* by Sovran and Klomp (1967)), is

$$\eta_D = C_p/C_{pi}. \tag{2.51}$$

(2) A *total pressure recovery factor*, p_{02}/p_{01}, is sometimes used as an indicator of the performance of diffusers. From eqn. (2.45a), the diffuser efficiency can be written

$$\eta_D = (T_{2s}/T_1 - 1)/(T_2 T_1 - 1). \tag{2.52}$$

For the isentropic process 1–2s:

$$\frac{T_{2s}}{T_1} = \left[\frac{p_2}{p_1}\right]^{(\gamma-1)/\gamma}.$$

For the constant temperature process 01–02, $Tds = -dp/\rho$ which, when combined with the gas law, $p/\rho = RT$, gives $ds = -Rdp/p$:

$$\therefore \Delta s = R \ln\left(\frac{p_{01}}{p_{02}}\right).$$

For the constant pressure process 2s–2, $Tds = dh = C_p dT$,

$$\therefore \Delta s = C_p \ln\left(\frac{T_2}{T_{2s}}\right).$$

Equating these expressions for the entropy increase and using $R/C_p = (\gamma - 1)/\gamma$, then

$$\frac{T_2}{T_{2s}} = \left(\frac{p_{01}}{p_{02}}\right)^{(\gamma-1)/\gamma},$$

$$\therefore \frac{T_2}{T_1} = \left(\frac{T_2}{T_{2s}}\right)\left(\frac{T_{2s}}{T_1}\right) = \left[\left(\frac{p_{01}}{p_{02}}\right)\left(\frac{p_2}{p_1}\right)\right]^{(\gamma-1)/\gamma}.$$

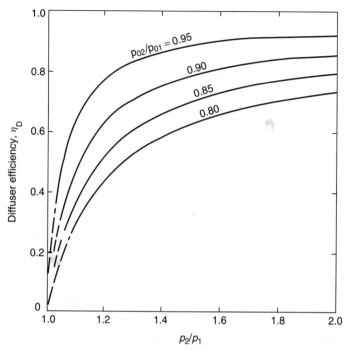

FIG. 2.14. Variation of diffuser efficiency with static pressure ratio for constant values of total pressure recovery factor ($\gamma = 1.4$).

Substituting these two expressions into eqn. (2.52):

$$\eta_D = \frac{(p_2/p_1)^{(\gamma-1)/\gamma} - 1}{[(p_{01}/p_{02})(p_2/p_1)]^{(\gamma-1)/\gamma} - 1}. \tag{2.53}$$

The variation of η_D as a function of the static pressure ratio, p_2/p_1, for specific values of the total pressure recovery factor, p_{02}/p_{01}, is shown in Figure 2.14.

Some remarks on diffuser performance

It was pointed out by Sovran and Klomp (1967) that the uniformity or steadiness of the flow at the diffuser exit is as important as the reduction in flow velocity (or the static pressure rise) produced. This is particularly so in the case of a compressor located at the diffuser exit since the compressor performance is sensitive to non-uniformities in velocity in its inlet flow. Figure 2.15, from Sovran and Klomp (1967), shows the occurrence of flow unsteadiness and/or non-uniform flow at the exit from two-dimensional diffusers (correlated originally by Kline, Abbott and Fox 1959). Four different flow regimes exist, three of which have steady or reasonably steady flow. The region of "no appreciable stall" is steady and uniform. The region marked "large transitory stall" is unsteady and non-uniform, while the "fully-developed" and "jet flow" regions are reasonably steady but very non-uniform.

The line marked a–a will be of interest in turbomachinery applications. However, a sharply marked transition does not exist and the definition of an appropriate line

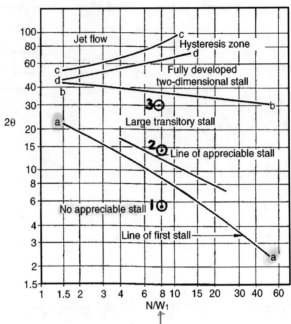

FIG. 2.15. Flow regime chart for two-dimensional diffusers (adapted from Sovran and Klomp 1967).

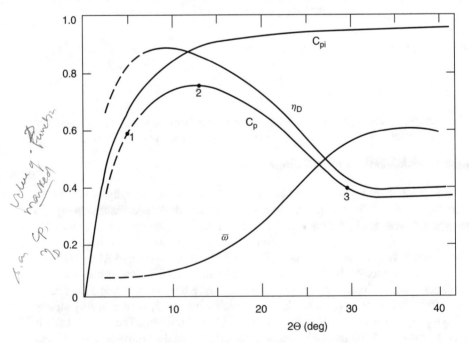

FIG. 2.16. Typical diffuser performance curves for a two-dimensional diffuser, with $L/W_1 = 8.0$ (adapted from Kline *et al.* 1959).

involves a certain degree of arbitrariness and subjectivity on the occurrence of "first stall".

Figure 2.16 shows typical performance curves for a rectangular diffuser with a fixed sidewall to length ratio, $L/W_1 = 8.0$, given in Kline *et al.* (1959). On the line labelled C_p, points numbered 1, 2 and 3 are shown. These same numbered points are redrawn onto Figure 2.15 to show where they lie in relation to the various flow regimes. Inspection of the location of point 2 shows that optimum recovery at constant length occurs slightly above the line marked "no appreciable stall". The performance of the diffuser between points 2 and 3 in Figure 2.16 is shows a very significant deterioration and is in the regime of large amplitude, very unsteady flow.

Maximum pressure recovery

From an inspection of eqn. (2.46) it will be observed that when diffuser efficiency η_D is a maximum, the total pressure loss is a minimum for a given rise in static pressure. Another optimum problem is the requirement of *maximum pressure recovery* for a given diffuser length in the flow direction *regardless of the area ratio $A_r = A_2/A_1$*. This may seem surprising but, in general, this optimum condition produces a different diffuser geometry from that needed for optimum efficiency. This can be demonstrated by means of the following considerations.

From eqn. (2.51), taking logs of both sides and differentiating, we get:

$$\frac{\partial}{\partial \theta}(\ln \eta_D) = \frac{\partial}{\partial \theta}(\ln C_p) - \frac{\partial}{\partial \theta}(\ln C_{pi}).$$

Setting the L.H.S to zero for the condition of maximum η_D, then

$$\frac{1}{C_p}\frac{\partial C_p}{\partial \theta} = \frac{1}{C_{pi}}\frac{\partial C_{pi}}{\partial \theta}. \tag{2.54}$$

Thus, at the maximum efficiency the fractional rate of increase of C_p with a change in θ is equal to the fractional rate of increase of C_{pi} with a change in θ. At this point C_p is positive and, by definition, both C_{pi} and $\partial C_p/\partial \theta$ are also positive. Equation (2.54) shows that $\partial C_p/\partial \theta > 0$ at the maximum efficiency point. Clearly, C_p *cannot* be at its maximum when η_D is at its peak value! What happens is that C_p *continues to increase* until $\partial C_p/\partial \theta = 0$, as can be seen from the curves in Figure 2.16.

Now, upon differentiating eqn. (2.50) with respect to θ and setting the lhs to zero, the condition for maximum C_p is obtained, namely

$$\frac{\partial C_{pi}}{\partial \theta} = \frac{\partial}{\partial \theta}(\Delta p_0/q_1).$$

Thus, as the diffuser angle is increased beyond the divergence which gave maximum efficiency, the actual pressure rise will continue to rise until the additional losses in total pressure balance the theoretical gain in pressure recovery produced by the increased area ratio.

Diffuser design calculation

The performance of a conical diffuser has been chosen for this purpose using data presented by Sovran and Klomp (1967). This is shown in Figure 2.17 as contour

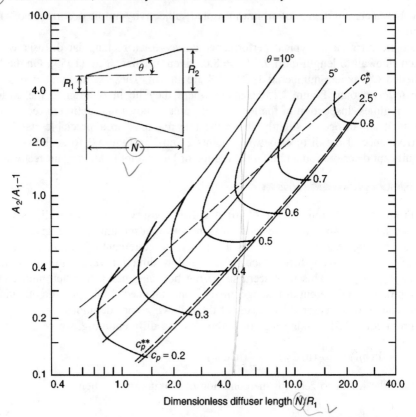

FIG. 2.17. Performance chart for conical diffusers, $B_1 \cong 0.02$. (adapted from Sovran and Klomp 1967).

plots of C_p in terms of the geometry of the diffuser, L/R_1 and the area ratio A_R. Two optimum diffuser lines, useful for design purposes, were added by the authors. The first is the line C_p^*, the locus of points which defines the diffuser area ratio A_R, producing the maximum pressure recovery for a prescribed non-dimensional length, L/R_1. The second is the line C_p^{**}, the locus of points defining the diffuser non-dimensional length, producing the maximum pressure recovery at a prescribed area ratio.

EXAMPLE 2.3. Design a conical diffuser to give maximum pressure recovery in a non-dimensional length $N/R_1 = 4.66$ using the data given in Figure 2.17.

Solution. From the graph, using log-linear scaling, the appropriate value of C_p is 0.6 and the corresponding value of A_R is 2.13. From eqn. (2.47b), $C_{pi} = 1 - (1/2 \cdot 13^2) = 0.78$. Hence, $\eta_D = 0.6/0.78 = \underline{0.77}$.

Transposing the expression given in Figure 2.13b, the included cone angle can be found:

$$2\theta = 2\tan^{-1}\{(A_R^{0.5} - 1)/(L/R_1)\} = \underline{11.26 \deg}.$$

EXAMPLE 2.4. Design a conical diffuser to give maximum pressure recovery at a prescribed area ratio $A_R = 1.8$ using the data given in Figure 2.17.

Solution. From the graph, $C_p = 0.6$ and $N/R_1 = 7.85$ (using log-linear scaling). Thus,

$$2\theta = 2\tan^{-1}\{(1.8^{0.5} - 1)/7.85\} = 5 \deg.$$

$$C_{pi} = 1 - (1/1.8^2) = 0.69 \text{ and } \eta_D = 0.6/0.69 = \underline{0.87}.$$

Analysis of a non-uniform diffuser flow

The actual pressure recovery produced by a diffuser of optimum geometry is known to be strongly affected by the shape of the velocity profile at inlet. A large reduction in the pressure rise which might be expected from a diffuser can result from inlet flow non-uniformities (e.g. wall boundary layers and, possibly, wakes from a preceding row of blades). Sovran and Klomp (1967) presented an incompressible flow analysis which helps to explain how this deterioration in performance occurs and some of the main details of their analysis are included in the following account.

The mass-averaged total pressure $\overline{p}_0$ at any cross-section of a diffuser can be obtained by integrating over the section area. For symmetrical ducts with straight centre lines the static pressure can be considered constant, as it is normally. Thus,

$$\overline{p}_0 = \int_A c(p + \tfrac{1}{2}\rho c^2)\mathrm{d}A \Big/ \int_A c\,\mathrm{d}A,$$

$$= p + \tfrac{1}{2}\rho \int_A c^3\mathrm{d}A \Big/ \int_A c\,\mathrm{d}A. \tag{2.55}$$

The average axial velocity U and the average dynamic pressure q at a section are

$$U = \frac{1}{A}\int_A c\,\mathrm{d}A \text{ and } q = \frac{1}{2}\rho U^2.$$

Substituting into eqn. (2.55),

$$\overline{p}_0 = p + \tfrac{1}{2}\rho U^3 \int_A \left(\frac{c}{U}\right)^3 \mathrm{d}A/UA$$

$$= p + \frac{q}{A}\int_A \left(\frac{c}{U}\right)^3 \mathrm{d}A = p + \alpha q, \tag{2.56}$$

where α is the kinetic energy flux coefficient of the velocity profile, i.e.

$$\alpha = \frac{1}{A}\int_A \left(\frac{c}{U}\right)^3 \mathrm{d}A = \frac{1}{Q}\int_A \left(\frac{c}{U}\right)^2 \mathrm{d}Q = \overline{c^2}/U^2, \tag{2.57}$$

where $\overline{c^2}$ is the mean square of the velocity in the cross-section and $Q = AU$, i.e.

$$\overline{c^2} = (1/Q)\int_A c^2\,\mathrm{d}Q.$$

From eqn. (2.56) the change in static pressure in found as

$$p_2 - p_1 = (\alpha_1 q_1 - \alpha_2 q_2) - (\overline{p}_{01} - \overline{p}_{02}). \tag{2.58}$$

From eqn. (2.51), with eqns. (2.47a) and (2.47b), the diffuser efficiency (or diffuser effectiveness) can now be written:

$$\eta_D = C_p / C_{pi} = \frac{(p_2 - p_1)/q_1}{1 - 1/A_R^2}.$$

Substituting eqn. (2.58) into the above expression,

$$\eta_D = \frac{\alpha_1 [1 - (\alpha_2/\alpha_1)/A_R^2]}{(1 - 1/A_R^2)} - \frac{\varpi}{(1 - 1/A_R^2)} \tag{2.59}$$

where ϖ is the total pressure loss coefficient for the whole diffuser, i.e.

$$\varpi = (\overline{p}_{01} - \overline{p}_{02})/q_1. \tag{2.60}$$

Equation (2.59) is particularly useful as it enables the separate effects due the changes in the velocity profile and total pressure losses on the diffuser effectiveness to be found. The first term in the equation gives the reduction in η_D caused by *insufficient flow diffusion*. The second term gives the reduction in η_D produced by viscous effects and represents *inefficient flow diffusion*. An assessment of the relative proportion of these effects on the effectiveness requires the accurate measurement of both the inlet and exit velocity profiles as well as the static pressure rise. Such complete data is seldom derived by experiments. However, Sovran and Klomp (1967) made the observation that there is a widely held belief that fluid mechanical losses are the primary cause of poor performance in diffusers. One of the important conclusions they drew from their work was that it is the thickening of the inlet boundary layer which is primarily responsible for the reduction in η_D. Thus, it is *insufficient* flow diffusion rather than *inefficient* flow diffusion which is often the cause of poor performance.

Some of the most comprehensive tests made of diffuser performance were those of Stevens and Williams (1980) who included traverses of the flow at inlet and at exit as well as careful measurements of the static pressure increase and total pressure loss in low speed tests on annular diffusers. In the following worked example, to illustrate the preceding theoretical analysis, data from this source has been used.

EXAMPLE 2.5. An annular diffuser with an area ratio, $A_R = 2.0$ is tested at low speed and the results obtained give the following data:

at entry, $\alpha_1 = 1.059, B_1 = 0.109$

at exit, $\alpha_2 = 1.543, B_2 = 0.364, C_p = 0.577$

Determine the diffuser efficiency.

NB B_1 and B_2 are the fractions of the area blocked by the wall boundary layers at inlet and exit (displacement thicknesses) and are included only to illustrate the profound effect the diffusion process has on boundary layer thickening.

Solution. From eqns. (2.47a) and (2.58):

$$C_p = \frac{p_2 - p_1}{q_1} = (\alpha_1 - \alpha_2/A_R^2) - \varpi$$

$$= 1.059 - 1.543/4 - 0.09 = \underline{0.583}.$$

Using eqn. (2.59) directly,

$$\eta_D = C_p/C_{pi} = C_p/(1 - 1/A_R^2) = 0.583/0.75$$

$$\therefore \eta_D = \underline{0.7777}.$$

Stevens and Williams observed that an incipient transitory stall was in evidence on the diffuser outer wall which affected the accuracy of the results. So, it is not surprising that a slight mismatch is evident between the above calculated result and the measured result.

References

Çengel, Y. A. and Boles, M. A. (1994). *Thermodynamics: An Engineering Approach.* (2nd edn). McGraw-Hill.

Japikse, D. (1984). *Turbomachinery Diffuser Design Technology*, DTS-1. Concepts ETI.

Kline, S. J. and Johnson, J. P. (1986). Diffusers – flow phenomena and design. In *Advanced Topics in Turbomachinery Technology. Principal Lecture Series, No. 2.* (D. Japikse, ed.) pp. 6–1 to 6–44, Concepts, ETI.

Kline, S. J., Abbott, D. E. and Fox, R. W. (1959). Optimum design of straight-walled diffusers. *Trans. Am. Soc. Mech. Engrs.*, Series D, **81**.

Horlock, J. H. (1966). *Axial Flow Turbines.* Butterworths. (1973 Reprint with corrections, Huntington, New York: Krieger.)

Reynolds, William C. and Perkins, Henry C. (1977). *Engineering Thermodynamics.* (2nd edn). McGraw-Hill.

Rogers, G. F. C. and Mayhew, Y. R. (1992). *Engineering Thermodynamics, Work and Heat Transfer.* (4th edn). Longman.

Rogers, G. F. C. and Mayhew, Y. R. (1995). *Thermodynamic and Transport Properties of Fluids (SI Units).* (5th edn). Blackwell.

Runstadler, P. W., Dolan, F. X. and Dean, R. C. (1975). *Diffuser Data Book.* Creare TN186.

Sovran, G. and Klomp, E. (1967). Experimentally determined optimum geometries for rectilinear diffusers with rectangular, conical and annular cross-sections. *Fluid Mechanics of Internal Flow*, Elsevier, pp. 270–319.

Stevens, S. J. and Williams, G. J. (1980). The influence of inlet conditions on the performance of annular diffusers. *J. Fluids Engineering, Trans. Am. Soc. Mech. Engrs.*, **102**, 357–63.

Problems

1. For the adiabatic expansion of a perfect gas through a turbine, show that the overall efficiency η_t and small stage efficiency η_p are related by

$$\eta_t = (1 - \varepsilon^{\eta_p})/(1 - \epsilon),$$

where $\epsilon = r^{(1-\gamma)/\gamma}$, and r is the expansion pressure ratio, γ is the ratio of specific heats.

An axial flow turbine has a small stage efficiency of 86%, an overall pressure ratio of 4.5 to 1 and a mean value of γ equal to 1.333. Calculate the overall turbine efficiency.

2. Air is expanded in a multi-stage axial flow turbine, the pressure drop across each stage being very small. Assuming that air behaves as a perfect gas with ratio of specific heats γ, derive pressure-temperature relationships for the following processes:

(i) reversible adiabatic expansion;

(ii) irreversible adiabatic expansion, with small stage efficiency η_p;

(iii) reversible expansion in which the heat loss in each stage is a constant fraction k of the enthalpy drop in that stage;

(iv) reversible expansion in which the heat loss is proportional to the absolute temperature T.

Sketch the first three processes on a T, s diagram.

If the entry temperature is 1100 K, and the pressure ratio across the turbine is 6 to 1, calculate the exhaust temperatures in each of these three cases. Assume that γ is 1.333, that $\eta_p = 0.85$, and that $k = 0.1$.

3. A multi-stage high-pressure steam turbine is supplied with steam at a stagnation pressure of 7 MPa. and a stagnation temperature of 500°C. The corresponding specific enthalpy is 3410 kJ/kg. The steam exhausts from the turbine at a stagnation pressure of 0.7 MPa, the steam having been in a superheated condition throughout the expansion. It can be assumed that the steam behaves like a perfect gas over the range of the expansion and that $\gamma = 1.3$. Given that the turbine flow process has a small-stage efficiency of 0.82, determine.

(i) the temperature and specific volume at the end of the expansion;

(ii) the reheat factor.

The specific volume of superheated steam is represented by $pv = 0.231(h - 1943)$, where p is in kPa, v is in m^3/kg and h is in kJ/kg.

4. A 20 MW back-pressure turbine receives steam at 4 MPa and 300°C, exhausting from the last stage at 0.35 MPa. The stage efficiency is 0.85, reheat factor 1.04 and external losses 2% of the actual isentropic enthalpy drop. Determine the rate of steam flow.

At the exit from the first stage nozzles the steam velocity is 244 m/s, specific volume 68.6 dm^3/kg, mean diameter 762 mm and steam exit angle 76 deg measured from the axial direction. Determine the nozzle exit height of this stage.

5. Steam is supplied to the first stage of a five stage pressure-compounded steam turbine at a stagnation pressure of 1.5 MPa and a stagnation temperature of 350°C. The steam leaves the last stage at a stagnation pressure of 7.0 kPa with a corresponding dryness fraction of 0.95. By using a Mollier chart for steam and assuming that the stagnation state point locus is a straight line joining the initial and final states, determine

(i) the stagnation conditions between each stage assuming that each stage does the same amount of work;

(ii) the total-to-total efficiency of each stage;

(iii) the overall total-to-total efficiency and total-to-static efficiency assuming the steam enters the condenser with a velocity of 200 m/s;

(iv) the reheat factor based upon stagnation conditions.

CHAPTER 3

Two-dimensional Cascades

Let us first understand the facts and then we may seek the causes. (ARISTOTLE.)

Introduction

The operation of any turbomachine is directly dependent upon changes in the working fluid's angular momentum as it crosses individual blade rows. A deeper insight of turbomachinery mechanics may be gained from consideration of the flow changes and forces exerted within these individual blade rows. In this chapter the flow past two-dimensional blade cascades is examined.

A review of the many different types of cascade tunnel, which includes low-speed, high-speed, intermittent blowdown and suction tunnels, etc. is given by Sieverding (1985). The range of Mach number in axial-flow turbomachines can be considered to extend from $M = 0.2$ to 2.5 (of course, if we also include fans then the lower end of the range is very low). Two main types of cascade tunnel are:

(1) low-speed, operating in the range 20–60 m/s; and
(2) high-speed, for the compressible flow range of testing.

A typical low-speed, continuous running, cascade tunnel is shown in Figure 3.1(a). The linear cascade of blades comprises a number of identical blades, equally spaced and parallel to one another. A suction slot is situated on the ceiling of the tunnel just before the cascade to allow the controlled removal of the tunnel boundary layer. Carefully controlled suction is usually provided on the tunnel sidewalls immediately upstream of the cascade so that two-dimensional, constant axial velocity flow can be achieved.

Figure 3.1b shows the test section of a cascade facility for transonic and moderate supersonic inlet velocities. The upper wall is slotted and equipped for suction, allowing operation in the transonic regime. The flexible section of the upper wall allows for a change of geometry so that a convergent–divergent nozzle is formed, thus allowing the flow to expand supersonically upstream of the cascade.

To obtain truly two-dimensional flow would require a cascade of infinite extent. Of necessity cascades must be limited in size, and careful design is needed to ensure that at least the central regions (where flow measurements are made) operate with approximately two-dimensional flow.

For axial flow machines of high hub-tip ratio, radial velocities are negligible and, to a close approximation, the flow may be described as two-dimensional. The flow in a cascade is then a reasonable model of the flow in the machine. With lower hub-tip radius ratios, the blades of a turbomachine will normally have an appreciable amount

55

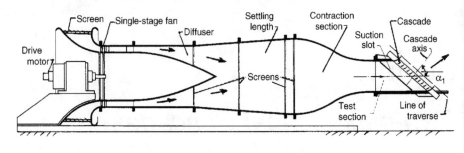

(a)

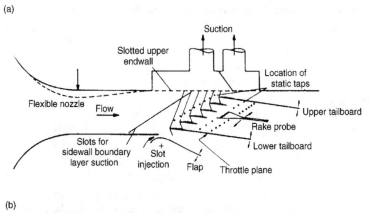

(b)

FIG. 3.1. Compressor cascade wind tunnels. (a) Conventional low-speed, continuous running cascade tunnel (adapted from Carter *et al.* 1950). (b) Transonic/supersonic cascade tunnel (adapted from Sieverding 1985).

of twist along their length, the amount depending upon the sort of "vortex design" chosen (see Chapter 6). However, data obtained from two-dimensional cascades can still be of value to a designer requiring the performance at discrete blade sections of such blade rows.

Cascade nomenclature

A cascade blade profile can be conceived as a curved *camber line* upon which a *profile thickness distribution* is symmetrically superimposed. Referring to Figure 3.2 the camber line $y(x)$ and profile thickness $t(x)$ are shown as functions of the distance x along the *blade chord l*. In British practice the shape of the camber line is usually either a circular arc or a parabolic arc defined by the maximum camber b located at distance a from the leading edge of the blade. The profile thickness distribution may be that of a standard aerofoil section but, more usually, is one of the sections specifically developed by the various research establishments for compressor or turbine applications. Blade camber and thickness distributions are generally presented as tables of y/l and t/l against x/l. Some examples of these tables are quoted by Horlock (1958, 1966). Summarising, the useful parameters for describing a cascade blade are: camber line shape, b/l, a/l, type of thickness distribution and maximum thickness to chord ratio, t_{max}/l.

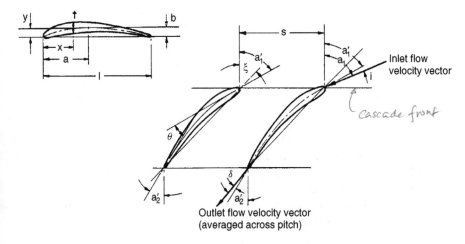

FIG. 3.2. Compressor cascade and blade notation.

With the blades arranged in cascade, two important additional geometric variables which define the cascade are the *space-chord ratio* s/l and the *stagger angle* ξ, which is the angle between the chord line and a reference direction *perpendicular to the cascade front*. Throughout the remainder of this book, all fluid and blade angles are referred to this perpendicular so as to avoid the needless complication arising from the use of other reference directions. However, custom dies hard; in steam turbine practice, blade and flow angles are conventionally measured from the *tangential* direction (i.e. parallel to the cascade front). Despite this, it is better to avoid ambiguity of meaning by adopting the single reference direction already given.

The blades angles at entry to and at exit from the cascade are denoted by α_1' and α_2' respectively. A most useful blade parameter is the *camber angle* θ which is the change in angle of the camber line between the leading and trailing edges and equals $\alpha_1' - \alpha_2'$ in the notation of Figure 3.2. For circular arc camber lines the stagger angle is $\xi = \frac{1}{2}(\alpha_1' + \alpha_2')$. For parabolic arc camber lines of low camber (i.e. small b/l) as used in some compressor cascades, the inlet and outlet blade angles are

$$\alpha_1' = \xi + \tan^{-1}\frac{b/l}{(a/l)^2} \quad \alpha_2' = \xi - \tan^{-1}\frac{b/l}{(1 - a/l)^2}$$

the equation approximating for the parabolic arc being $Y = X\{A(X - 1) + BY\}$ where $X = x/l$, $Y = y/l$. A, B are two arbitrary constants which can be solved with the conditions that at $x = a$, $y = b$ and $dy/dx = 0$. The exact general equation of a parabolic arc camber line which has been used in the design of highly cambered turbine blades is dealt with by Dunham (1974).

Analysis of cascade forces

The fluid approaches the cascade from far upstream with velocity c_1 at an angle α_1 and leaves far downstream of the cascade with velocity c_2 at an angle α_2. In the following analysis the fluid is assumed to be incompressible and the flow to be steady. The assumption of steady flow is valid for an isolated cascade row but, in a turbomachine, relative motion between successive blade rows gives rise to unsteady flow effects. As regards the assumption of incompressible flow, the majority of cascade tests are conducted at fairly low Mach numbers (e.g. 0.3 on compressor cascades) when compressibility effects are negligible. Various techniques are available for correlating incompressible and compressible cascades; a brief review is given by Csanady (1964).

A portion of an isolated blade cascade (for a compressor) is shown in Figure 3.3. The forces X and Y are exerted by unit depth of blade upon the fluid, exactly equal and opposite to the forces exerted by the fluid upon unit depth of blade. A control surface is drawn with end boundaries far upstream and downstream of the cascade and with side boundaries coinciding with the median stream lines.

Applying the principle of continuity to a unit depth of span and noting the assumption of incompressibility, yields

$$c_1 \cos \alpha_1 = c_2 \cos \alpha_2 = c_x. \tag{3.1}$$

The momentum equation applied in the x and y directions with constant axial velocity gives,

$$X = (p_2 - p_1)s, \tag{3.2}$$

$$Y = \rho s c_x (c_{y1} - c_{y2}), \tag{3.3}$$

or

$$Y = \rho s c_x^2 (\tan \alpha_1 - \tan \alpha_2) \tag{3.3a}$$

Equations (3.1) and (3.3) are completely valid for a flow incurring total pressure losses in the cascade.

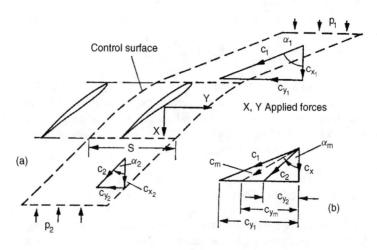

FIG. 3.3. Forces and velocities in a blade cascade.

Energy losses

A real fluid crossing the cascade experiences a loss in total pressure Δp_0 due to skin friction and related effects. Thus

$$\frac{\Delta p_0}{\rho} = \frac{p_1 - p_2}{\rho} + \frac{1}{2}(c_1^2 - c_2^2). \tag{3.4}$$

Noting that $c_1^2 - c_2^2 = (c_{y1}^2 + c_x^2) - (c_{y2}^2 + c_x^2) = (c_{y1} + c_{y2})(c_{y1} - c_{y2})$, substitute eqns. (3.2) and (3.3) into eqn. (3.4) to derive the relation,

$$\frac{\Delta p_0}{\rho} = \frac{1}{\rho s}(-X + Y \tan \alpha_m), \tag{3.5}$$

where

$$\tan \alpha_m = \frac{1}{2}(\tan \alpha_1 + \tan \alpha_2). \tag{3.6}$$

A non-dimensional form of eqn. (3.5) is often useful in presenting the results of cascade tests. Several forms of total pressure-loss coefficient can be defined of which the most popular are,

$$\zeta = \Delta p_0 / (\tfrac{1}{2} \rho c_x^2) \tag{3.7a}$$

and

$$\overline{\omega} = \Delta p_0 / (\tfrac{1}{2} \rho c_1^2). \tag{3.7b}$$

Using again the same reference parameter, a pressure rise coefficient C_p and a tangential force coefficient C_f may be defined

$$C_p = \frac{p_2 - p_1}{\frac{1}{2} \rho c_x^2} = \frac{X}{\frac{1}{2} \rho s c_x^2}, \tag{3.8}$$

$$C_f = \frac{Y}{\frac{1}{2} \rho s c_x^2} = 2(\tan \alpha_1 - \tan \alpha_2), \tag{3.9}$$

using eqns. (3.2) and (3.3a).

Substituting these coefficients into eqn. (3.5) to give, after some rearrangement,

$$C_p = C_f \tan \alpha_m - \zeta. \tag{3.10}$$

Lift and drag

A mean velocity c_m is defined as

$$c_m = c_x / \cos \alpha_m, \tag{3.11}$$

where α_m is itself defined by eqn. (3.6). Considering unit depth of a cascade blade, a lift force L acts in a direction perpendicular to c_m and a drag force D in a direction parallel to c_m. Figure 3.4 shows L and D as the reaction forces exerted *by* the blade *upon* the fluid.

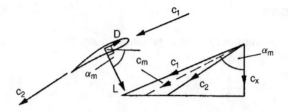

FIG. 3.4. Lift and drag forces exerted by a cascade blade (of unit span) upon the fluid.

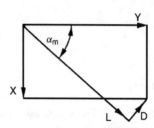

FIG. 3.5. Axial and tangential forces exerted by unit span of a blade upon the fluid.

Experimental data are often presented in terms of lift and drag when the data may be of greater use in the form of tangential force and total pressure loss. The lift and drag forces can be resolved in terms of the axial and tangential forces. Referring to Figure 3.5,

$$L = X \sin \alpha_m + Y \cos \alpha_m, \tag{3.12}$$

$$D = Y \sin \alpha_m - X \cos \alpha_m. \tag{3.13}$$

From eqn. (3.5)

$$D = \cos \alpha_m (Y \tan \alpha_m - X) = s\Delta p_0 \cos \alpha_m. \tag{3.14}$$

Rearranging eqn. (3.14) for X and substituting into eqn. (3.12) gives,

$$\begin{aligned}
L &= (Y \tan \alpha_m - s\Delta p_0) \sin \alpha_m + Y \cos \alpha_m \\
&= Y \sec \alpha_m - s\Delta p_0 \sin \alpha_m \\
&= \rho s c_x^2 (\tan \alpha_1 - \tan \alpha_2) \sec \alpha_m - s\Delta p_0 \sin \alpha_m, \tag{3.15}
\end{aligned}$$

after using eqn. (3.9).

Lift and drag coefficients may be introduced as

$$C_L = \frac{L}{\frac{1}{2}\rho c_m^2 l}, \tag{3.16a}$$

$$C_D = \frac{D}{\frac{1}{2}\rho c_m^2 l}. \tag{3.16b}$$

Using eqn. (3.14) together with eqn. (3.7),

$$C_D = \frac{s\Delta p_0 \cos\alpha_m}{\frac{1}{2}\rho c_m^2 l} = \zeta\frac{s}{l}\cos^3\alpha_m. \tag{3.17}$$

With eqn. (3.15)

$$C_L = \frac{\rho s c_x^2(\tan\alpha_1 - \tan\alpha_2)\sec\alpha_m - s\Delta p_0 \sin\alpha_m}{\frac{1}{2}\rho c_m^2 l}$$

$$= 2\frac{s}{l}\cos\alpha_m(\tan\alpha_1 - \tan\alpha_2) - C_D\tan\alpha_m. \tag{3.18}$$

Alternatively, employing eqns. (3.9) and (3.17),

$$C_L = \frac{s}{l}\cos\alpha_m\left(C_f - \zeta\frac{\sin 2\alpha_m}{2}\right). \tag{3.19}$$

Within the normal range of operation in a cascade, values of C_D are very much less than C_L. As α_m is unlikely to exceed 60 deg, the quantity $C_D\tan\alpha_m$ in eqn. (3.18) can be dropped, resulting in the approximation,

$$\frac{L}{D} = \frac{C_L}{C_D} \doteqdot \frac{2\sec^2\alpha_m}{\zeta}(\tan\alpha_1 - \tan\alpha_2) = \frac{C_f}{\zeta}\sec^2\alpha_m. \tag{3.20}$$

Circulation and lift

The lift of a single isolated aerofoil for the ideal case when $D = 0$ is given by the Kutta–Joukowski theorem

$$L = \rho\Gamma c, \tag{3.21}$$

where c is the relative velocity between the aerofoil and the fluid at infinity and Γ is the circulation about the aerofoil. This theorem is of fundamental importance in the development of the theory of aerofoils (for further information see Glauert (1959).

In the absence of total pressure losses, the lift force per unit span of a blade *in cascade*, using eqn. (3.15), is

$$L = \rho s c_x^2(\tan\alpha_1 - \tan\alpha_2)\sec\alpha_m$$

$$= \rho s c_m(c_{y1} - c_{y2}). \tag{3.22}$$

Now the *circulation* is the contour integral of velocity around a closed curve. For the cascade blade the circulation is

$$\Gamma = s(c_{y1} - c_{y2}). \tag{3.23}$$

Combining eqns. (3.22) and (3.23),

$$L = \rho\Gamma c_m. \tag{3.24}$$

As the spacing between the cascade blades is increased without limit (i.e. $s \to \infty$), the inlet and outlet velocities to the cascade, c_1 and c_2, becomes equal in magnitude and direction. Thus $c_1 = c_2 = c$ and eqn. (3.24) becomes identical with the Kutta–Joukowski theorem obtained for an isolated aerofoil.

Efficiency of a compressor cascade

The efficiency η_D of a compressor blade cascade can be defined in the same way as diffuser efficiency; this is the ratio of the actual static pressure rise in the cascade to the maximum possible theoretical pressure rise (i.e. with $\Delta p_0 = 0$). Thus,

$$\eta_D = \frac{p_2 - p_1}{\frac{1}{2}\rho(c_1^2 - c_2^2)}$$

$$= 1 - \frac{\Delta p_0}{\rho c_x^2 \tan \alpha_m (\tan \alpha_1 - \tan \alpha_2)}.$$

Inserting eqns. (3.7) and (3.9) into the above equation,

$$\eta_D = 1 - \frac{\zeta}{C_f \tan \alpha_m}. \tag{3.25}$$

Equation (3.20) can be written as $\zeta/C_f \doteq (\sec^2 \alpha_m)C_D/C_L$ which when substituted into eqn. (3.25) gives

$$\eta_D = 1 - \frac{2C_D}{C_L \sin 2\alpha_m}. \tag{3.26}$$

Assuming a constant lift–drag ratio, eqn. (3.26) can be differentiated with respect to α_m to give the optimum mean flow angle for maximum efficiency. Thus,

$$\frac{\partial \eta_D}{\partial \alpha_m} = \frac{4C_D \cos 2\alpha_m}{C_L \sin^2 2\alpha_m} = 0,$$

so that

$$\alpha_{m\,opt} = 45 \deg,$$

therefore

$$\eta_{D\,\max} = 1 - \frac{2C_D}{C_L}. \tag{3.27}$$

This simple analysis suggests that maximum efficiency of a compressor cascade is obtained when the mean flow angle is 45 deg, but ignores changes in the ratio C_D/C_L with varying α_m. Howell (1945) calculated the effect of having a specified variation of C_D/C_L upon cascade efficiency, comparing it with the case when C_D/C_L is constant. Figure 3.6 shows the results of this calculation as well as the variation of C_D/C_L with α_m. The graph shows that $\eta_{D\,\max}$ is at an optimum angle only a little less than 45 deg but that the curve is rather flat for a rather wide change in α_m. Howell suggested that value of α_m *rather less* than the optimum could well be chosen with little sacrifice in efficiency, and with some benefit with regard to power–weight ratio of compressors. In Howell's calculations, the drag is an estimate based on cascade

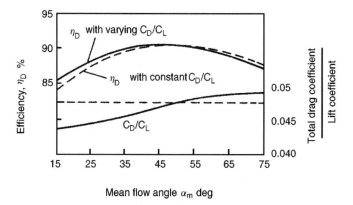

FIG. 3.6. Efficiency variation with average flow angle (adapted from Howell 1945).

experimental data together with an allowance for wall boundary-layer losses and "secondary-flow" losses.

Performance of two-dimensional cascades

From the relationships developed earlier in this chapter it is apparent that the effects of a cascade may be completely deduced if the flow angles at inlet and outlet together with the pressure loss coefficient are known. However, for a given cascade only one of these quantities may be arbitrarily specified, the other two being fixed by the cascade geometry and, to a lesser extent, by the Mach number and Reynolds number of the flow. For a given family of geometrically similar cascades the performance may be expressed functionally as,

$$\zeta, a_2 = (a_1, M_1, Re), \tag{3.28}$$

where ζ is the pressure loss coefficient, eqn. (3.7), M_1 is the inlet Mach number $= c_1/(\gamma RT_1)^{1/2}$, Re is the inlet Reynolds number $= \rho_1 c_1 l/\mu$ based on blade chord length.

Despite numerous attempts it has not been found possible to determine, accurately, cascade performance characteristics by theoretical means alone and the experimental method still remains the most reliable technique. An account of the theoretical approach to the problem lies outside the scope of this book, however, a useful summary of the subject is given by Horlock (1958).

The cascade wind tunnel

The basis of much turbomachinery research and development derives from the cascade wind tunnel, e.g. Figure 3.1 (or one of its numerous variants), and a brief description of the basic aerodynamic design is given below. A more complete description of the cascade tunnel is given by Carter *et al.* (1950) including many of the research techniques developed.

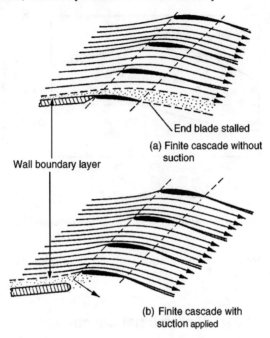

Wall boundary layer

End blade stalled

(a) Finite cascade without suction

(b) Finite cascade with suction applied

FIG. 3.7. Streamline flow through cascades (adapted from Carter *et al.* 1950).

In a well-designed cascade tunnel it is most important that the flow near the central region of the cascade blades (where the flow measurements are made) is approximately two-dimensional. This effect could be achieved by employing a large number of long blades, but an excessive amount of power would be required to operate the tunnel. With a tunnel of more reasonable size, aerodynamic difficulties become apparent and arise from the tunnel wall boundary layers interacting with the blades. In particular, and as illustrated in Figure 3.7a, the tunnel wall boundary layer mingles with the end blade boundary layer and, as a consequence, this blade stalls resulting in a non-uniform flow field.

Stalling of the end blade may be delayed by applying a controlled amount of suction to a slit just upstream of the blade, and sufficient to remove the tunnel wall boundary layer (Figure 3.7b). Without such boundary-layer removal the effects of flow interference can be quite pronounced. They become most pronounced near the cascade "stalling point" (defined later) when any small disturbance of the upstream flow field precipitates stall on blades adjacent to the end blade. Instability of this type has been observed in compressor cascades and can affect every blade of the cascade. It is usually characterised by regular, periodic "cells" of stall crossing rapidly from blade to blade; the term *propagating stall* is often applied to the phenomenon. Some discussion of the mechanism of propagating stall is given in Chapter 6.

The boundary layers on the walls to which the blade roots are attached, generate *secondary vorticity* in passing through the blades which may produce substantial *secondary flows*. The mechanism of this phenomenon has been discussed at some length by Carter (1948), Horlock (1958) and many others and a brief explanation is included in Chapter 6.

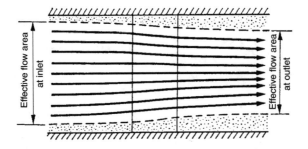

FIG. 3.8. Contraction of streamlines due to boundary layer thickening (adapted from Carter *et al.* 1950).

In a compressor cascade the rapid increase in pressure across the blades causes a marked thickening of the wall boundary layers and produces an effective contraction of the flow, as depicted in Figure 3.8. A *contraction coefficient*, used as a measure of the boundary-layer growth through the cascade, is defined by $\rho_1 c_1 \cos a_1 / (\rho_2 c_2 \cos a_2)$. Carter *et al.* (1950) quotes values of 0.9 for a good tunnel dropping to 0.8 in normal high-speed tunnels and even less in bad cases. These are values for compressor cascades; with turbine cascades slightly higher values can be expected.

Because of the contraction of the main through-flow, the theoretical pressure rise across a compressor cascade, even allowing for losses, is never achieved. This will be evident since a contraction (in a subsonic flow) accelerates the fluid, which is in conflict with the diffuser action of the cascade.

To counteract these effects it is customary (in Great Britain) to use *at least* seven blades in a compressor cascade, each blade having a minimum aspect ratio (blade span–chord length) of 3. With seven blades, suction is desirable in a compressor cascade but it is not usual in a turbine cascade. In the United States much lower aspect ratios are commonly employed in compressor cascade testing, the technique being the almost complete removal of tunnel wall boundary layers from all four walls using a combination of suction slots and perforated end walls to which suction is applied.

Cascade test results

The basic cascade performance data in low-speed flows are obtained from measurements of total pressure, flow angle and velocity taken across one or more complete pitches of the cascade, the plane of measurement being about half a chord downstream of the trailing edge plane. The literature on instrumentation is very extensive as are the various measurement techniques employed and the student is referred to the works of Horlock (1958), Bryer and Pankhurst (1971), Sieverding (1975, 1985). The publication by Bryer and Pankhurst for deriving air speed and flow direction is particularly instructive and recommended, containing as it does details of the design and construction of various instruments used in cascade tunnel measurements as well as their general principles and performance details.

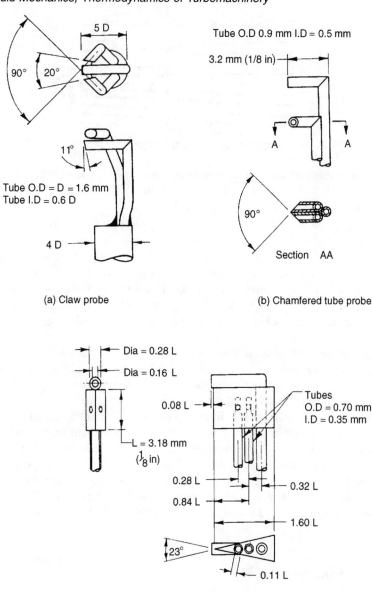

(a) Claw probe

(b) Chamfered tube probe

(c) Wedge probe

FIG. 3.9. Some combination pressure probes (adapted from Bryer and Pankhurst 1971):
(a) claw probe; (b) chamfered tube probe; (c) wedge probe.

Some representative combination pressure probes are shown in Figure 3.9. These types are frequently used for pitchwise traversing across blade cascades but, because of their small size, they are also used for interstage (radial) flow traversing in compressors. For the measurement of flow direction in conditions of severe transverse total pressure gradients, as would be experienced during the measurement of

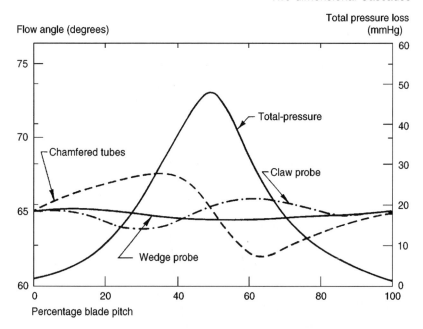

Fɪɢ. 3.10. Apparent flow angle variation measured by three different combination probes traversed across a transverse variation of total pressure (adapted from Bryer and Pankhurst 1971).

blade cascade flows, quite substantial errors in the measurement of flow direction do arise. Figure 3.10 indicates the *apparent* flow angle variation measured by these same three types of pressure probe when traversed across a transverse gradient of total pressure caused by a compressor stator blade. It is clear that the wedge probe is the least affected by the total pressure gradient. An investigation by Dixon (1978) did confirm that all pressure probe instruments are subject to this type of directional error when traversed across a total pressure variation such as a blade wake.

An extensive bibliography on all types of measurement in fluid flow is given by Dowden (1972). Figure 3.11 shows a typical cascade test result from a traverse across 2 blade pitches taken by Todd (1947) at an inlet Mach number of 0.6. It is observed that a total pressure deficit occurs across the blade row arising from the fluid friction on the blades. The fluid deflection is not uniform and is a maximum at each blade trailing edge on the pressure side of the blades. From such test results, average values of total pressure loss and fluid outlet angle are found (usually on a mass flow basis). The use of terms like "total pressure loss" and "fluid outlet angle" in the subsequent discussion will signify these *average* values.

Similar tests performed for a range of fluid inlet angles, at the same inlet Mach number M_1 and Reynolds number Re, enables the complete performance of the cascade to be determined (at that M_1 and Re). So as to minimise the amount of testing required, much cascade work is performed at low inlet velocities, but at a Reynolds number greater than the "critical" value. This critical Reynolds number Re_c is approximately 2×10^5 based on inlet velocity and blade chord. With $Re > Re_c$, total pressure losses and fluid deflections are only slightly dependent on changes

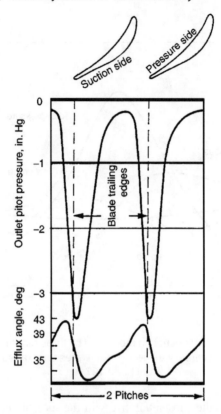

FIG. 3.11. A sample plot of inlet and outlet stagnation pressures and fluid outlet angle (adapted from Todd 1947).

in Re. Mach number effects are negligible when $M_1 < 0.3$. Thus, the performance laws, eqn. (3.28), for this flow simplify to,

$$\zeta, a_2 = f(a_1).\qquad(3.28a)$$

There is a fundamental difference between the flows in turbine cascades and those in compressor cascades which needs emphasising. A fluid flowing through a channel in which the mean pressure is falling (mean flow is accelerating) experiences a relatively small total pressure loss in contrast with the mean flow through a channel in which the pressure is rising (diffusing flow) when losses may be high. This characteristic difference in flow is reflected in turbine cascades by a wide range of low loss performance and in compressor cascades by a rather narrow range.

Compressor cascade performance

A typical set of low-speed compressor cascade results (Howell 1942) for a blade cascade of specified geometry, is shown in Figure 3.12. These results are presented in the form of a pressure loss coefficient $\Delta p_0/(\frac{1}{2}\rho c_1^2)$ and fluid deflection $\epsilon = a_1 - a_2$ against incidence $i = a_1 - a_1'$ (refer to Figure. 3.2 for nomenclature). Note

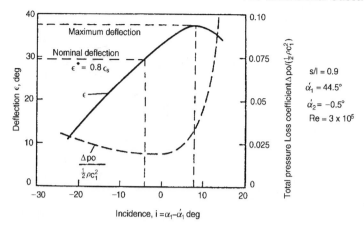

Fig. 3.12. Compressor cascade characteristics (Howell 1942). (By courtesy of the Controller of H.M.S.O., Crown copyright reserved).

that from eqn. (3.7), $\Delta p_0/(\frac{1}{2}\rho c_1^2) = \zeta \cos^2 \alpha_1$. There is a pronounced increase in total pressure loss as the incidence rises beyond a certain value and the cascade is *stalled* in this region. The precise incidence at which stalling occurs is difficult to define and a *stall point* is arbitrarily specified as the incidence at which the total pressure loss is *twice* the minimum loss in total pressure. Physically, stall is characterised (at positive incidence) by the flow separating from the suction side of the blade surfaces. With decreasing incidence, total pressure losses again rise and a "negative incidence" stall point can also be defined as above. The *working range* is conventionally defined as the incidence range between these two limits at which the losses are twice the minimum loss. Accurate knowledge of the extent of the working range, obtained from two-dimensional cascade tests, is of great importance when attempting to assess the suitability of blading for changing conditions of operation. A *reference incidence* angle can be most conveniently defined either at the mid-point of the working range or, less precisely, at the minimum loss condition. These two conditions do not necessarily give the same reference incidence.

From such cascade test results the *profile losses* through compressor blading of the same geometry may be estimated. To these losses estimates of the annulus skin friction losses and other secondary losses must be added, and from which the efficiency of the compressor blade row may be determined. Howell (1945) suggested that these losses could be estimated using the following drag coefficients. For the annulus walls loss,

$$C_{Da} = 0.02s/H \tag{3.29a}$$

and for the so-called "secondary" loss,

$$C_{Ds} = 0.018C_L^2 \tag{3.29b}$$

where s, H are the blade pitch and blade length respectively, and C_L the blade lift coefficient. Calculations of this type were made by Howell and others to estimate the efficiency of a complete compressor stage. A worked example to illustrate the

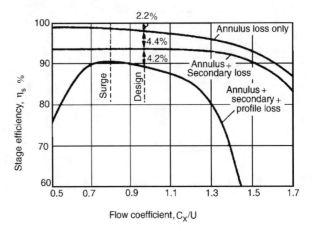

FIG. 3.13. Losses in a compressor stage (Howell 1945). (Courtesy of the Institution of Mechanical Engineers).

details of the method is given in Chapter 5. Figure 3.13 shows the variation of stage efficiency with flow coefficient and it is of particular interest to note the relative magnitude of the profile losses in comparison with the overall losses, especially at the design point.

Cascade performance data to be easily used, are best presented in some condensed form. Several methods of empirically correlating low-speed performance data have been developed in Great Britain. Howell's correlation (1942) relates the performance of a cascade to its performance at a "nominal" condition defined at 80% of the stalling deflection. Carter (1950) has referred performance to an optimum incidence given by the highest lift–drag ratio of the cascade. In the United States, the National Advisory Committee for Aeronautics (NACA), now called the National Aeronautics and Space Administration (NASA), systematically tested whole families of different cascade geometries, in particular, the widely used NACA 65 Series (Herrig *et al.* 1957). The data on the NACA 65 Series has been usefully summarised by Felix (1957) where the performance of a fixed geometry cascade can be more readily found. A concise summary is also given by Horlock (1958).

Turbine cascade performance

Figure 3.14 shows results obtained by Ainley (1948) from two sets of turbine cascade blades, impulse and "reaction". The term "reaction" is used here to denote, in a qualitative sense, that the fluid accelerates through the blade row and thus experiences a *pressure drop* during its passage. There is no pressure change across an impulse blade row. The performance is expressed in the form $\lambda = \Delta p_o/(p_{o2} - p_2)$ and α_2 against incidence.

From these results it is observed that:

(a) the reaction blades have a much wider range of low loss performance than the impulse blades, a result to be expected as the blade boundary layers are subjected to a favourable pressure gradient,

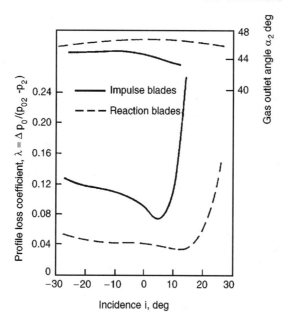

FIG. 3.14. Variation in profile loss with incidence for typical turbine blades (adapted from Ainley 1948).

(b) the fluid outlet angle α_2 remains relatively constant over the whole range of incidence in contrast with the compressor cascade results.

For turbine cascade blades, a method of correlation is given by Ainley and Mathieson (1951) which enables the performance of a gas turbine to be predicted with an estimated tolerance of within 2% on peak efficiency. In Chapter 4 a rather different approach, using a method attributed to Soderberg, is outlined. While being possibly slightly less accurate than Ainley's correlation, Soderberg's method employs fewer parameters and is rather easier to apply.

Compressor cascade correlations

Many experimental investigations have confirmed that the efficient performance of compressor cascade blades is limited by the growth and separation of the blade surface boundary layers. One of the aims of cascade research is to establish the generalised loss characteristics and stall limits of conventional blades. This task is made difficult because of the large number of factors which can influence the growth of the blade surface boundary layers, viz. surface velocity distribution, blade Reynolds number, inlet Mach number, free-stream turbulence and unsteadiness, and surface roughness. From the analysis of experimental data several correlation methods have been evolved which enable the first-order behaviour of the blade losses and limiting fluid deflection to be predicted with sufficient accuracy for engineering purposes.

LIEBLEIN. The correlation of Lieblein (1959), NASA (1965) is based on the experimental observation that a large amount of velocity diffusion on blade surfaces tends to produce thick boundary layers and eventual flow separation. Lieblein states the general hypothesis that in the region of minimum loss, the wake thickness and consequently the magnitude of the loss in total pressure, is proportional to the diffusion in velocity on the suction-surface of the blade in that region. The hypothesis is based on the consideration that the boundary layer on the suction-surface of conventional compressor blades contributes the largest share of the blade wake. Therefore, the suction-surface velocity distribution becomes the main factor in determining the total pressure loss.

Figure 3.15 shows a typical velocity distribution derived from surface pressure measurements on a compressor cascade blade in the region of minimum loss. The diffusion in velocity may be expressed as the ratio of maximum suction-surface

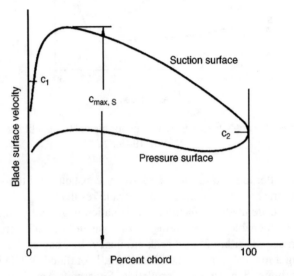

FIG. 3.15. Compressor cascade blade surface velocity distribution.

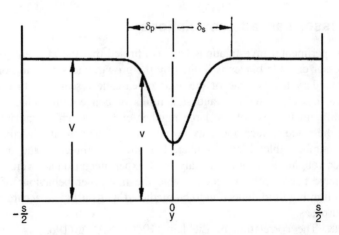

FIG. 3.16. Model variation in velocity in a plane normal to axial direction.

velocity to outlet velocity, $c_{max,s}/c_2$. Lieblein found a correlation between the diffusion ratio $c_{max,s}/c_2$ and the wake momentum-thickness to chord ratio, θ_2/l at the reference incidence (mid-point of working range) for American NACA 65-(A_{10}) and British C.4 circular-arc blades. The wake momentum-thickness, with the parameters of the flow model in Figure 3.16 is defined as

$$\theta_2 = \int_{\delta_p}^{\delta_s} \frac{v}{V}\left(1 - \frac{v}{V}\right) dy. \tag{3.30}$$

The Lieblein correlation, with his data points removed for clarity, is closely fitted by the mean curve in Figure 3.17. This curve represents the equation

$$\frac{\theta_2}{l} = 0.004 \left/ \left\{1 - 1.17 \ln\left(\frac{c_{max,s}}{c_2}\right)\right\}\right. \tag{3.31}$$

which may be more convenient to use in calculating results. It will be noticed that for the limiting case when $(\theta_2/l) \to \infty$, the corresponding *upper* limit for the diffusion ratio $c_{max,s}/c_2$ is 2.35. The *practical* limit of efficient operation would correspond to a diffusion ratio of between 1.9 and 2.0.

Losses are usually expressed in terms of the stagnation pressure loss coefficient $\overline{\omega} = \Delta p_0 / \left(\frac{1}{2}\rho c_1^2\right)$ or $\zeta = \Delta p_0 / \left(\frac{1}{2}\rho c_x^2\right)$ as well as the drag coefficient C_D. Lieblein and Roudebush (1956) have demonstrated the simplified relationship between momentum–thickness ratio and total pressure loss coefficient, valid for unstalled blades,

$$\overline{\omega} = 2\left(\frac{\theta_2}{l}\right)\left(\frac{l}{s}\right)\frac{\cos^2 \alpha_1}{\cos^3 \alpha_2}. \tag{3.32}$$

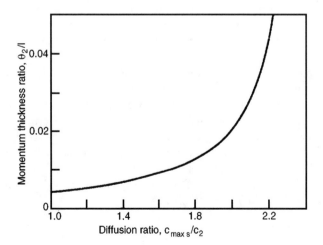

FIG. 3.17. Mean variation of wake momentum–thickness/chord ratio with suction-surface diffusion ratio at reference incidence condition for NACA 65 − ($C_{10}A_{10}$)10 blades and British C.4 circular-arc blades (adapted from Lieblein (1959)).

Combining this relation with eqns. (3.7) and (3.17) the following useful results can be obtained:

$$C_D = \bar{\omega} \left(\frac{s}{l}\right) \frac{\cos^3 \alpha_m}{\cos^2 \alpha_1} = 2 \left(\frac{\theta_2}{l}\right) \left(\frac{\cos \alpha_m}{\cos \alpha_2}\right)^3 = \zeta \left(\frac{s}{l}\right) \cos^3 \alpha_m. \qquad (3.33)$$

The correlation given above *assumes* a knowledge of suction-surface velocities in order that total pressure loss and stall limits can be estimated. As this data may be unavailable it is necessary to establish an *equivalent diffusion ratio*, approximately equal to $c_{max,s}/c_2$, that can be easily calculated from the inlet and outlet conditions of the cascade. An empirical correlation was established by Lieblein (1959) between a circulation parameter defined by $f(\Gamma) = \Gamma \cos \alpha_1/(lc_1)$ and $c_{max,s}/c_1$ at the reference incidence, where the ideal circulation $\Gamma = s(c_{y1} - c_{y2})$, using eqn. (3.23). The correlation obtained is the simple *linear* relation.

$$c_{max,s}/c_1 = 1.12 + 0.61 f(\Gamma) \qquad (3.34)$$

which applies to both NACA 65-(A_{10}) and C.4 circular arc blades. Hence, the equivalent diffusion ratio, after substituting for Γ and simplifying, is

$$D_{eq} = \frac{c_{max,s}}{c_2} = \frac{\cos \alpha_2}{\cos \alpha_1} \left\{ 1.12 + 0.61 \left(\frac{s}{l}\right) \cos^2 \alpha_1 (\tan \alpha_1 - \tan \alpha_2) \right\} \qquad (3.35)$$

At incidence angles greater than reference incidence, Lieblein found that the following correlation was adequate:

$$D_{eq} = \frac{\cos \alpha_2}{\cos \alpha_1} \left\{ 1.12 + k(i - i_{ref})^{1.43} + 0.61 \left(\frac{s}{l}\right) \cos^2 \alpha_1 (\tan \alpha_1 - \tan \alpha_2) \right\} \qquad (3.36)$$

where $k = 0.0117$ for the NACA 65-(A_{10}) blades and $k = 0.007$ for the C.4 circular arc blades.

The expressions given above are still very widely used as a means of estimating total pressure loss and the unstalled range of operation of blades commonly employed in subsonic axial compressors. The method has been modified and extended by Swann to include the additional losses caused by shock waves in transonic compressors. The discussion of transonic compressors is outside the scope of this text and is not included.

HOWELL. The low-speed correlation of Howell (1942) has been widely used by designers of axial compressors and is based on a nominal condition such that the deflection ε^* is 80% of the stalling deflection, ε_s (Figure 3.12). Choosing $\varepsilon^* = 0.8\varepsilon_s$ as the *design condition* represents a compromise between the ultraconservative and the overoptimistic! Howell found that the nominal deflections of various compressor cascades are, primarily, a function of the space–chord ratio s/l, the nominal fluid outlet angle α_2^* and the Reynolds number Re

$$\varepsilon^* = f(s/l, \alpha_2^*, Re). \qquad (3.37)$$

It is important to note that the correlation (which is really a correlation of stalling deflection, $\varepsilon_s = 1.25\varepsilon^*$) is virtually independent of blade camber θ in the normal range of choice of this parameter ($20° < \theta < 40°$). Figure 3.18 shows the variation of

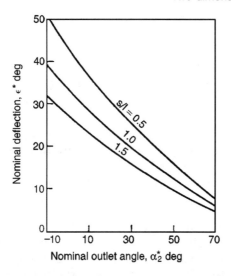

FIG. 3.18. Variation of nominal deflection with nominal outlet angle for several space/chord ratios (adapted from Howell 1945).

ε^* found by Howell (1945) against α_2^* for several space–chord ratios. The dependence on Reynolds number is small for $Re > 3 \times 10^5$, based on blade chord.

An approximating formula to the data given in Figure 3.18, which was quoted by Howell and frequently found to be useful in preliminary performance estimation, is the tangent-difference rule:

$$\tan \alpha_1^* - \tan \alpha_2^* = \frac{1.55}{1 + 1.5s/l} \tag{3.38}$$

which is applicable in the range $0 \leq \alpha_2^* \leq 40°$.

Fluid deviation

The *difference* between the fluid and blade inlet angles at cascade inlet is under the arbitrary control of the designer. At cascade outlet however, the difference between the fluid and blade angles, called the *deviation* δ, is a function of blade camber, blade shape, space–chord ratio and stagger angle. Referring to Figure 3.2, the deviation $\delta = \alpha_2 - \alpha_2'$ is drawn as positive; almost without exception it is in such a direction that the deflection of the fluid is reduced. The deviation may be of considerable magnitude and it is important that an accurate estimate is made of it. Re-examining Figure 3.11 again, it will be observed that the fluid receives its maximum guidance on the pressure side of the cascade channel and that this diminishes almost linearly towards the suction side of the channel.

Howell used an empirical rule to relate nominal deviation δ^* to the camber and space–chord ratio,

$$\delta^* = m\theta(s/l)^n, \tag{3.39}$$

where $n \doteq \frac{1}{2}$ for compressor cascades and $n \doteq 1$ for compressor *inlet guide vanes*. The value of m depends upon the shape of the camber line and the blade setting.

For a compressor cascade (i.e. diffusing flow),

$$m = 0.23(2a/l)^2 + \alpha_2^*/500, \tag{3.40a}$$

where α is the distance of maximum camber from the leading edge. For the inlet guide vanes, which are essentially *turbine* nozzles (i.e. accelerating flow),

$$m = \text{constant} = 0.19 \tag{3.40b}$$

EXAMPLE 3.1. A compressor cascade has a space–chord ratio of unity and blade inlet and outlet angles of 50 deg and 20 deg respectively. If the blade camber line is a circular arc (i.e. $a/l = 50\%$) and the cascade is designed to operate at Howell's nominal condition, determine the fluid deflection, incidence and ideal lift coefficient at the design point.

Solution. The camber, $\theta = \alpha_1' - \alpha_2' = 30 \deg$. As a first approximation put $\alpha_2^* = 20 \deg$ in eqn. (3.40) to give $m = 0.27$ and, using eqn. (3.39), $\delta^* = 0.27 \times 30 = 8.1 \deg$. As a better approximation put $\alpha_2^* = 28.1 \deg$ in eqn. (3.40) giving $m = 0.2862$ and $\delta^* = 8.6 \deg$. Thus, $\alpha_2^* = 28.6 \deg$ is sufficiently accurate.

From Figure 3.16, with $s/l = 1.0$ and $\alpha_2^* = 28.6 \deg$ obtain $\varepsilon^* = \alpha_1^* - \alpha_2^* = 21 \deg$. Hence $\alpha_1^* = 49.6 \deg$ and the nominal incidence $i^* = \alpha_1^* - \alpha_1' = -0.4 \deg$.

The *ideal* lift coefficient is found by setting $C_D = 0$ in eqn. (3.18),

$$C_L = 2(s/l) \cos \alpha_m (\tan \alpha_1 - \tan \alpha_2).$$

Putting $\alpha_1 = \alpha_1^*$, $\alpha_2 = \alpha_2^*$ and noting $\tan \alpha_m = \frac{1}{2}(\tan \alpha_1^* + \tan \alpha_2^*)$ obtain $\alpha_m = 40.75 \deg$ and $C_L^* = 2(1.172 - 0.545)0.758 \doteqdot 0.95$.

In conclusion it will be noted that the estimated deviation is one of the most important quantities for design purposes, as small errors in it are reflected in large changes in deflection and thus, in predicted performance.

Off-design performance

To obtain the performance of a given cascade at conditions removed from the design point, generalised performance curves of Howell (1942) shown in Figure 3.19 may be used. If the nominal deflection ε^* and nominal incidence i^* are known the off-design performance (deflection, total pressure loss coefficient) of the cascade at any other incidence is readily calculated.

EXAMPLE 3.2. In the previous exercise, with a cascade of $s/l = 1.0$, $a_1' = 50 \deg$ and $a_2' = 20 \deg$ the nominal conditions were $\varepsilon^* = 21 \deg$ and $i^* = -0.4 \deg$.

Determine the off-design performance of this cascade at an incidence $i = 3.8 \deg$.

Solution. Referring to Figure 3.19 and with $(i - i^*)/\varepsilon^* = 0.2$ obtain $C_D \doteqdot 0.017$, $\varepsilon/\varepsilon^* = 1.15$. Thus, the off-design deflection, $\varepsilon = 24.1 \deg$.

From eqn. (3.17), the total pressure loss coefficient is,

$$\zeta = \Delta p_0 / \left(\tfrac{1}{2} \rho c_x^2\right) = C_D / [(s/l) \cos^3 \alpha_m].$$

Now $\alpha_1 = \alpha_1' + i = 53.8 \deg$, also $\alpha_2 = \alpha_1 - \varepsilon = 29.7 \deg$, therefore,

$$\alpha_m = \tan^{-1}\{\tfrac{1}{2}(\tan \alpha_1 + \tan \alpha_2)\} = \tan^{-1}\{0.969\} = 44.1 \deg,$$

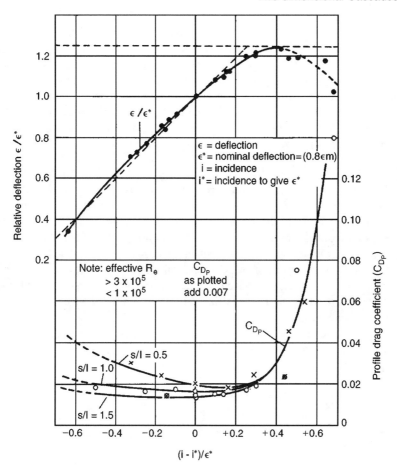

Fig. 3.19. The off-design performance of a compressor cascade (Howell 1942). (By courtesy of the Controller of H.M.S.O., Crown copyright reserved).

hence

$$\zeta = 0.017/0.719^3 = 0.0458.$$

The tangential lift force coefficient, eqn. (3.9), is

$$C_f = (p_2 - p_1)/(\tfrac{1}{2}\rho c_x^2) = 2(\tan\alpha_1 - \tan\alpha_2) = 1.596.$$

The diffuser efficiency, eqn. (3.25), is

$$\eta_D = 1 - \zeta/(C_f \tan\alpha_m) = 1 - 0.0458/(1.596 \times 0.969) = 97\%.$$

It is worth nothing, from the representative data contained in the above exercise, that the validity of the approximation in eqn. (3.20) is amply justified.

Howell's correlation, clearly, is a simple and fairly direct method of assessing the performance of a given cascade for a range of inlet flow angles. The data can also be used for solving the more complex *inverse problem*, namely, the selection of a suitable cascade geometry when the fluid deflection is given. For this case, if the previous method of a nominal design condition is used, mechanically unsuitable space–chord ratios are a possibility. The space–chord ratio may, however, be determined to some extent by the mechanical layout of the compressor, the design incidence then only fortuitously coinciding with the nominal incidence. The design incidence is therefore somewhat arbitrary and some designers, ignoring nominal design conditions, may select an incidence best suited to the operating conditions under which the compressor will run. For instance, a negative design incidence may be chosen so that at reduced flow rates a *positive* incidence condition is approached.

Mach number effects

High-speed cascade characteristics are similar to those at low speed until the *critical* Mach number M_c is reached, after which the performance declines. Figure 3.20, taken from Howell (1942) illustrates for a particular cascade tested at varying Mach number and fixed incidence, the drastic decline in pressure rise coefficient up to the *maximum* Mach number at entry M_m, when the cascade is fully choked. When the cascade is choked, no further increase in mass flow through the cascade is possible. The definition of inlet critical Mach number is less precise, but one fairly satisfactory definition (Horlock 1958) is that the maximum *local* Mach number in the cascade has reached unity.

Howell attempted to correlate the decrease in both efficiency and deflection in the range of inlet Mach numbers, $M_c \leq M \leq M_m$ and these are shown in Figure 3.21. By employing this correlation, curves similar to that in Figure 3.20 may be found for each incidence.

One of the principal aims of high-speed cascade testing is to obtain data for determining the values of M_c and M_m. Howell (1945) indicates how, for a typical cascade, M_c and M_m vary with incidence (Figure 3.22)

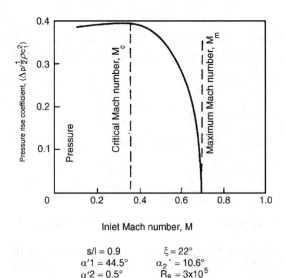

$$s/l = 0.9 \qquad \xi = 22°$$
$$\alpha'_1 = 44.5° \qquad \alpha_2{}^* = 10.6°$$
$$\alpha'_2 = 0.5° \qquad R_e = 3 \times 10^5$$

FIG. 3.20. Variation of cascade pressure rise coefficient with inlet Mach number (Howell 1942). (By courtesy of the Controller of H.M.S.O., Crown copyright reserved).

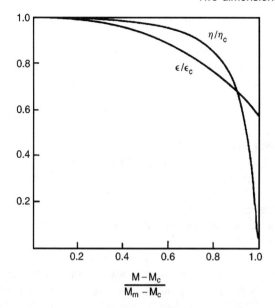

FIG. 3.21. Variation of efficiency and deflection with Mach number (adapted from Howell 1942).

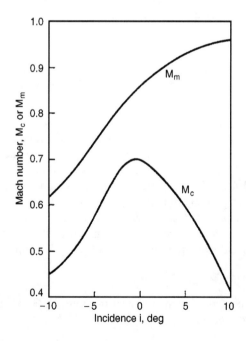

FIG. 3.22. Dependence of critical and maximum Mach numbers upon incidence (Howell 1945). (By courtesy of the Institution of Mechanical Engineers).

Fan blade design (McKenzie)

The cascade tests and design methods evolved by Howell, Carter and others, which were described earlier, established the basis of British axial compressor design. However, a number of empirical factors had to be introduced into the methods in order to correlate actual compressor performance with the performance predicted from cascade data. The system has been in use for many years and has been gradually modified and improved during this time.

McKenzie (1980) has described work done at Rolls-Royce to further develop the correlation of cascade and compressor performance. The work was done on a low-speed four-stage compressor with 50 per cent reaction blading of constant section. The compressor hub to tip radius ratio was 0.8 and a large number of combinations of stagger and camber was tested.

McKenzie pointed out that the deviation rule originated by Howell (1945), i.e. eqns. (3.39) and (3.40a) with $n = 0.5$, was developed from cascade tests performed *without* sidewall suction. Earlier in this chapter it was explained that the consequent thickening of the sidewall boundary layers caused a contraction of the main through-flow (Figure 3.8), resulting in a reduced static pressure rise across the cascade and an increased air deflection. Rolls-Royce conducted a series of tests on C5 profiles with circular arc camber lines using a number of wall suction slots to control the axial velocity ratio (AVR). The deviation angles at mid-span with an AVR of unity were found to be significantly greater than those given by eqn. (3.39).

From cascade tests McKenzie derived the following rule for the deviation angle:

$$\delta = (1.1 + 0.31\theta)(s/l)^{1/3} \tag{3.41}$$

where δ and θ are in degrees. From the results a relationship between the blade stagger angle ξ and the vector mean flow angle α_m was obtained:

$$\tan \xi = \tan \alpha_m - 0.213, \tag{3.42}$$

where $\tan \alpha_m$ is defined by eqn. (3.6). The significance of eqn. (3.42) is, that if the air inlet and outlet angles (α_1 and α_2 respectively) are specified, then the stagger angle *for maximum efficiency* can be determined, assuming that a C5 profile (or a similar profile such as C4) on a circular arc camber line is being considered. Of course, the camber angle θ and the pitch/chord ratio s/l still need to be determined.

In a subsequent paper McKenzie (1988) gave a graph of efficiency in terms of Cp_i and s/l, which was an improved presentation of the correlation given in his earlier paper. The ideal static pressure rise coefficient is defined as

$$Cp_i = 1 - (c_2/c_1)^2. \tag{3.43}$$

McKenzie's efficiency correlation is shown in Figure 3.23, where the ridge line of optimum efficiency is given by

$$s/l = 9 \times (0.567 - Cp_i) \tag{3.44}$$

EXAMPLE 3.3. At the midspan of a proposed fan stator blade the inlet and outlet air angles are to be $\alpha_1 = 58^0$ and $\alpha_2 = 44^0$. Using the data and correlation of McKenzie, determine a suitable blade camber and space–chord ratio.

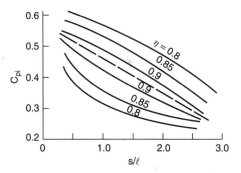

FIG. 3.23. Efficiency correlation (adapted from McKenzie 1988).

Solution. From eqn. (3.6) the vector mean flow angle is found,

$$\tan \alpha_m = \tfrac{1}{2}(\tan \alpha_1 + \tan \alpha_2) = 1.2830.$$

From eqn. (3.42) we get the stagger angle,

$$\tan \xi = \tan \alpha_m - 0.213 = 1.0700.$$

Thus, $\alpha_m = 52.066^0$ and $\xi = 46.937^0$.
  From eqn. (3.43), assuming that AVR $= 1.0$, we find

$$Cp_i = 1 - \left(\frac{\cos \alpha_1}{\cos \alpha_2} \right)^2 = 0.4573.$$

Using the optimum efficiency correlation, eqn. (3.44),

$$s/l = 9 \times (0.567 - 0.4573),$$

$$\therefore s/l = \underline{0.9872}.$$

To determine the blade camber we combine

$$\delta = \alpha_2 - \alpha_2' = \alpha_2 - \xi - \theta/2$$

with eqn. (3.41), to get

$$\theta = \frac{\xi - \alpha_2 + 1.1(s/l)^{1/3}}{0.5 - 0.31(s/l)^{1/3}} = \frac{46.937 - 44 + 1.1 \times 0.9957}{0.5 - 0.31 \times 0.9957}$$

$$\therefore \theta = \underline{21.08^0}.$$

According to McKenzie the correlation gives, for high stagger designs, peak efficiency conditions well removed from stall and is in good agreement with earlier fan blade design methods.

Turbine cascade correlation (Ainley)

Ainley and Mathieson (1951) published a method of estimating the performance of an axial flow turbine and the method has been widely used ever since. In essence the total pressure loss and gas efflux angle for each row of a turbine stage is determined at a single *reference* diameter and under a wide range of inlet conditions. This reference diameter was taken as the arithmetic mean of the rotor and stator rows inner and outer diameters. Dunham and Came (1970) gathered together details of several improvements to the method of Ainley and Mathieson which gave better performance prediction for *small* turbines than did the original method. When the blading is *competently* designed the revised method appears to give reliable predictions of efficiency to within 2% over a wide range of designs, sizes and operating conditions.

Total pressure loss correlations

The overall total pressure loss is composed of three parts, viz. (i) profile loss, (ii) secondary loss, and (iii) tip clearance loss.

(i) A profile loss coefficient is defined as the loss in stagnation pressure across the blade row or cascade, divided by the difference between stagnation and static pressures at blade outlet; i.e.

$$Y_p = \frac{p_{01} - p_{02}}{p_{02} - p_2}. \tag{3.45}$$

In the Ainley method, profile loss is determined initially at zero incidence ($i = 0$). At any other incidence the profile loss ratio $Y_p/Y_{p(i=0)}$ is assumed to be defined by a unique function of the incidence ratio i/i_s (Figure 3.24), where i_s is the stalling incidence. This is defined as the incidence at which $Y_p/Y_{p(i=0)} = 2.0$.

Ainley and Mathieson correlated the profile losses of turbine blade rows against space/chord ratio s/l, fluid outlet angle α_2, blade maximum thickness/chord ratio t/l

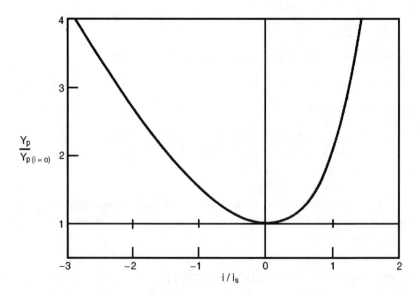

FIG. 3.24. Variation of profile loss with incidence for typical turbine blading (adapted from Ainley and Mathieson 1951).

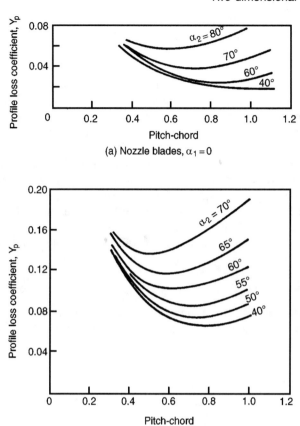

FIG. 3.25. Profile loss coefficients of turbine nozzle and impulse blades at zero incidence ($t/l = 20\%$; $Re = 2 \times 10^5$; $M < 0.6$) (adapted from Ainley and Mathieson 1951).

and blade inlet angle. The variation of $Y_{p(i=0)}$ against s/l is shown in Figure 3.25 for nozzles and impulse blading at various flow outlet angles. The sign convention used for flow angles in a turbine cascade is indicated in Figure 3.27. For other types of blading intermediate between nozzle blades and impulse blades the following expression is employed:

$$Y_{p(i=0)} = \left\{ Y_{p(\alpha1=0)} + \left(\frac{\alpha_1}{\alpha_2}\right)^2 \left[Y_{p(\alpha1=\alpha2)} - Y_{p(\alpha1=0)} \right] \right\} \left(\frac{t/l}{0.2}\right)^{\alpha_1/\alpha_2} \quad (3.46)$$

where all the Y_p's are taken at the same space/chord ratio and flow outlet angle. If rotor blades are being considered, put $3\beta_2$ for α_1 and β_3 for α_2. Equation (3.46) includes a correction for the effect of thickness–chord ratio and is valid in the range $0.15 \leq t/l \leq 0.25$. If the actual blade has a t/l greater or less than the limits quoted, Ainley recommends that the loss should be taken as equal to a blade having

t/l either 0.25 or 0.15. By substituting $\alpha_1 = \alpha_2$ and $t/l = 0.2$ in eqn. (3.46), the zero incidence loss coefficient for the impulse blades $Y_{p(\alpha1=\alpha2)}$ given in Figure 3.25 is recovered. Similarly, with $\alpha_1 = 0$ at $t/l = 0.2$ in eqn. (3.46) gives $Y_{p(\alpha1=0)}$ of Figure 3.25.

A feature of the losses given in Figure 3.25 is that, compared with the impulse blades, the nozzle blades have a much lower loss coefficient. This trend confirms the results shown in Figure 3.14, that flow in which the mean pressure is falling has a lower loss coefficient than a flow in which the mean pressure is constant or increasing.

(ii) The secondary losses arise from complex three-dimensional flows set up as a result of the end wall boundary layers passing through the cascade. There is substantial evidence that the end wall boundary layers are convected inwards along the suction-surface of the blades as the main flow passes through the blade row, resulting in a serious mal-distribution of the flow, with losses in stagnation pressure often a significant fraction of the total loss. Ainley found that secondary losses could be represented by

$$C_{Ds} = \lambda C_L^2 / (s/l) \tag{3.47}$$

where λ is parameter which is a function of the flow acceleration through the blade row. From eqn. (3.17), together with the definition of Y, eqn. (3.45) for incompressible flow, $C_D = Y(s/l) \cos^3 \alpha_m / \cos^2 \alpha_2$, hence

$$Y_s = \frac{C_{Ds} \cos^2 \alpha_2}{(s/l) \cos^3 \alpha_m} = \lambda \left(\frac{C_L}{s/l}\right)^2 \frac{\cos^2 \alpha_2}{\cos^3 \alpha_m} = \lambda Z \tag{3.48}$$

where Z is the blade aerodynamic loading coefficient. Dunham (1970) subsequently found that this equation was not correct for blades of low aspect ratio, as in small turbines. He modified Ainley's result to include a better correlation with aspect ratio and at the same time simplified the flow acceleration parameter. The correlation, given by Dunham and Came (1970), is

$$Y_s = 0.0334 \left(\frac{l}{H}\right) \left(\frac{\cos \alpha_2}{\cos \alpha_1'}\right) Z \tag{3.49}$$

and this represents a significant improvement in the prediction of secondary losses using Ainley's method.

Recently, more advanced methods of predicting losses in turbine blade rows have been suggested which take into account the *thickness* of the entering boundary layers on the annulus walls. Came (1973) measured the secondary flow losses on *one* end wall of several turbine cascades for various thicknesses of inlet boundary layer. He correlated his own results, and those of several other investigators, and obtained a modified form of Dunham's earlier result, viz.,

$$Y_s = \left(0.25Y_1 \frac{\cos^2 \alpha_1}{\cos^2 \alpha_2} + 0.009 \frac{l}{H}\right) \left(\frac{\cos \alpha_2}{\cos \alpha_1'}\right) Z - Y_1 \tag{3.50}$$

which is the net secondary loss coefficient for *one* end wall only and where Y_1 is a mass-averaged inlet boundary layer total pressure loss coefficient. It is evident that the increased accuracy obtained by use of eqn. (3.50) requires the additional effort of calculating the wall boundary layer development. In *initial* calculations of performance it is probably sufficient to use the earlier result of Dunham and Came, eqn. (3.49), to achieve a reasonably accurate result.

(iii) The tip clearance loss coefficient Y_k depends upon the blade loading Z and the size and nature of the clearance gap k. Dunham and Came presented an amended version of Ainley's original result for Y_k:

$$Y_k = B \left(\frac{l}{H} \right) \left(\frac{k}{l} \right)^{0.78} Z \qquad (3.51)$$

where $B = 0.5$ for a plain tip clearance, 0.25 for shrouded tips.

Reynolds number correction

Ainley and Mathieson (1951) obtained their data for a mean Reynolds number of 2×10^5 based on the mean chord and exit flow conditions from the turbine state. They recommended for lower Reynolds numbers, down to 5×10^4, that a correction be made to stage efficiency according to the rough rule:

$$(1 - \eta_{tt}) \propto Re^{-1/5}.$$

Dunham and Came (1970) gave an optional correction which is applied directly to the sum of the profile and secondary loss coefficients for a blade row using the Reynolds number *appropriate* to that row. The rule is:

$$Y_p + Y_s \propto Re^{-1/5}.$$

Flow outlet angle from a turbine cascade

It was pointed out by Ainley (1948) that the method of defining deviation angle as adopted in several well-known compressor cascade correlations had proved to be impracticable for turbine blade cascade. In order to predict fluid outlet angle α_2, steam turbine designers had made much use of the simple empirical rule that

$$\alpha_2 = \cos^{-1} \Theta / s \qquad (3.52a)$$

where Θ is the opening at the throat, depicted in Figure 3.26, and s is the pitch. This widely used rule gives a very good approximation to measured pitchwise averaged flow angles when the outlet Mach number is at or close to unity. However, at low Mach numbers substantial variations have been found between the rule and observed flow angles. Ainley and Mathieson (1951) recommended that for low outlet Mach numbers $0 < M_2 \leq 0.5$, the following rule be used:

$$\alpha_2 = f(\cos^{-1} \Theta / s) + 4s/e \text{ (deg)} \qquad (3.52b)$$

where $f(\cos^{-1} \Theta / s) = -11.15 + 1.154 \cos^{-1} \Theta / s$ and $e = j^2/(8z)$ is the mean radius of curvature of the blade suction surface between the throat and the trailing edge. At a gas outlet Mach number of unity Ainley and Mathieson assumed, for a turbine blade row, that

$$\alpha_2 = \cos^{-1} A_t / A_{n2} \qquad (3.52c)$$

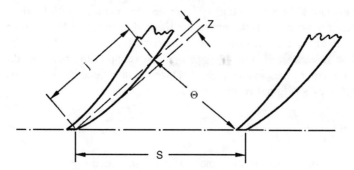

FIG. 3.26. Details near turbine cascade exit showing "throat" and suction-surface curvature parameters.

where A_t is the passage throat area and A_{n2} is the annulus area in the reference plane downstream of the blades. If the annulus walls at the ends of the cascade are not flared then eqn. (3.52c) is the same as eqn. (3.52a). Between $M_2 = 0.5$ and $M_2 = 1.0$ a linear variation of α_2 can be reasonably assumed in the absence of any other data.

Comparison of the profile loss in a cascade and in a turbine stage

The aerodynamic efficiency of an axial-flow turbine is significantly less than that predicted from measurements made on equivalent cascades operating under steady flow conditions. The importance of flow unsteadiness originating from the wakes of a preceding blade row was studied by Lopatitiskii *et al.* (1969) who reported that the rotor blade profile loss was (depending on blade geometry and Reynolds number) between two and four times greater than that for an equivalent cascade operating with the same flow. Hodson (1984) made an experimental investigation of the rotor to stator interaction using a large-scale, low-speed turbine, comparing the results with those of a rectilinear cascade of identical geometry. Both tunnels were operated at a Reynolds number of 3.15×10^5. Hodson reported that the turbine rotor midspan profile loss was approximately 50 per cent higher than that of the rectilinear cascade. Measurements of the shear stress showed that as a stator wake is convected through a rotor blade passage, the *laminar* boundary layer on the suction surface undergoes transition in the vicinity of the wake. The 50 per cent increase in profile loss was caused by the time-dependent transitional nature of the boundary layers. The loss increase was largely independent of spacing between the rotor and the stator.

In a turbine stage the interaction between the two rows can be split into two parts: (a) the effects of the potential flow; and (b) the effects due to wake interactions. The effects of the potential influence extend upstream and downstream and decay exponentially with a length scale typically of the order of the blade chord or pitch. Some aspects of these decay effects are studied in Chapter 6 under the heading "Actuator Disc Approach". In contrast, blade wakes are convected downstream of the blade row with very little mixing with the mainstream flow. The wakes tend to persist even where the blade rows of a turbomachine are very widely spaced.

A designer usually assumes that the blade rows of an axial-flow turbomachine are sufficiently far apart that the flow is steady in both the stationary and rotating frames of reference. The flow in a real machine, however, is unsteady both as a result of the relative motion of the blade wakes between the blade rows and the potential influence. In modern turbomachines, the spacing between the blade rows is typically of the order of 1/4 to 1/2 of a blade chord. As attempts are made to make turbomachines more compact and blade loadings are increased, then the levels of unsteadiness will increase.

The earlier Russian results showed that the losses due to flow unsteadiness were greater in turbomachines of high reaction and low Reynolds number. With such designs, a larger proportion of the blade suction surface would have a laminar boundary layer and would then exhibit a correspondingly greater profile loss as a result of the wake-induced boundary layer transition.

Optimum space–chord ratio of turbine blades (Zweifel)

It is worth pondering a little upon the effect of the space–chord ratio in turbine blade rows as this is a factor strongly affecting efficiency. Now if the spacing between blades is made small, the fluid then tends to receive the maximum amount of guidance from the blades, but the friction losses will be very large. On the other hand, with the same blades spaced well apart, friction losses are small but, because of poor fluid guidance, the losses resulting from flow separation are high. These considerations led Zweifel (1945) to formulate his criterion for the optimum space–chord ratio of blading having large deflection angles. Essentially, *Zweifel's criterion* is simply that the ratio (ψ_T) of the actual to an "ideal" tangential blade loading, has a certain constant value for minimum losses. The tangential blade loads are obtained from the real and ideal pressure distributions on both blade surfaces, as described below.

Figure 3.27 indicates a typical pressure distribution around one blade in a turbine cascade, curves P and S corresponding to the pressure (or concave) side and suction (convex) side respectively. The pressures are projected parallel to the cascade front so that the area enclosed between the curves S and P represents the *actual tangential blade load per unit span*,

$$Y = \rho s c_x (c_{y2} + c_{y1}),\tag{3.53}$$

cf. eqn. (3.3) for a compressor cascade.

It is instructive to examine the pressures along the blade surfaces. Assuming incompressible flow the static inlet pressure is $p_1 = p_0 - \frac{1}{2}\rho c_1^2$; if losses are also ignored the outlet static pressure $p_2 = p_0 - \frac{1}{2}\rho c_2^2$. The pressure on the P side remains high at first (p_0 being the maximum, attained only at the stagnation point), then falls sharply to p_2. On the S side there is a rapid decrease in static pressure from the leading edge, but it may even rise again towards the trailing edge. The closer the blade spacing s the smaller the load Y becomes (eqn. (3.53)). Conversely, wide spacing implies an increased load with pressure rising on the P side and falling

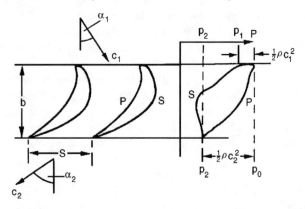

FIG. 3.27. Pressure distribution around a turbine cascade blade (after Zweifel 1945).

on the S side. Now, whereas the static pressure can never rise above p_0 on the P surface, very low pressures are possible, at least in theory on the S surface. However, the pressure rise towards the trailing edge is limited in practice if flow separation is to be avoided, which implies that the load carried by the blade is restricted.

To give some idea of blade load capacity, the real pressure distribution is compared with an ideal pressure distribution giving a maximum load Y_{id} without risk of fluid separation on the S surface. Upon reflection, one sees that these conditions for the ideal load are fulfilled by p_0 acting over the *whole* P surface and p_2 acting over the *whole* S surface. With this ideal pressure distribution (which cannot, of course, be realised), the tangential load per unit span is,

$$Y_{id} = \tfrac{1}{2}\rho c_2^2 b \tag{3.54}$$

and, therefore,

$$\psi_T = Y/Y_{id} = 2(s/b)\cos^2\alpha_2(\tan\alpha_1 + \tan\alpha_2) \tag{3.55}$$

after combining eqns. (3.53) and (3.54) together with angles defined by the geometry of Figure 3.27.

Zweifel found from a number of experiments on turbine cascades that for minimum losses the value of ψ_T was approximately 0.8. Thus, for specified inlet and outlet angles the optimum space–chord ratio can be estimated. However, according to Horlock (1966). Zweifel's criterion predicts optimum space–chord ratio for the data of Ainley *only for outlet angles of 60 to 70 deg*. At other outlet angles it does not give an accurate estimate of optimum space–chord ratio.

References

Ainley, D. G. (1948). Performance of axial flow turbines. *Proc. Instn. Mech. Engrs.*, **159**.
Ainley, D. G. and Mathieson, G. C. R. (1951). A method of performance estimation for axial flow turbines. *ARC. R. and M. 2974.*

Bryer, D. W. and Pankhurst, R. C. (1971). *Pressure-probe Methods for Determining Wind Speed and Flow Direction*. National Physical Laboratory, HMSO.

Came, P. M. (1973). Secondary loss measurements in a cascade of turbine blades. *Proc. Instn. Mech. Engrs*. Conference Publication 3.

Carter, A. D. S. (1950). Low-speed performance of related aerofoils in cascade. ARC. Current Paper, No. 29.

Carter, A. D. S., Andrews, S. J. and Shaw, H. (1950). Some fluid dynamic research techniques. *Proc. Instn. Mech. Engrs.*, **163**.

Carter, A. D. S. (1948). Three-dimensional flow theories for axial compressors and turbines. *Proc. Instn. Mech. Engrs.*, **159**.

Csanady, G. T. (1964). *Theory of Turbomachines*. McGraw-Hill, New York.

Dixon, S. L. (1978). Measurement of flow direction in a shear flow. *J. Physics E: Scientific Instruments*, **2**, 31–4.

Dowden, R. R. (1972). *Fluid Flow Measurement – a Bibliography*. BHRA.

Dunham, J. (1970). A review of cascade data on secondary losses in turbines. *J. Mech. Eng. Sci.*, **12**.

Dunham, J. (1974). A parametric method of turbine blade profile design. *Am. Soc. Mech. Engrs*. Paper 74-GT-119.

Dunham, J. and Came, P. (1970). Improvements to the Ainley-Mathieson method of turbine performance prediction. *Trans. Am. Soc. Mech. Engrs.*, Series A, **92**.

Felix, A. R. (1957). Summary of 65-Series compressor blade low-speed cascade data by use of the carpet-plotting technique. *NACA T.N.* 3913.

Glauert, H. (1959). *Aerofoil and Airscrew Theory*. (2nd edn). Cambridge University Press.

Hay, N., Metcalfe, R. and Reizes, J. A. (1978). A simple method for the selection of axial fan blade profiles. *Proc. Instn Mech Engrs.*, **192**, (25) 269–75.

Herrig, L. J., Emery, J. C. and Erwin, J. R. (1957). Systematic two-dimensional cascade tests of NACA 65-Series compressor blades at low speeds. *NACA T.N.* 3916.

Hodson, H. P. (1984). Boundary layer and loss measurements on the rotor of an axial-flow turbine. *J. Eng. for Gas Turbines and Power. Trans Am. Soc. Mech. Engrs.*, **106**, 391–9.

Horlock, J. H. (1958). *Axial Flow Compressors*. Butterworths. (1973 reprint with supplemental material, Huntington, New York: Krieger).

Horlock, J. H. (1966). *Axial-flow Turbines*. Butterworths. (1973 reprint with corrections, Huntington, New York: Krieger).

Howell, A. R. (1942). The present basis of axial flow compressor design: Part I, Cascade theory and performance. *ARC R and M*. 2095.

Howell, A. R. (1945). Design of axial compressors. *Proc. Instn. Mech. Engrs.*, **153**.

Howell, A. R. (1945). Fluid dynamics of axial compressors. *Proc. Instn. Mech. Engrs.*, **153**.

Lieblein, S. (1959). Loss and stall analysis of compressor cascades. *Trans. Am. Soc. Mech. Engrs.*, Series D, **81**.

Lieblein, S. and Roudebush, W. H. (1956). Theoretical loss relation for low-speed 2D cascade flow. *NACA T.N.* 3662.

Lopatitskii, A. O. *et al.* (1969). Energy losses in the transient state of an incident flow on the moving blades of turbine stages. *Energomashinostroenie*, **15**.

McKenzie, A. B. (1980). The design of axial compressor blading based on tests of a low speed compressor. *Proc. Instn. Mech. Engrs.*, **194**, 6.

McKenzie, A. B. (1988). The selection of fan blade geometry for optimum efficiency. *Proc. Instn. Mech. Engrs.*, **202**, A1, 39–44.

National Aeronautics and Space Administration, (1965). Aerodynamic design of axial-flow compressors. NASA SP 36.

Sieverding, C. H. (1975). Pressure probe measurements in cascades. In *Modern Methods of Testing Rotating Components of Turbomachines*, AGARDograph 207.

Sieverding, C. H. (1985). Aerodynamic development of axial turbomachinery blading. In *Thermodynamics and Fluid Mechanics of Turbomachinery*, Vol. 1 (A.S. Ücer, P. Stow and Ch. Hirsch, eds) pp. 513–65. Martinus Nijhoff.

Swann, W. C. (1961). A practical method of predicting transonic compressor performance. *Trans. Am. Soc. Mech. Engrs.*, Series A, **83**.

Todd, K. W. (1947). Practical aspects of cascade wind tunnel research. *Proc. Instn. Mech. Engrs.*, **157**.

Zweifel, O. (1945). The spacing of turbomachine blading, especially with large angular deflection. *Brown Boveri Rev.*, **32**, 12.

Problems

1. Experimental compressor cascade results suggest that the stalling lift coefficient of a cascade blade may be expressed as

$$C_L \left(\frac{c_1}{c_2} \right)^3 = 2.2$$

where c_1 and c_2 are the entry and exit velocities. Find the stalling inlet angle for a compressor cascade of space–chord ratio unity if the outlet air angle is 30 deg.

2. Show, for a turbine cascade, using the angle notation of Figure 3.27, that the lift coefficient is

$$C_L = 2(s/l)(\tan \alpha_1 + \tan \alpha_2) \cos \alpha_m + C_D \tan \alpha_m$$

where $\tan \alpha_m = \frac{1}{2}(\tan \alpha_2 - \tan \alpha_1)$ and $C_D = \mathrm{Drag}/(\frac{1}{2}\rho c_m{}^2 l)$.

A cascade of turbine nozzle vanes has a blade inlet angle $\alpha_1' = 0$ deg, a blade outlet angle α_2' of 65.5 deg, a chord length l of 45 mm and an axial chord b of 32 mm. The flow entering the blades is to have zero incidence and an estimate of the deviation angle based upon similar cascades is that δ will be about 1.5 deg at low outlet Mach number. If the blade load ratio ψ_T defined by eqn. (3.55) is to be 0.85, estimate a suitable space–chord ratio for the cascade.

Determine the drag and lift coefficients for the cascade given that the profile loss coefficient

$$\lambda = \Delta p_0 / (\tfrac{1}{2}\rho c_2^2) = 0.035.$$

3. A compressor cascade is to be designed for the following conditions:

Nominal fluid outlet angle	α_2^*	=	30 deg
Cascade camber angle	θ	=	30 deg
Pitch/chord ratio	s/l	=	1.0
Circular arc camberline	a/l	=	0.5

Using Howell's curves and his formula for nominal deviation, determine the nominal incidence, the actual deviation for an incidence of $+2.7$ deg and the approximate lift coefficient at this incidence.

4. A compressor cascade is built with blades of circular arc camber line, a space/chord ratio of 1.1 and blade angles of 48 and 21 deg at inlet and outlet. Test data taken from the cascade shows that at zero incidence ($i = 0$) the deviation $\delta = 8.2$ deg and the total pressure loss coefficient $\bar{\omega} = \Delta p_0/(\frac{1}{2}\rho c_1{}^2) = 0.015$. At positive incidence over a limited range ($0 \leqslant i \leqslant 6°$) the variation of both δ and $\bar{\omega}$ for this particular cascade can be represented with sufficient accuracy by linear approximations, viz.

$$\frac{d\delta}{di} = 0.06, \qquad \frac{d\bar{\omega}}{di} = 0.001$$

where i is in degrees.

For a flow incidence of 5.0 deg determine

(i) the flow angles at inlet and outlet;
(ii) the diffuser efficiency of the cascade;
(iii) the static pressure rise of air with a velocity 50 m/s normal to the plane of the cascade.

Assume density of air is 1.2 kg/m³.

5. (a) A cascade of compressor blades is to be designed to give an outlet air angle α_2 of 30 deg for an inlet air angle α_1 of 50 deg measured from the normal to the plane of the cascade. The blades are to have a *parabolic arc* camber line with $a/l = 0.4$ (i.e. the fractional distance along the chord to the point of maximum camber). Determine the space/chord ratio and blade outlet angle if the cascade is to operate at zero incidence and nominal conditions. You may assume the linear approximation for nominal deflection of Howell's cascade correlation:

$$\epsilon^* = (16 - 0.2\alpha_2^*)(3 - s/l)\,\text{deg}$$

as well as the formula for nominal deviation:

$$\delta^* = \left[0.23\left(\frac{2a}{l}\right)^2 + \frac{\alpha_2{}^*}{500}\right]\theta\sqrt{\frac{s}{l}}\,\text{deg.}$$

(b) The space/chord ratio is now changed to 0.8, but the blade angles remain as they are in part (a) above. Determine the lift coefficient when the incidence of the flow is 2.0 deg. Assume that there is a linear relationship between ϵ/ϵ^* and $(i - i^*)/\epsilon^*$ over a limited region, viz. at $(i - i^*)/\epsilon^* = 0.2$, $\epsilon/\epsilon^* = 1.15$ and at $i = i^*$, $\epsilon/\epsilon^* = 1$. In this region take $C_D = 0.02$.

6. (a) Show that the pressure rise coefficient $C_p = \Delta p/(\frac{1}{2}\rho c_x^2)$ of a compressor cascade is related to the diffuser efficiency η_D and the total pressure loss coefficient ζ by the following expressions:

$$C_p = \eta_D(1 - \sec^2\alpha_2/\sec^2\alpha_1) = 1 - (\sec^2\alpha_2 + \zeta)/\sec^2\alpha_1$$

where
$$\eta_d = \Delta p/\{\tfrac{1}{2}\rho(c_1^2 - c_2^2)\}$$

$$\zeta = \Delta p_0/(\tfrac{1}{2}\rho c_x^2)$$

$\alpha_1, \alpha_2 = $ flow angles at cascade inlet and outlet.

(b) Determine a suitable *maximum* inlet flow angle of a compressor cascade having a space/chord ratio 0.8 and $\alpha_2 = 30$ deg when the diffusion factor D is to be limited to 0.6. The definition of diffusion factor which should be used is the early Lieblein formula (1956),

$$D = \left(1 - \frac{\cos\alpha_1}{\cos\alpha_2}\right) + \left(\frac{s}{l}\right)\frac{\cos\alpha_1}{2}(\tan\alpha_1 - \tan\alpha_2).$$

(c) The stagnation pressure loss derived from flow measurements on the above cascade is 149 Pa when the inlet velocity c_1 is 100 m/s at an air density ρ of 1.2 kg/m³. Determine the values of

 (i) pressure rise;
 (ii) diffuser efficiency;
 (iii) drag and lift coefficients.

7 (a) A set of circular arc fan blades, camber $\theta = 8$ deg, are to be tested in a cascade wind tunnel at a space/chord ratio, $s/l = 1.5$, with a stagger angle $\xi = 68$ deg. Using McKenzie's method of correlation and assuming optimum conditions at an axial velocity ratio of unity, obtain values for the air inlet and outlet angles.

(b) Assuming the values of the derived air angles are correct and that the cascade has an effective lift/drag ratio of 18, determine

 (i) the coefficient of lift of the blades;
 (ii) the efficiency of the cascade (treating it as a diffuser).

CHAPTER 4

Axial-flow Turbines: Two-dimensional Theory

Power is more certainly retained by wary measures than by daring counsels.
(TACITUS, *Annals.*)

Introduction

The simplest approach to the study of axial-flow turbines (and also axial-flow compressors) is to assume that the flow conditions prevailing at the mean radius fully represent the flow at all other radii. This two-dimensional analysis at the *pitch-line* can provide a reasonable approximation to the actual flow, if the ratio of blade height to mean radius is small. When this ratio is large, however, as in the final stages of a steam turbine or, in the first stages of an axial compressor, a three-dimensional analysis is required. Some important aspects of three-dimensional flows in axial turbomachines are discussed in Chapter 6. Two further assumptions are, that radial velocities are zero, and that the flow is invariant along the circumferential direction (i.e. there are no "blade-to-blade" flow variations).

In this chapter the presentation of the analysis has been devised with compressible flow effects in mind. This approach is then applicable to both steam and gas turbines provided that, in the former case, the steam condition remains wholly within the *vapour* phase (i.e. superheat region). Much early work concerning flows in steam turbine nozzles and blade rows are reported in Stodola (1945), Kearton (1958) and Horlock (1960).

Velocity diagrams of the axial turbine stage

The axial turbine stage comprises a row of fixed guide vanes or nozzles (often called a *stator* row) and a row of moving blades or buckets (a *rotor* row). Fluid enters the stator with absolute velocity c_1 at angle α_1 and accelerates to an absolute velocity c_2 at angle α_2 (Figure 4.1). All angles are measured from the axial (x) direction. The *sign convention* is such that angles and velocities as drawn in Figure 4.1 will be taken as positive throughout this chapter. From the velocity diagram, the rotor inlet *relative* velocity w_2, at an angle β_2, is found by subtracting, vectorially, the blade speed U from the absolute velocity c_2. The relative flow within the rotor accelerates to velocity w_3 at an angle β_3 at rotor outlet; the corresponding absolute flow (c_3, α_3) is obtained by adding, vectorially, the blade speed U to the relative velocity w_3.

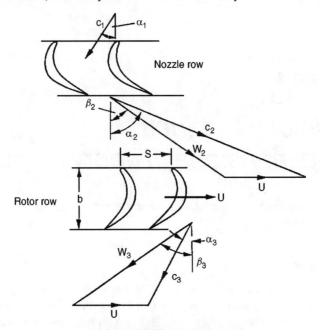

FIG. 4.1. Turbine stage velocity diagrams.

The continuity equation for uniform, steady flow is,

$$\rho_1 A_1 c_{x1} = \rho_2 A_2 c_{x2} = \rho_3 A_3 c_{x3}.$$

In two-dimensional theory of turbomachines it is usually assumed, for simplicity, that the axial velocity remains constant i.e. $c_{x1} = c_{x2} = c_{x3} = c_x$.

This must imply that,

$$\rho_1 A_1 = \rho_2 A_2 = \rho_3 A_3 = \text{constant}. \tag{4.1}$$

Thermodynamics of the axial turbine stage

The work done on the rotor by unit mass of fluid, the specific work, equals the stagnation enthalpy drop incurred by the fluid passing through the stage (assuming adiabatic flow), or,

$$\Delta W = \dot{W}/\dot{m} = h_{01} - h_{03} = U(c_{y2} + c_{y3}). \tag{4.2}$$

In eqn. (4.2) the absolute tangential velocity components (c_y) are *added*, so as to adhere to the agreed sign convention of Figure 4.1. As no work is done in the nozzle row, the stagnation enthalpy across it remains constant and

$$h_{01} = h_{02}. \tag{4.3}$$

Writing $h_0 = h + \frac{1}{2}(c_x^2 + c_y^2)$ and using eqn. (4.3) in eqn. (4.2) we obtain,

$$h_{02} - h_{03} = (h_2 - h_3) + \frac{1}{2}(c_{y2}^2 - c_{y3}^2) = U(c_{y2} + c_{y3}),$$

hence,

$$(h_2 - h_3) + \tfrac{1}{2}(c_{y2} + c_{y3})[(c_{y2} - U) - (c_{y3} + U)] = 0.$$

It is observed from the velocity triangles of Figure 4.1 that $c_{y2} - U = w_{y2}, c_{y3} + U = w_{y3}$ and $c_{y2} + c_{y3} = w_{y2} + w_{y3}$. Thus,

$$(h_2 - h_3) + \tfrac{1}{2}(w_{y2}^2 - w_{y3}^2) = 0.$$

Add and subtract $\tfrac{1}{2}c_x^2$ to the above equation

$$h_2 + \tfrac{1}{2}w_2^2 = h_3 + \tfrac{1}{2}w_3^2 \text{ or } h_{02\text{rel}} = h_{03\text{rel}}. \qquad (4.4)$$

Thus, we have proved that the *relative* stagnation enthalpy, $h_{0\text{rel}} = h + \tfrac{1}{2}w^2$, remains unchanged through the rotor of an axial turbomachine. It is implicitly assumed that no radial shift of the streamlines occurs in this flow. In a *radial flow* machine a more general analysis is necessary (see Chapter 7) which takes account of the blade speed change between rotor inlet and outlet.

A Mollier diagram showing the change of state through a complete turbine stage, including the effects of irreversibility, is given in Figure 4.2.

Through the nozzles, the state point moves from 1 to 2 and the static pressure decreases from p_1 to p_2. In the rotor row, the absolute static pressure reduces (in general) from p_2 to p_3. It is important to note that the conditions contained in eqns. (4.2)–(4.4) are all satisfied in the figure.

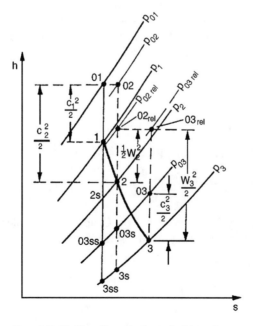

FIG. 4.2. Mollier diagram for a turbine stage.

Stage losses and efficiency

In Chapter 2 various definitions of efficiency for complete turbomachines were given. For a *turbine stage* the total-to-total efficiency is,

$$\eta_{tt} = \frac{\text{actual work output}}{\text{ideal work output when operating to same back pressure}}$$

$$= (h_{01} - h_{03})/(h_{01} - h_{03ss}).$$

At the entry and exit of a *normal* stage the flow conditions (absolute velocity and flow angle) are identical, i.e. $c_1 = c_3$ and $a_1 = a_3$. If it is assumed that $c_{3ss} = c_3$, which is a reasonable approximation, the total-to-total efficiency becomes.

$$\eta_{tt} = (h_1 - h_3)/(h_1 - h_{3ss})$$

$$\doteq (h_1 - h_3)/\{(h_1 - h_3) + (h_3 - h_{3s}) + (h_{3s} - h_{3ss})\}. \tag{4.5}$$

Now the slope of a constant pressure line on a Mollier diagram is $(\partial h/\partial s)_p = T$, obtained from eqn. (2.18). Thus, for a finite change of enthaply in a constant pressure process, $\Delta h \doteq T \Delta s$ and, therefore,

$$h_{3s} - h_{3ss} \doteq T_3(s_{3s} - s_{3ss}), \tag{4.6a}$$

$$h_2 - h_{2s} \doteq T_2(s_2 - s_{2s}). \tag{4.6b}$$

Noting, from Figure 4.2, that $s_{3s} - s_{3ss} = s_2 - s_{2s}$, the last two equations can be combined to give

$$h_{3s} - h_{3ss} = (T_3/T_2)(h_2 - h_{2s}). \tag{4.7}$$

The effects of irreversibility through the stator and rotor are expressed by the differences in static enthalpies, $(h_2 - h_{2s})$ and $(h_3 - h_{3s})$ respectively. Non-dimensional enthalpy "loss" coefficients can be defined in terms of the exit kinetic energy from each blade row. Thus, for the nozzle row,

$$h_2 - h_{2s} = \tfrac{1}{2}c_2^2\zeta_N. \tag{4.8a}$$

For the rotor row,

$$h_3 - h_{3s} = \tfrac{1}{2}w_3^2\zeta_R. \tag{4.8b}$$

Combining eqns. (4.7) and (4.8) with eqn. (4.5) gives

$$\eta_{tt} = \left[1 + \frac{\zeta_R w_3^2 + \zeta_N c_2^2 T_3/T_2}{2(h_1 - h_3)}\right]^{-1}. \tag{4.9}$$

When the exit velocity is not recovered (in Chapter 2, examples of such cases are quoted) a total-to-static efficiency for the stage is used.

$$\eta_{ts} = (h_{01} - h_{03})/(h_{01} - h_{3ss})$$

$$= \left[1 + \frac{\zeta_R w_3^2 + \zeta_N c_2^2 T_3/T_2 + c_1^2}{2(h_1 - h_3)}\right]^{-1}, \tag{4.10}$$

where, as before, it is assumed that $c_1 = c_3$.

In initial calculations or, in cases where the static temperature drop through the rotor is not large, the temperature ratio T_3/T_2 is set equal to unity, resulting in the more convenient approximations,

$$\eta_{tt} = \left[1 + \frac{\zeta_R w_3^2 + \zeta_N c_2^2}{2(h_1 - h_3)}\right]^{-1}, \tag{4.9a}$$

$$\eta_{ts} = \left[1 + \frac{\zeta_R w_3^2 + \zeta_N c_2^2 + c_1^2}{2(h_1 - h_3)}\right]^{-1}. \tag{4.10a}$$

So that estimates can be made of the efficiency of a proposed turbine stage as part of the preliminary design process, some means of determining the loss coefficients is required. Several methods for doing this are available with varying degrees of complexity. The blade row method proposed by Soderberg (1949) and reported by Horlock (1966), although old, is still remarkably valid despite its simplicity. Ainley and Mathieson (1952) correlated the profile loss coefficients for nozzle blades (which give 100% expansion) and impulse blades (which give 0% expansion) against flow deflection and pitch/chord ratio for stated values of Reynolds number and Mach number. Details of their method are given in Chapter 3. For blading between impulse and reaction the profile loss is derived from a combination of the impulse and reaction profile losses (see eqn. (3.42)). Horlock (1966) has given a detailed comparison between these two methods of loss prediction. A refinement of the Ainley and Mathieson prediction method was later published by Dunham and Came (1970).

Various other methods of predicting the efficiency of axial flow turbiness have been devised such as those of Craig and Cox (1971), Kacker and Okapuu (1982) and Wilson (1987). It was Wilson who, tellingly, remarked that despite the emergence of "computer programs of great power and sophistication", and "generally incorporating computational fluid dynamics", that these have not yet replaced the preliminary design methods mentioned above. It is, clearly, essential for a design to converge as closely as possible to an optimum configuration using preliminary design methods before carrying out the final design refinements using computational fluid dynamics.

Soderberg's correlation

One method of obtaining design data on turbine blade losses is to assemble information on the overall efficiencies of a wide variety of turbines, and from this calculate the individual blade row losses. This system was developed by Soderberg (1949) from a large number of tests performed on steam turbines and on cascades, and extended to fit data obtained from small turbines with very low aspect ratio blading (small height–chord). Soderberg's method was intended only for turbines conforming to the standards of "good design", as discussed below. The method was used by Stenning (1953) to whom reference can be made.

A paper by Horlock (1960) has critically reviewed several different and widely used methods of obtaining design data for turbines. His paper confirms the claim

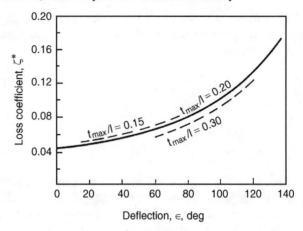

FIG. 4.3. Soderberg's correlation of turbine blade loss coefficient with fluid deflection
(adapted from Horlock (1960)).

made for Soderberg's correlation that, although based on relatively few parameters, it is of comparable accuracy with the best of the other methods.

Soderberg found that for the *optimum* space–chord ratio, turbine blade losses (with "full admission" to the complete annulus) could be correlated with space-chord ratio, blade aspect ratio, blade thickness–chord ratio and Reynolds number. Soderberg used *Zweifel's criterion* (see Chapter 3) to obtain the optimum space-chord ratio of turbine cascades based upon the cascade geometry. Zweifel suggested that the aerodynamic load coefficient ψ_T should be approximately 0.8. Following the notation of Figure 4.1

$$\psi_T = 0.8 = 2(s/b)(\tan\alpha_1 + \tan\alpha_2)\cos^2\alpha_2. \tag{4.11}$$

The optimum space–chord ratio may be obtained from eqn. (4.11) for specified values of α_1 and α_2.

For turbine blade rows operating at this load coefficient, with a Reynolds number of 10^5 and aspect ratio $H/b =$ blade height/axial chord) of 3, the "nominal" loss coefficient ζ^* is a simple function of the fluid deflection angle $\epsilon = \alpha_1 + \alpha_2$, for a given thickness–chord ratio (t_{max}/l). Values of ζ^* are drawn in Figure 4.3 as a function of deflection ϵ, for several ratios of t_{max}/l. A frequently used analytical simplification of this correlation (for $t_{max}/l = 0.2$), which is useful in initial performance calculations, is

$$\zeta^* = 0.04 + 0.06\left(\frac{\epsilon}{100}\right)^2. \tag{4.12}$$

This expression fits Soderberg's curve (for $t_{max}/l = 0.2$) quite well for $\epsilon \leq 120°$, but is less accurate at higher deflections. For turbine rows operating at zero incidence, which is the basis of Soderberg's correlation, the fluid deflection is little different from the blading deflection since, for *turbine cascades*, deviations are usually small. Thus, for a nozzle row, $\epsilon = \epsilon_N = \alpha_2' + \alpha_1'$ and for a rotor row, $\epsilon = \epsilon_R = \beta_2' + \beta_3'$ can be used (the prime referring to the actual blade angles).

If the aspect ratio H/b is other than 3, a correction to the nominal loss coefficient ζ^* is made as follows:

for nozzles,

$$1 + \zeta_1 = (1 + \zeta^*)(0.993 + 0.021b/H), \tag{4.13a}$$

for rotors,

$$1 + \zeta_1 = (1 + \zeta^*)(0.975 + 0.075b/H), \tag{4.13b}$$

where ζ_1 is the loss coefficient at a Reynolds number of 10^5.

A further correction can be made if the Reynolds number is different from 10^5. As used in this section, Reynolds number is based upon exit velocity c_2 and the hydraulic mean diameter D_h at the throat section.

$$Re = \rho_2 c_2 D_h / \mu, \tag{4.14}$$

where

$$D_h = 2sH \cos \alpha_2 / (s \cos \alpha_2 + H).$$

(N.B. Hydraulic mean diameter $= 4 \times$ flow area $\div$ wetted perimeter.)

The Reynolds number correction is

$$\zeta_2 = \left(\frac{10^5}{Re} \right)^{1/4} \zeta_1. \tag{4.15}$$

Soderberg's method of loss prediction gives turbine efficiencies with an error of less than 3% over a wide range of Reynolds number and aspect ratio when additional corrections are included to allow for tip leakage and disc friction. An approximate correction for tip clearance may be incorporated by the simple expedient of multiplying the final calculated stage efficiency by the ratio of "blade" area to *total* area (i.e. "blade" area + clearance area).

Types of axial turbine design

The process of choosing the best turbine design for a given application usually involves juggling several parameters which may be of equal importance, for instance, rotor angular velocity, weight, outside diameter, efficiency, so that the final design lies within acceptable limits for each parameter. In consequence, a simple presentation can hardly do justice to the real problem. However, a consideration of the factors affecting turbine efficiency for a simplified case can provide a useful guide to the designer.

Consider the problem of selecting an axial turbine design for which the mean blade speed U, the specific work ΔW, and the axial velocity c_x, have already been selected. The upper limit of blade speed is limited by stress; the limit on blade tip speed is roughly 450 m/s although some experimental turbines have been operated at higher speeds. The axial velocity is limited by flow area considerations. It is assumed that the blades are sufficiently short to treat the flow as two-dimensional.

The specific work done is

$$\Delta W = U(c_{y2} + c_{y3}).$$

With ΔW, U and c_x fixed the only remaining parameter required to completely define the velocity triangles is c_{y2}, since

$$c_{y3} = \Delta W/U - c_{y2}. \tag{4.16}$$

For different values of c_{y2} the velocity triangles can be constructed, the loss coefficients determined and η_{tt}, η_{ts} calculated. In Shapiro *et al.* (1957) Stenning considered a family of turbines each having a flow coefficient $c_x/U = 0.4$, blade aspect ratio $H/b = 3$ and Reynolds number $Re = 10^5$, and calculated η_{tt}, η_{ts} for stage loading factors $\Delta W/U^2$ of 1, 2 and 3 using Soderberg's correlation. The results of this calculation are shown in Figure 4.4. It will be noted that these results relate to blading efficiency and make no allowance for losses due to tip clearance and disc friction.

EXAMPLE 4.1. Verify the peak value of the total to static efficiency η_{ts} shown in Figure 4.4 for the curve marked $\Delta W/U^2 = 1$, using Soderberg's correlation and the same data used by Stenning in Shapiro *et al.* (1957).

Solution. From eqn. (4.10a):

$$\frac{1}{\eta_{ts}} = 1 + \frac{\zeta_R w_3^2 + \zeta_N c_2^2 + c_1^2}{2\Delta W}.$$

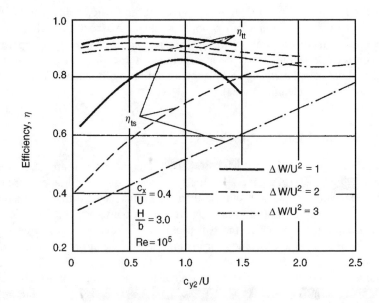

FIG. 4.4. Variation of efficiency with (c_{y2}/U) for several values of stage loading factor $\Delta W/U^2$ (adapted from Shapiro *et al.* 1957).

As $\Delta W = U^2 = U(c_{y2} + C_{y3})$ then as $c_{y2} = U, c_{y3} = 0$,

$$\phi = c_x/U = \cot \alpha_2 = 0.4, \text{ hence } \alpha_2 = 68.2 \text{ deg.}$$

The velocity triangles are symmetrical, so that $\alpha_2 = \beta_3$. Also, $\epsilon_R = \epsilon_N = \alpha_2 = 68.2°$,

$$\therefore \zeta = 0.04 \times (1 + 1.5 \times 0.682^2) = 0.0679,$$

$$\frac{1}{\eta_{ts}} = 1 + \frac{2\zeta w_3^2 + c_x^2}{2U^2} = 1 + \zeta \phi^2 \sec^2 \beta_3 + \frac{1}{2}\phi^2$$

$$= 1 + \phi^2 (\zeta \sec^2 \beta_3 + 0.5)$$

$$= 1 + 0.4^2 \times (0.0679 \times 2.6928^2 + 0.5)$$

$$= 1 + 0.16 \times (0.49235 + 0.5),$$

$$\therefore \eta_{ts} = \underline{0.863.}$$

This value appears to be the same as the peak value of the efficiency curve $\Delta W/U^2 = 1.0$, in Figure 4.4.

Stage reaction

The classification of different types of axial turbine is more conveniently described by the *degree of reaction* or *reaction ratio R*, of each stage rather than by the ratio c_{y2}/U. As a means of description the term reaction has certain inherent advantages which become apparent later. Several definitions of reaction are available; the classical definition is given as the ratio of the static pressure drop in the rotor to the static pressure drop in the stage. However, it is more useful to define the reaction ratio as the static *enthalpy* drop in the rotor to the static *enthalpy* drop in the stage because it then becomes, in effect, a statement of the stage *flow geometry* Thus,

$$R = (h_2 - h_3)/(h_1 - h_3). \tag{4.17}$$

If the stage is normal (i.e. $c_1 = c_3$) then,

$$R = (h_2 - h_3)/(h_{01} - h_{03}). \tag{4.18}$$

Using eqn. (4.4), $h_2 - h_3 = \frac{1}{2}(w_3^2 - w_2^2)$ and eqn. (4.18) becomes,

$$R = \frac{w_3^2 - w_2^2}{2U(c_{y2} + c_{y3})}. \tag{4.19}$$

Assuming constant axial velocity through the stage

$$R = \frac{(w_{y3} - w_{y2})(w_{y3} + w_{y2})}{2U(c_{y2} + c_{y3})} = \frac{w_{y3} - w_{y2}}{2U}, \tag{4.20}$$

since, upon referring to Figure 4.1, it is seen that

$$c_{y2} = w_{y2} + U \text{ and } c_{y3} = w_{y3} - U. \tag{4.21}$$

Thus,

$$R = \frac{c_x}{2U}(\tan\beta_3 - \tan\beta_2) \tag{4.22a}$$

or

$$R = \frac{1}{2} + \frac{c_x}{2U}(\tan\beta_3 - \tan\alpha_2), \tag{4.22b}$$

after using eqn. (4.21).

If $\beta_3 = \beta_2$, the reaction is zero; if $\beta_3 = \alpha_2$ the reaction is 50%. These two special cases are discussed below in more detail.

Zero reaction stage

From the definition of reaction, when $R = 0$, eqn. (4.18) indicates that $h_2 = h_3$ and eqn. (4.22a) that $\beta_2 = \beta_3$. The Mollier diagram and velocity triangles corresponding to these conditions are sketched in Figure 4.5. Now as $h_{02rel} = h_{03rel}$ and $h_2 = h_3$ for $R = 0$ it must follow, therefore, that $w_2 = w_3$. It will be observed from Figure 4.5 that, because of irreversibility, there is a *pressure drop* through the rotor row. The zero reaction stage is *not* the same thing as an *impulse* stage; in the latter case there is, by definition, no pressure drop through the rotor. The Mollier diagram for an impulse stage is shown in Figure 4.6 where it is seen that the enthalpy *increases*

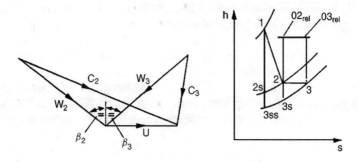

FIG. 4.5. Velocity diagram and Mollier diagram for a zero reaction turbine stage.

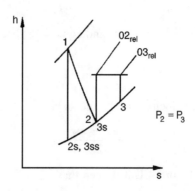

FIG. 4.6. Mollier diagram for an impulse turbine stage.

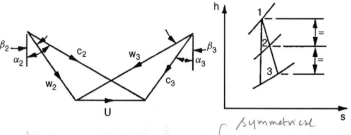

FIG. 4.7. Velocity diagram and Mollier diagram for a 50% reaction turbine stage.

through the rotor. The implication is clear from eqn. (4.18); the reaction is negative for the impulse turbine stage when account is taken of the irreversibility.

50 per cent reaction stage

The combined velocity diagram for this case is symmetrical as can be seen from Figure 4.7, since $\beta_3 = \alpha_2$. Because of the symmetry it is at once obvious that $\beta_2 = \alpha_3$, also. Now with $R = \frac{1}{2}$, eqn. (4.18) implies that the enthalpy drop in the nozzle row equals the enthalpy drop in the rotor, or

$$h_1 - h_2 = h_2 - h_3. \tag{4.23}$$

Figure 4.7 has been drawn with the same values of c_x, U and ΔW, as in Figure 4.5 (zero reaction case), to emphasise the difference in flow geometry between the 50% reaction and zero reaction stages.

Diffusion within blade rows

Any diffusion of the flow through turbine blade rows is particularly undesirable and must, at the design stage, be avoided at all costs. This is because the adverse pressure gradient (arising from the flow diffusion) coupled with large amounts of fluid deflection (usual in turbine blade rows), makes boundary-layer separation more than merely possible, with the result that large scale losses arise. A compressor blade row, on the other hand, is designed to cause the fluid pressure to rise in the direction of flow, i.e. an *adverse* pressure gradient. The magnitude of this gradient is strictly controlled in a compressor, mainly by having a fairly limited amount of fluid deflection in each blade row.

The comparison of the profile losses given in Figure 3.14 is illustrative of the undesirable result of negative "reaction" in a turbine blade row. The use of the term reaction here needs qualifying as it was only defined with respect to a complete stage. From eqn. (4.22a) the ratio R/ϕ can be expressed for a single row of blades if the flow angles are known. The original data provided with Figure 3.14 gives the blade inlet angles for impulse and reaction blades as 45.5 and 18.9 deg respectively. Thus, the flow angles can be found from Figure 3.14 for the range of incidence given, and R/ϕ can be calculated. For the reaction blades R/ϕ decreases as incidence increases going from 0.36 to 0.25 as i changes from 0 to 10 deg. The impulse blades, which it

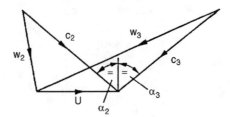

FIG. 4.8. Velocity diagram for 100% reaction turbine stage.

will be observed have a dramatic increase in blade profile loss, has R/ϕ decreasing from zero to -0.25 in the same range of incidence.

It was shown above that negative values of reaction indicated diffusion of the rotor relative velocity (i.e. for $R < 0$, $w_3 < w_2$). A similar condition which holds for diffusion of the nozzle absolute velocity, is that if $R > 1$, $c_2 < c_1$.

Substituting $\tan \beta_3 = \tan \alpha_3 + U/c_x$ into eqn. (4.22b) gives

$$R = 1 + \frac{c_x}{2U}(\tan \alpha_3 - \tan \alpha_2). \tag{4.22c}$$

Thus, when $\alpha_3 = \alpha_2$, the reaction is unity (also $c_2 = c_3$). The velocity diagram for $R = 1$ is shown in Figure 4.8 with the same values of c_x, U and ΔW used for $R = 0$ and $R = \frac{1}{2}$. It will be apparent that if R exceeds unity, then $c_2 < c_1$ (i.e. nozzle flow *diffusion*).

EXAMPLE 4.2. A single-stage gas turbine operates at its design condition with an axial absolute flow at entry and exit from the stage. The absolute flow angle at nozzle exit is 70 deg. At stage entry the total pressure and temperature are 311 kPa and 850°C respectively. The exhaust static pressure is 100 kPa, the total-to-static efficiency is 0.87 and the mean blade speed is 500 m/s.

Assuming constant axial velocity through the stage, determine

(i) the specific work done;
(ii) the Mach number leaving the nozzle;
(iii) the axial velocity;
(iv) the total-to-total efficiency;
(v) the stage reaction.

Take $C_p = 1.148$ kJ/(kg°C) and $\gamma = 1.33$ for the gas.

Solution. (i) From eqn. (4.10), total-to-static efficiency is

$$\eta_{ts} = \frac{h_{01} - h_{03}}{h_{01} - h_{3ss}} = \frac{\Delta W}{h_{01}\{1 - (p_3/p_{01})^{(\gamma-1)/\gamma}\}}.$$

Thus, the specific work is

$$\Delta W = \eta_{ts} C_p T_{01}\{1 - (p_3/p_{01})^{(\gamma-1)/\gamma}\}$$

$$= 0.87 \times 1148 \times 1123 \times \{1 - (1/3.11)^{0.248}\}$$

$$= 276 \text{ kJ/kg}.$$

(ii) At nozzle exit the Mach number is

$$M_2 = c_2/(\gamma R T_2)^{1/2}$$

and it is necessary to solve the velocity diagram to find c_2 and hence to determine T_2.

As $c_{y3} = 0$, $\quad \Delta W = U c_{y2}$

$$c_{y2} = \frac{\Delta W}{U} = \frac{276 \times 10^3}{500} = 552 \, \text{m/s}$$

$$c_2 = c_{y2}/\sin \alpha_2 = 588 \, \text{m/s}.$$

Referring to Figure 4.2, across the nozzle $h_{01} = h_{02} = h_2 + \frac{1}{2}c_2^2$, thus

$$T_2 = T_{01} - \frac{1}{2}c_2^2/C_p = 973 \, \text{K}.$$

Hence, $M_2 = 0.97$ with $\gamma R = (\gamma - 1)C_p$.

(iii) The axial velocity, $c_x = c_2 \cos \alpha_2 = 200 \, \text{m/s}$.

(iv) $\eta_{tt} = \Delta W/(h_{01} - h_{3ss} - \frac{1}{2}c_3^2)$.

After some rearrangement,

$$\frac{1}{\eta_{tt}} = \frac{1}{\eta_{ts}} - \frac{c_3^2}{2\Delta W} = \frac{1}{0.87} - \frac{200^2}{2 \times 276 \times 10^3} = 1.0775.$$

Therefore $\eta_{tt} = 0.93$.

(v) Using eqn. (4.22a), the reaction is

$$R = \frac{1}{2}(c_x/U)(\tan \beta_3 - \tan \beta_2).$$

From the velocity diagram, $\tan \beta_3 = U/c_x$ and $\tan \beta_2 = \tan \alpha_2 - U/c_x$

$$R = 1 - \frac{1}{2}(c_x/U)\tan \alpha_2 = 1 - 200 \times 0.2745/1000$$

$$= 0.451.$$

EXAMPLE 4.3. Verify the assumed value of total-to-static efficiency in the above example using Soderberg's correlation method. The average blade aspect ratio for the stage $H/b = 5.0$, the maximum blade thickness–chord ratio is 0.2 and the average Reynolds number, defined by eqn. (4.14), is 10^5.

Solution. The approximation for total-to-static efficiency, eqn. (4.10a), is used and can be rewritten as

$$\frac{1}{\eta_{ts}} = 1 + \frac{\zeta_R(w_3/U)^2 + \zeta_N(c_2/U)^2 + (c_x/U)^2}{2\Delta W/U^2}.$$

The loss coefficients ζ_R and ζ_N, uncorrected for the effects of blade aspect ratio, are determined using eqn. (4.12) which requires a knowledge of flow turning angle ϵ for each blade row.

For the nozzles, $\alpha_1 = 0$ and $\alpha_2 = 70 \, \text{deg}$, thus $\epsilon_N = 70 \, \text{deg}$.

$$\zeta_N^* = 0.04(1 + 1.5 \times 0.7^2) = 0.0694.$$

Correcting for aspect ratio with eqn. (4.13a),

$$\zeta_{N1} = 1.0694(0.993 + 0.021/5) - 1 = 0.0666.$$

For the rotor, $\tan \beta_2 = (c_{y2} - U)/c_x = (552 - 500)/200 = 0.26$,

$$\therefore \beta_2 = 14.55 \deg.$$

Therefore,

$$\tan \beta_3 = U/c_x = 2.5,$$

and $\qquad \beta_3 = 68.2 \deg.$

Therefore $\qquad \varepsilon_R = \beta_2 + \beta_3 = 82.75 \deg,$

$$\zeta_R^* = 0.04(1 + 1.5 \times 0.8275^2) = 0.0812.$$

Correcting for aspect ratio with eqn. (4.13b)

$$\zeta_{R1} = 1.0812(0.975 + 0.075/5) - 1 = 0.0712.$$

The velocity ratios are:

$$\left(\frac{w_3}{U}\right)^2 = 1 + \left(\frac{c_x}{U}\right)^2 = 1.16$$

$$\left(\frac{c_2}{U}\right)^2 = \left(\frac{588}{500}\right)^2 = 1.382; \left(\frac{c_x}{U}\right)^2 = 0.16$$

and the stage loading factor is,

$$\frac{\Delta W}{U^2} = \frac{c_{y2}}{U} = \frac{552}{500} = 1.104$$

Therefore

$$\frac{1}{\eta_{ts}} = 1 + \frac{0.0712 \times 1.16 + 0.0666 \times 1.382 + 0.16}{2 \times 1.104}$$

$$= 1 + 0.1515$$

$$\therefore \eta_{ts} = 0.869.$$

This result is very close to the value assumed in the first example.

It is not too difficult to include the temperature ratio T_3/T_2 implicit in the more exact eqn. (4.10) in order to see how little effect the correction will have. To calculate T_3

$$T_3 = T_{01} - \frac{\Delta W + \frac{1}{2}c_3^2}{C_p} = 1123 - \frac{276,000 + 20,000}{1148}$$

$$= 865 \,\text{K}.$$

$$T_3/T_2 = 865/973 = 0.89.$$

Therefore

$$\frac{1}{\eta_{ts}} = 1 + \frac{0.0712 \times 1.16 + 0.89 \times 0.0666 \times 1.382 + 0.16}{2 \times 1.104}$$

$$= 1 + 0.1468.$$

$$\therefore \eta_{ts} = 0.872.$$

Choice of reaction and effect on efficiency

In Figure 4.4 the total-to-total and total-to-static efficiencies are shown plotted against c_{y2}/U for several values of stage loading factor $\Delta W/U^2$. These curves can now easily be replotted against the degree of reaction R instead of c_{y2}/U. Equation (4.22c) can be rewritten as $R = 1 + (c_{y3} - c_{y2})/(2U)$ and c_{y3} eliminated using eqn. (4.16) to give

$$R = 1 + \frac{\Delta W}{2U^2} - \frac{c_{y2}}{U}. \tag{4.24}$$

The replotted curves are shown in Figure 4.9 as presented by Shapiro *et al.* (1957). In the case of total-to-static efficiency, it is at once apparent that this is optimised, at a given blade loading, by a suitable choice of reaction. When $\Delta W/U^2 = 2$, the maximum value of η_{ts} occurs with approximately zero reaction. With lighter blade loading, the optimum η_{ts} is obtained with higher reaction ratios. When $\Delta W/U^2 > 2$, the highest value of η_{ts} attainable without rotor *relative* flow diffusion occurring, is obtained with $R = 0$.

From Figure 4.4, for a fixed value of $\Delta W/U^2$, there is evidently only a relatively small changes in total-to-total efficiency (compared with η_{ts}), for a wide range of

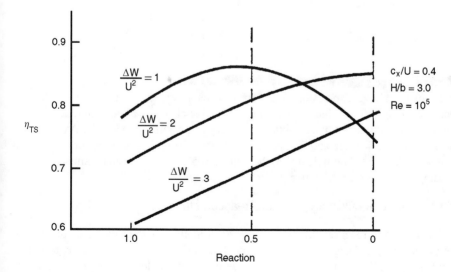

FIG. 4.9. Influence of reaction on total-to-static efficiency with fixed values of stage loading factor.

possible designs. Thus η_{tt} is not greatly affected by the choice of reaction. However, the maximum value of η_{tt} decreases as the stage loading factor increases. To obtain high total-to-total efficiency, it is therefore necessary to use the highest possible value of blade speed consistent with blade stress limitations (i.e. to reduce $\Delta W/U^2$).

Design point efficiency of a turbine stage

The performance of a turbine stage in terms of its efficiency is calculated for several types of design, i.e. 50 per cent reaction, zero reaction and zero exit flow angle, using the loss correlation method of Soderberg described earlier. These are most usefully presented in the form of carpet plots of stage loading coefficient, ψ, and flow coefficient, ϕ.

(1) Total-to-total efficiency of 50 per cent reaction stage

In a multistage turbine the total-to-total efficiency is the relevant performance criterion, the kinetic energy at stage exit being recovered in the next stage. After the last stage of a multistage turbine, or a single-stage turbine, the kinetic energy in the exit flow would be recovered in a diffuser or used for another purpose (e.g. as a contribution to the propulsive thrust).

From eqn. (4.9a), where it has already been assumed that $c_1 = c_3$ and $T_3 = T_2$, we have:

$$\frac{1}{\eta_{tt}} = 1 + \frac{(\zeta_R w_3^2 + \zeta_N c_2^2)}{2\Delta W},$$

where $\Delta W = \psi U^2$ and, for 50 per cent reaction, $w_3 = c_2$ and $\zeta_R = \zeta_N = \zeta$

$$w_3^2 = c_x^2 \sec^2 \beta_3 = c_x^2(1 + \tan^2 \beta_3)$$

$$\therefore \frac{1}{\eta_{tt}} = 1 + \frac{\zeta\phi^2}{\psi}(1 + \tan^2 \beta_3) = 1 + \frac{\zeta\phi^2}{\psi}\left[1 + \left(\frac{1 + \psi}{2\phi}\right)^2\right]$$

as $\tan \beta_3 = (\psi + 1)/(2\phi)$ and $\tan \beta_2 = (\psi - 1)/(2\phi)$.

From the above expressions the performance chart, shown in Figure 4.10, was derived for specified values of ψ and ϕ. From this chart it can be seen that the peak total-to-total efficiency, η_{tt}, is obtained at very low values of ϕ and ψ. As indicated in a survey by Kacker and Okapuu (1982), most aircraft gas turbine designs will operate with flow coefficients in the range, $0.5 \leqslant \psi \leqslant 1.5$, and values of stage loading coefficient in the range, $0.8 \leqslant \psi \leqslant 2.8$.

(2) Total-to-total efficiency of a zero reaction stage

The degree of reaction will normally vary along the length of the blade depending upon the type of design that is specified. The performance for $R = 0$ represents a limit, lower values of reaction are possible but undesirable as they would give rise to large losses in efficiency. For $R < 0$, $w_3 < w_2$, which means the relative flow decelerates across the rotor.

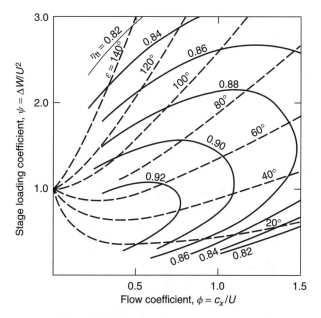

FIG. 4.10. Design point total-to-total efficiency and deflection angle contours for a turbine stage of 50 per cent reaction.

Referring to Figure 4.5, for zero reaction $\beta_2 = \beta_3$, and from eqn. (4.21)

$$\tan \alpha_2 = 1/\phi + \tan \beta_2 \text{ and } \tan \alpha_3 = \tan \beta_3 - 1/\phi.$$

Also, $\psi = \Delta W/U^2 = \phi(\tan \alpha_2 + \tan \alpha_3) = \phi(\tan \beta_2 + \tan \beta_3) = 2\phi \tan \beta_2$,

$$\therefore \tan \beta_2 = \frac{\psi}{2\phi}.$$

Thus, using the above expressions:

$$\tan \alpha_2 = (\psi/2 + 1)/\phi \text{ and } \tan \alpha_3 = (\psi/2 - 1)/\phi.$$

From these expressions the flow angles can be calculated if values for ψ and ϕ are specified. From an inspection of the velocity diagram,

$$c_2 = c_x \sec \alpha_2, \text{ hence } c_2^2 = c_x^2(1 + \tan^2 \alpha_2) = c_x^2[1 + (\psi/2 + 1)^2/\phi^2],$$

$$w_3 = c_x \sec \beta_3, \text{ hence } w_3^2 = c_x^2(1 + \tan^2 \beta_3) = c_x^2[1 + (\psi/2\phi)^2].$$

Substituting the above expressions into eqn. (4.9a):

$$\frac{1}{\eta_{tt}} = 1 + \frac{\zeta_R w_3^2 + \zeta_N c_2^2}{2\psi U^2},$$

$$\frac{1}{\eta_{tt}} = 1 + \frac{1}{2\psi} \left\{ \zeta_R \left[\phi^2 + \left(\frac{\psi}{2} \right)^2 \right] + \zeta_N \left[\phi^2 + \left(1 + \frac{\psi}{2} \right)^2 \right] \right\}.$$

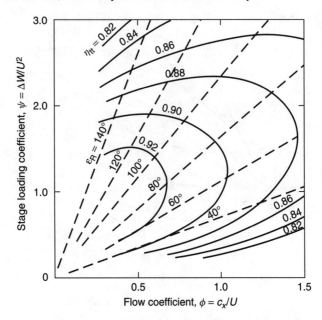

FIG. 4.11. Design point total-to-total efficiency and rotor flow deflection angle for a zero reaction turbine stage.

The performance chart shown in Figure 4.11 has been derived using the above expressions. This is similar in its general form to Figure 4.10 for 50 per cent reaction, with the highest efficiencies being obtained at the lowest values of ϕ and ψ, except that higher efficiencies are obtained at higher values of the stage loading but at reduced values of the flow coefficient.

(3) Total-to-static efficiency of stage with axial velocity at exit

A single-stage axial turbine will have axial flow at exit and the most appropriate efficiency is usually total-to-static. To calculate the performance, eqn. (4.10a) is used:

$$\frac{1}{\eta_{ts}} = 1 + \frac{(\zeta_R w_3^2 + \zeta_N c_2^2 + c_1^2)}{2\Delta W},$$

$$= 1 + \frac{\phi^2}{2\psi}(\zeta_R \sec^2 \beta_3 + \zeta_N \sec^2 \alpha_2 + 1).$$

With axial flow at exit, $c_1 = c_3 = c_x$, and from the velocity diagram, Figure 4.12,

$$\tan \beta_3 = U/c_x, \tan \beta_2 = \tan \alpha_2 - \tan \beta_3,$$

$$\sec^2 \beta_3 = 1 + \tan^2 \beta_3 = 1 + 1/\phi^2,$$

$$\sec^2 \alpha_2 = 1 + \tan^2 \alpha_2 = 1 + (\psi/\phi)^2,$$

$$\therefore \frac{1}{\eta_{ts}} = 1 + \frac{1}{2\phi}[\zeta_R(1 + \phi^2) + \zeta_N(\psi^2 + \phi^2) + \phi^2].$$

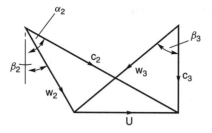

FIG. 4.12. Velocity diagram for turbine stage with axial exit flow.

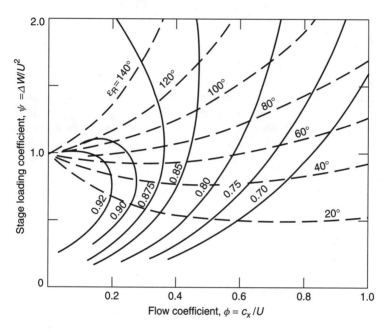

FIG. 4.13. Total-to-total efficiency contours for a stage with axial flow at exit.

Specifying ϕ, ψ, the unknown values of the loss coefficients, ζ_R and ζ_N, can be derived using Soderberg's correlation, eqn. (4.12), in which

$$\varepsilon_N = \alpha_2 = \tan^{-1}(\psi/\phi) \text{ and } \varepsilon_R = \beta_2 + \beta_3 = \tan^{-1}(1/\phi) + \tan^{-1}[(\psi - 1)/\phi].$$

From the above expressions the performance chart, Figure 4.13, has been derived.

An additional limitation is imposed on the performance chart because of the reaction which must remain greater than or, in the limit, equal to zero. From eqn. (4.22a),

$$R = \frac{\phi}{2}(\tan \beta_3 - \tan \beta_2) = 1 - \frac{\psi}{2}.$$

Thus at the limit, $R = 0$, the stage loading coefficient, $\psi = 2$.

Maximum total-to-static efficiency of a reversible turbine stage

When blade losses and exit kinetic energy loss are included in the definition of efficiency, we have shown, eqn. (4.10a), that the efficiency is

$$\eta_{ts} = \frac{h_{01} - h_{03}}{h_{01} - h_{3ss}} = \left[1 + \frac{w_3^2 \zeta_r + c_2^2 \zeta_n + c_3^2}{2(h_1 - h_3)}\right]^{-1}.$$

In the case of the *ideal* (or reversible) turbine stage the only loss is due to the exhaust kinetic energy and then the total-to-static efficiency is

$$\eta_{ts} = \frac{h_{01} - h_{03ss}}{h_{01} - h_{3ss}} = \left[1 + \frac{c_3^2}{2U(c_{y2} + c_{y3})}\right]^{-1} \tag{4.25a}$$

since $\Delta W = h_{01} - h_{03ss} = U(c_{y2} + c_{y3})$ and $h_{03ss} - h_{3ss} = \frac{1}{2}c_3^2$.

The maximum value of η_{ts} is obtained when the exit velocity c_3 is *nearly* a minimum for given turbine stage operating conditions (R, ϕ and α_2). On first thought it may appear obvious that maximum η_{ts} will be obtained when c_3 is absolutely axial (i.e. $\alpha_3 = 0°$) but this is incorrect. By allowing the exit flow to have some counterswirl (i.e. $\alpha_3 > 0 \deg$) the work done is increased for only a relatively small increase in the exit kinetic energy loss. Two analyses are now given to show how the total-to-static efficiency of the ideal turbine stage can be optimised for specified conditions.

Substituting $c_{y2} = c_x \tan \alpha_2$, $c_{y3} = c_x \tan \alpha_3$, $c_3 = c_x / \cos \alpha_3$ and $\phi = c_x / U$ into eqn. (4.25), leads to

$$\eta_{ts} = \left[1 + \frac{\phi(1 + \tan^2 a_3)}{2(\tan a_2 + \tan a_3)}\right]^{-1} \tag{4.25b}$$

i.e. $\eta_{ts} = \text{fn}(\phi, a_2, a_3)$.

(i) To find the optimum η_{ts} when R and ϕ are specified

From eqn. (4.22c) the nozzle flow outlet angle α_2 can be expressed in terms of R, ϕ and α_3 as

$$\tan \alpha_2 = \tan \alpha_3 + 2(1 - R)/\phi. \tag{4.26}$$

Substituting into eqn. (4.25b)

$$\eta_{ts} = \left[1 + \frac{\phi^2(1 + \tan^2 \alpha_3)}{4(\phi \tan a_3 + 1 - R)}\right]^{-1}.$$

Differentiating this expression with respect to $\tan \alpha_3$, and equating the result to zero,

$$\tan^2 \alpha_3 + 2k \tan \alpha_3 - 1 = 0$$

where $k = (1 - R)/\phi$. This quadratic equation has the solution

$$\tan \alpha_3 = -k + \sqrt{(k^2 + 1)} \tag{4.27}$$

the value of α_3 being the optimum flow outlet angle from the stage when R and ϕ are specified. From eqn. (4.26), $k = (\tan_2 - \tan \alpha_3)/2$ which when substituted into eqn. (4.27) and simplified gives

$$\tan \alpha_3 = \cot \alpha_2 = \tan(\pi/2 - \alpha_2).$$

Hence, the exact result that

$$\alpha_3 = \pi/2 - \alpha_2.$$

The corresponding idealised $\eta_{ts\,max}$ and R are

$$\eta_{ts\,max} = [1 + (\phi/2)\cot \alpha_2]^{-1} \tag{4.28a}$$
$$R = 1 - \phi(\tan \alpha_2 - \cot \alpha_2)/2.$$

(ii) To find the optimum η_{ts} when α_2 and ϕ are specified

Differentiating eqn. (4.25b) with respect to $\tan \alpha_3$ and equating the result to zero,

$$\tan^2 \alpha_3 + 2 \tan \alpha_2 \tan \alpha_3 - 1 = 0.$$

Solving this quadratic, the relevant root is

$$\tan \alpha_3 = \sec \alpha_2 - \tan \alpha_2.$$

Using simple trignometric relations this simplifies still further to

$$\alpha_3 = (\pi/2 - \alpha_2)/2.$$

Substituting this expression for α_3 into eqn. (4.25b) the idealised maximum η_{ts} is obtained

$$\eta_{ts\,max} = [1 + \phi(\sec \alpha_2 - \tan \alpha_2)]^{-1}. \tag{4.28b}$$

The corresponding expressions for the degree of reaction R and stage loading coefficient $\Delta W/U^2$ are

$$R = 1 - \phi(\tan \alpha_2 - \tfrac{1}{2}\sec \alpha_2)$$
$$\frac{\Delta W}{U^2} = \phi \sec \alpha_2 = \frac{c_2}{U}. \tag{4.29}$$

It is interesting that in this analysis the exit swirl angle α_3 is only half that of the constant reaction case. The difference is merely the outcome of the two different sets of constraints used for the two analyses.

For both analyses, as the flow coefficient is reduced towards zero, α_2 approaches $\pi/2$ and α_3 approaches zero. Thus, for such high nozzle exit angle turbine stages, the appropriate blade loading factor for maximum η_{ts} can be specified if the reaction is known (and conversely). For a turbine stage of 50% reaction (and with $\alpha_3 \rightarrow$ 0 deg) the appropriate velocity diagram shows that $\Delta W/U^2 \doteq 1$ for maximum η_{ts}. Similarly, a turbine stage of zero reaction (which is an *impulse* stage for ideal, reversible flow) has a blade loading factor $\Delta W/U^2 \doteq 2$ for maximum η_{ts}.

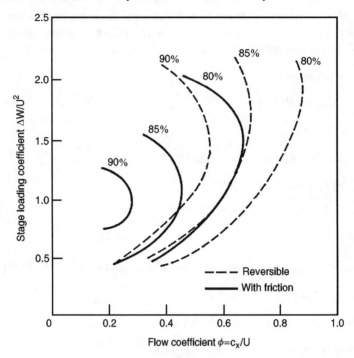

FIG. 4.14. Total-to-static efficiency of a 50% reaction axial flow turbine stage (adapted from Horlock 1966).

Calculations of turbine stage performance have been made by Horlock (1966) both for the reversible and irreversible cases with $R = 0$ and 50%. Figure 4.14 shows the effect of blade losses, determined with Soderberg's correlation, on the total-to-static efficiency of the turbine stage for the constant reaction of 50%. It is evident that exit losses become increasingly dominant as the flow coefficient is increased.

Stresses in turbine rotor blades

Although this chapter is primarily concerned with the fluid mechanics and thermodynamics of turbines, some consideration of stresses in rotor blades is needed as these can place restrictions on the allowable blade height and annulus flow area, particularly in high temperature, high stress situations. Only a very brief outline is attempted here of a very large subject which is treated at much greater length by Horlock (1966), in texts dealing with the mechanics of solids, e.g. Den Hartog (1952), Timoshenko (1957), and in specialised discourses, e.g. Japiske (1986) and Smith (1986). The stresses in turbine blades arise from centrifugal loads, from gas bending loads and from vibrational effects caused by non-constant gas loads. Although the centrifugal stress produces the biggest contribution to the total stress, the vibrational stress is very significant and thought to be responsible for fairly common vibratory fatigue failures (Smith 1986). The direct and simple approach to blade vibration is to "tune" the blades so that resonance does not occur in the operating range of the turbine. This means obtaining a blade design in which none of its natural frequencies coincides with any excitation frequency. The subject is complex, interesting but outside of the scope of the present text.

Centrifugal stresses

Consider a blade rotating about an axis O as shown in Figure 4.15. For an element of the blade of length dr at radius r, at a rotational speed Ω the elemetary centrifugal load dF_c is given by,

$$\mathrm{d}F_c = -\Omega^2 r\, \mathrm{d}m,$$

where d$m = \rho_m A\, \mathrm{d}r$ and the negative sign accounts for the direction of the stress gradient (i.e. zero stress at the blade tip to a maximum at the blade root).

$$\frac{\mathrm{d}\sigma_c}{\rho_m} = \frac{\mathrm{d}F_c}{\rho_m A} = -\Omega^2 r\, \mathrm{d}r.$$

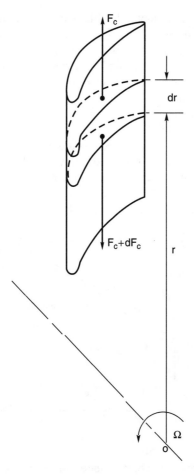

F$_c$

dr

F$_c$+dF$_c$

r

Ω

O

FIG. 4.15. Centrifugal forces acting on rotor blade element.

For blades with a constant cross-sectional area, we get

$$\frac{\sigma_c}{\rho_m} = \Omega^2 \int_{r_h}^{r_t} r \, dr = \frac{U_t^2}{2} \left[1 - \left(\frac{r_h}{r_t} \right)^2 \right].$$

(4.30a)

A rotor blade is usually tapered both in chord and in thickness from root to tip such that the area ratio A_t/A_h is between 1/3 and 1/4. For such a blade taper it is often assumed that the blade stress is reduced to 2/3 of the value obtained for an untapered blade. A blade stress taper factor can be defined as:

$$K = \frac{\text{stress at root of tapered blade}}{\text{stress at root of untapered blade}}$$

Thus, for tapered blades

$$\frac{\sigma_c}{\rho_m} = \frac{K U_t^2}{2} \left[1 - \left(\frac{r_h}{r_t} \right)^2 \right].$$

(4.30b)

Values of the taper factor K quoted by Emmert (1950), are shown in Figure 4.16 for various taper geometries.

Typical data for the allowable stresses of commonly used alloys are shown in Figure 4.17 for the "1000 hr rupture life" limit with maximum stress allowed plotted as a function of blade temperature. It can be seen that in the temperature range 900–1100 K, nickel or cobalt alloys are likely to be suitable and for temperatures up to about 1300 K molybdenum alloys would be needed.

By means of blade cooling techniques it is possible to operate with turbine entry temperatures up to 1650–1700 K, according to Le Grivès (1986). Further detailed information on one of the many alloys used for gas turbines blades is shown in Figure 4.18. This material is *Inconel*, a nickel-based alloy containing 13% chromium, 6% iron, with a little manganese, silicon and copper. Figure 4.18 shows the influence of the "rupture life" and also the "percentage creep", which is the

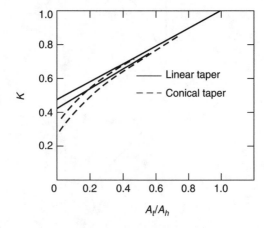

FIG. 4.16. Effect of tapering on centrifugal stress at blade root (adapted from Emmert 1950).

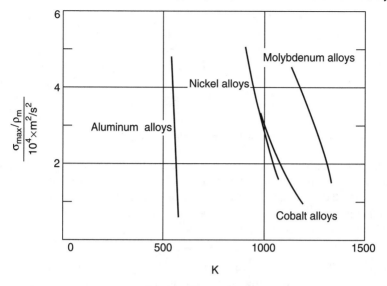

FIG. 4.17. Maximum allowable stress for various alloys (1000 hr rupture life) (adapted from Freeman 1955).

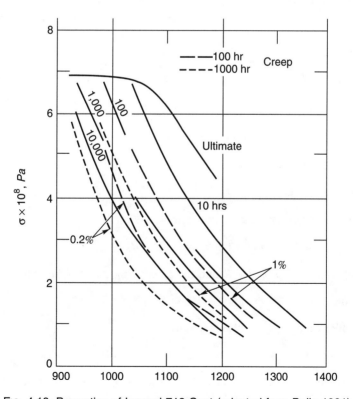

FIG. 4.18. Properties of Inconel 713 Cast (adapted from Balje 1981).

elongation strain at the allowable stress and temperature of the blade. To enable operation at high temperatures and for long life of the blades, the creep strength criterion is the one usually applied by designers.

An estimate of the average rotor blade temperature T_b can be made using the approximation,

$$T_b = T_2 + 0.85 w_2^2/(2C_p), \qquad (4.31)$$

i.e. 85% temperature recovery of the inlet *relative kinetic energy*.

EXAMPLE 4.4. Combustion gases enter the first stage of a gas turbine at a stagnation temperature and pressure of 1200 K and 4.0 bar. The rotor blade tip diameter is 0.75 m, the blade height is 0.12 m and the shaft speed is 10 500 rev/min. At the mean radius the stage operates with a reaction of 50 per cent, a flow coefficient of 0.7 and a stage loading coefficient of 2.5.

Determine:

(1) the relative and absolute flow angles for the stage;
(2) the velocity at nozzle exit;
(3) the static temperature and pressure at nozzle exit assuming a nozzle efficiency of 0.96 and the mass flow;
(4) the rotor blade root stress assuming the blade is tapered with a stress taper factor K of 2/3 and the blade material density is $8000 \, kg/m^2$;
(5) the approximate mean blade temperature;
(6) taking only the centrifugal stress into account suggest a suitable alloy from the information provided which could be used to withstand 1000 hr of operation.

Solution. (1) The stage loading is

$$\psi = \Delta h_0/U^2 = (w_{y3} + w_{y2})/U = \phi(\tan \beta_3 + \tan \beta_2).$$

From eqn. (4.20) the reaction is

$$R = \phi(\tan \beta_3 - \tan \beta_2)/2.$$

Adding and subtracting these two expressions, we get

$$\tan \beta_3 = (\psi/2 + R)/\phi \text{ and } \tan \beta_2 = (\psi/2 - R)/\phi.$$

Substituting values of ψ, ϕ and R into the preceding equations we obtain

$$\beta_3 = 68.2°, \beta_2 = 46.98°$$

and for similar triangles (i.e. 50% reaction)

$$\alpha_2 = \beta_3 \text{ and } \alpha_3 = \beta_2$$

(2) At the mean radius, $r_m = (0.75 - 0.12)/2 = 0.315 \, m$, the blade speed is $U_m = \Omega r_m = (10500/30) \times \pi \times 0.315 = 1099.6 \times 0.315 = 346.36 \, m/s$. The axial velocity $c_x = \phi U_m = 0.5 \times 346.36 = 242.45 \, m/s$ and the velocity of the gas at nozzle exit is, $c_2 = c_x/\cos \alpha_2 = 242.45/\cos 68.2 = 652.86 \, m/s$.

(3) To determine the conditions at nozzle exit, we have

$$T_2 = T_{02} - \tfrac{1}{2}c_2^2/C_p = 1200 - 652.86^2/(2 \times 1160) = 1016.3 \text{ K}$$

From eqn. (2.40), the nozzle efficiency is

$$\eta_N = \frac{h_{01} - h_2}{h_{01} - h_{2s}} = \frac{1 - T_2/T_{01}}{1 - (p_2/p_{01})^{(\gamma-1)/\gamma}}$$

$$\therefore \left(\frac{p_2}{p_{01}}\right)^{(\gamma-1)/\gamma} = 1 - \frac{1 - T_2/T_{01}}{\eta_N} = 1 - (1 - 1016.3/1200)/0.96 = 0.84052$$

$$\therefore p_2 = 4 \times 0.84052^{4.0303} = 1.986 \text{ bar}$$

The mass flow is found from the continuity equation:

$$\dot{m} = \rho_2 A_2 c_{x2} = \left(\frac{p_2}{RT_2}\right) A_2 c_{x2}$$

$$\therefore \dot{m} = \left(\frac{1.986 \times 10^5}{287.8 \times 1016.3}\right) \times 0.2375 \times 242.45 = 39.1 \text{ kg/s}$$

(4) For a tapered blade, eqn. (4.30b) gives

$$\frac{\sigma_c}{\rho_m} = \frac{2}{3} \times \frac{412.3^2}{2}\left[1 - \left(\frac{0.51}{0.75}\right)^2\right] = 30463.5 \text{ m}^2/\text{s}^2$$

where $U_t = 1099.6 \times 0.375 = 412.3$ m/s.

The density of the blade material is taken to be 8000 kg/m^3 and so the root stress is

$$\sigma_c = 8000 \times 30463.5 = 2.437 \times 10^8 \text{ N/m}^2 = 243.7 \text{ MPa}$$

(5) The approximate average mean blade temperature is

$$T_b = 1016.3 + 0.85 \times (242.45/\cos 46.975)^2/(2 \times 1160)$$

$$= 1016.3 + 46.26 = 1062.6 \text{ K}$$

(6) The data in Figure 4.17 suggests that for this moderate root stress, cobalt or nickel alloys would not withstand a lifespan of 1000 hr to rupture and the use of molybdenum would be necessary. However, it would be necessary to take account of bending and vibratory stresses and the decision about the choice of a suitable blade material would be decided on the outcome of these calculations.

Inspection of the data for Inconel 713 cast alloy, Figure 4.18, suggests that it might be a better choice of blade material as the temperature–stress point of the above calculation is to the left of the line marked creep strain of 0.2% in 1000 hr. Again, account must be taken of the additional stresses due to bending and vibration.

Design is a process of trial and error; changes in the values of some of the parameters can lead to a more viable solution. In the above case (with bending and vibrational stresses included) it might be necessary to reduce one or more of the values chosen, e.g.

(1) the rotational speed,
(2) the inlet stagnation temperature,
(3) the flow area.

NB. The combination of values for ψ and ϕ at $R = 0.5$ used in this example were selected from data given by Wilson (1987) and correspond to an optimum total-to-total efficiency of 91.9%.

Turbine blade cooling

In the gas turbine industry there has been a continuing trend towards higher turbine inlet temperatures, either to give increased specific thrust (thrust per unit air mass flow) or to reduce the specific fuel consumption. The highest allowable gas temperature at entry to a turbine with uncooled blades is 1250–1300 K while, with blade cooling systems, a range of gas temperatures up to 1800 K or so may be employed, depending on the nature of the cooling system.

Various types of cooling system for gas turbines have been considered in the past and a number of these are now in use. Wilde (1977) reviewed the progress in blade cooling techniques. He also considered the broader issues involving the various technical and design factors influencing the best choice of turbine inlet temperature for future turbofan engines. Le Grivès (1986) reviewed types of cooling system, outlining their respective advantages and drawbacks, and summarising important analytical considerations concerning their aerodynamics and heat transfer.

The system of blade cooling most commonly employed in aircraft gas turbines is where some cooling air is bled off from the exit stage of the high-pressure compressor and carried by ducts to the guide vanes and rotor of the high-pressure turbine. It was observed by Le Grivès that the cooling air leaving the compressor might be at a temperature of only 400 to 450 K less than the maximum allowable blade temperature of the turbine. Figure 4.19 illustrates a high-pressure turbine rotor blade, cut away to show the intricate labyrinth of passages through which the cooling air passes before it is vented to the blade surface via the rows of tiny holes along and around the leading edge of the blade. Ideally, the air emerges with little velocity and form a film of cool air around the blade surface (hence the term "film cooling"), insulating it from the hot gases. This type of cooling system enables turbine entry temperatures up to 1800 K to be used.

There is a rising thermodynamic penalty incurred with blade cooling systems as the turbine entry temperature rises, e.g. energy must be supplied to pressurise the air bled off from the compressor. Figure 4.21 is taken from Wilde (1977) showing how the net turbine efficiency decreases with increasing turbine entry temperature. Several in-service gas turbine engines are included in the graph. Wilde did question whether turbine entry temperatures greater than 1600 K could really be justified in turbofan engines because of the effect on the internal aerodynamic efficiency and specific fuel consumption.

Turbine flow characteristics

An accurate knowledge of the flow characteristics of a turbine is of considerable practical importance as, for instance, in the matching of flows between a compressor

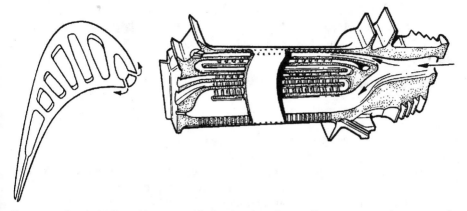

FIG. 4.19. Cooled HP turbine rotor blade showing the cooling passages (courtesy of Rolls-Royce plc).

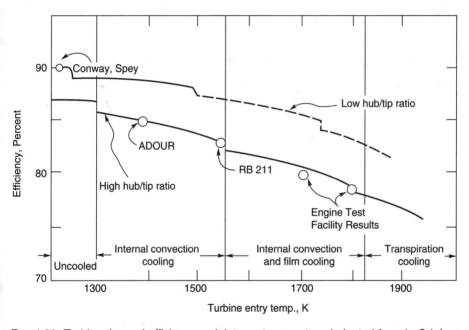

FIG. 4.20. Turbine thermal efficiency vs inlet gas temperature (adapted from le Grivès 1986).

and turbine of a jet engine. When a turbine can be expected to operate close to its design incidence (i.e. in the low loss region) the turbine characteristics can be reduced to a single curve. Figure 4.21, due to Mallinson and Lewis (1948), shows a comparison of typical characteristics for one, two and three stages plotted as turbine overall pressure ratio p_{0II}/p_{0I} against a mass flow coefficient $\dot{m}(\sqrt{T_{01}})/p_{0I}$. There is a noticeable tendency for the characteristic to become more ellipsoidal as the number of stages is increased. At a given pressure ratio the mass flow coefficient, or "swallowing capacity" tends to decrease with the addition of further stages to the turbine. One of the earliest attempts to assess the flow variation of a multistage turbine is credited to Stodola (1945), who formulated the much used "ellipse law". The curve labelled "multistage" in Figure 4.21 is in agreement with the "ellipse law" expression

$$m(\sqrt{T_{01}})/p_{0I} = k[1 - (p_{0II}/p_{0I})^2]^{1/2}, \tag{4.32}$$

where k is a constant.

This expression has been used for many years in steam turbine practice, but an accurate estimate of the variation in swallowing capacity with pressure ratio is of even greater importance in gas turbine technology. Whereas the average condensing steam turbine, even at part-load, operates at very high pressure ratios, some gas turbines may work at rather low pressure ratios, making flow matching with a compressor a difficult problem. The constant value of swallowing capacity, reached by the single-stage turbine at a pressure ratio a little above 2, and the other turbines at progressively higher pressure ratios, is associated with choking (sonic) conditions in the turbine stator blades.

Flow characteristics of a multistage turbine

Several derivations of the ellipse law are available in the literature. The derivation given below is a slightly amplified version of the proof given by Horlock (1958). A more general method has been given by Egli (1936) which takes into consideration the effects when operating outside the normal low loss region of the blade rows.

Consider a turbine comprising a large number of normal stages, each of 50% reaction; then, referring to the velocity diagram of Figure 4.22a, $c_1 = c_3 = w_2$ and $c_2 = w_3$. If the blade speed is maintained constant and the mass flow is reduced, the

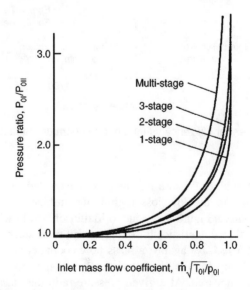

FIG. 4.21. Turbine flow characteristics (after Mallinson and Lewis 1948).

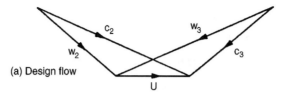

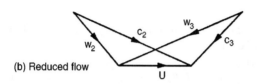

FIG. 4.22. Change in turbine stage velocity diagram with mass flow at constant blade speed.

fluid angles at exit from the rotor (β_3) and nozzles (α_2) will remain constant and the velocity diagram then assumes the form shown in Figure 4.22b. The turbine, if operated in this manner, will be of low efficiency, as the fluid direction at inlet to each blade row is likely to produce a negative incidence stall. To maintain high efficiency the fluid inlet angles must remain fairly close to the design values. It is therefore assumed that the turbine operates at its highest efficiency at *all off-design conditions* and, by implication, the blade speed is changed in direct proportion to the axial velocity. The velocity triangles are similar at off-design flows but of different scale.

Now the work done by unit mass of fluid through one stage is $U(c_{y2} + c_{y3})$ so that, assuming a perfect gas,

$$C_p \Delta T_0 = C_p \Delta T = U c_x (\tan \alpha_2 + \tan \alpha_3)$$

and, therefore,

$$\Delta T \propto c_x^2.$$

Denoting design conditions by subscript d, then

$$\frac{\Delta T}{\Delta T_d} = \left(\frac{c_x}{c_{xd}} \right)^2 \tag{4.33}$$

for equal values of c_x/U.

From the continuity equation, at off-design, $\dot{m} = \rho A c_x = \rho_1 A_1 c_{x1}$, and at design, $\dot{m}_d = \rho_d A c_{xd} = \rho_1 A_1 c_{x1}$, hence

$$\frac{c_x}{c_{xd}} = \frac{\rho_d}{\rho} \frac{c_{x1}}{c_{x1d}} = \frac{\rho_d}{\rho} \frac{\dot{m}}{\dot{m}_d}. \tag{4.34}$$

Consistent with the assumed mode of turbine operation, the polytropic efficiency is taken to be constant at off-design conditions and, from eqn. (2.37), the relationship between temperature and pressure is therefore,

$$T/p^{\eta_p(\gamma-1)/\gamma} = \text{constant.}$$

Combined with $p/\rho = RT$ the above expression gives, on eliminating p, $\rho/T^n = $ constant, hence

$$\frac{\rho}{\rho_d} = \left(\frac{T}{T_d}\right)^n, \tag{4.35}$$

where $n = \gamma/\{\eta_p(\gamma - 1)\} - 1$.

For an infinitesimal temperature drop eqn. (4.33) combined with eqns. (4.34) and (4.35) gives, with little error,

$$\frac{dT}{dT_d} = \left(\frac{c_x}{c_{xd}}\right)^2 = \left(\frac{T_d}{T}\right)^{2n}\left(\frac{\dot{m}}{\dot{m}_d}\right)^2. \tag{4.36}$$

Integrating eqn. (4.36),

$$T^{2n+1} = \left(\frac{\dot{m}}{\dot{m}_d}\right)^2 T_d^{2n+1} + K,$$

where K is an arbitrary constant.

To establish a value for K it is noted that if the turbine entry temperature is constant $T_d = T_1$ and $T = T_1$ also.

Thus, $K = [1 - (\dot{m}/\dot{m}_d)^2]T_I^{2n+1}$ and

$$\left(\frac{T}{T_I}\right)^{2n+1} - 1 = \left(\frac{\dot{m}}{\dot{m}_d}\right)^2\left[\left(\frac{T_d}{T_I}\right)^{2n+1} - 1\right]. \tag{4.37}$$

Equation (4.37) can be rewritten in terms of pressure ratio since $T/T_I = (p/p_1)^{\eta_p(\gamma-1)/\gamma}$. As $2n + 1 = 2\gamma/[\eta_p(\gamma - 1)] - 1$ then,

$$\frac{\dot{m}}{\dot{m}_d} = \left\{\frac{1 - (p/p_1)^{2-\eta_p(\gamma-1)/\gamma}}{1 - (p_d/p_1)^{2-\eta_p(\gamma-1)/\gamma}}\right\}^{1/2} \tag{4.38a}$$

With $\eta_p = 0.9$ and $\gamma = 1.3$ the pressure ratio index is about 1.8; thus the approximation is often used

$$\frac{\dot{m}}{\dot{m}_d} = \left\{\frac{1 - (p/p_1)^2}{1 - (p_d/p_1)^2}\right\}^{1/2}, \tag{4.38b}$$

which is ellipse law of a multistage turbine.

The Wells turbine

Introduction

Numerous methods for extracting energy from the motion of sea-waves have been proposed and investigated since the late 1970s. The problem is in finding an

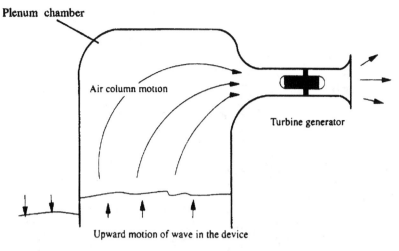

Plenum chamber

Air column motion

Turbine generator

Upward motion of wave in the device

FIG. 4.23. Arrangement of Wells turbine and oscillating water column (adapted from Raghunathan *et al.* 1995).

efficient and economical means of converting an oscillating flow of energy into a unidirectional rotary motion for driving an electrical generator. A novel solution of this problem is the Wells turbine (Wells 1976), a version of the axial-flow turbine. For countries surrounded by the sea, such as the British Isles and Japan to mention just two, or with extensive shorelines, wave energy conversion is an attractive proposition. Energy conversion systems based on the oscillating water column and the Wells turbine have been installed at several locations (Islay in Scotland and at Trivandrum in India). Figure 4.23 shows the arrangement of a turbine and generator together with the oscillating column of sea-water. The cross-sectional area of the plenum chamber is made very large compared to the flow area of the turbine so that a substantial air velocity through the turbine is attained.

One version of the Wells turbine consists of a rotor with about eight *uncambered* aerofoil section blades set at a stagger angle of ninety degrees (i.e. with their chord lines lying in the plane of rotation). A schematic diagram of such a Wells turbine is shown in Figure 4.24. At first sight the arrangement might seem to be a highly improbable means of energy conversion. However, once the blades have attained design speed the turbine is capable of producing a time-averaged positive power output from the cyclically reversing airflow with a fairly high efficiency. According to Raghunathan *et al.* (1995) *peak* efficiencies of 65% have been measured at the experimental wave power station on Islay. The results obtained from a theoretical analysis by Gato and Falcào (1984) showed that fairly high values of the mean efficiency, of the order 70–80%, may be attained in an oscillating flow "with properly designed Wells turbines".

Principle of operation

Figure 4.25(a) shows a blade in motion at the design speed U in a flow with an upward, absolute axial velocity c_1. It can be seen that the *relative velocity* w_1 is inclined to the chordline of the blade at an angle α. According to classical aerofoil

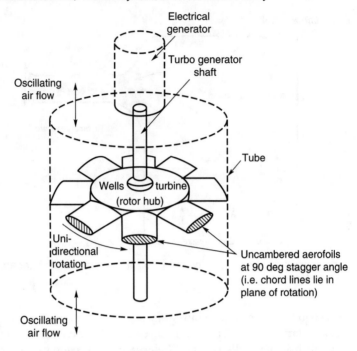

FIG. 4.24. Schematic of a Wells turbine (adapted from Raghunathan *et al.* 1995).

theory, an *isolated* aerofoil at an angle of incidence α to a free stream will generate a lift force L normal to the direction of the free stream. In a viscous fluid the aerofoil will also experience a drag force D in the direction of the free stream. These lift and drag forces can be resolved into the components of force X and Y as indicated in Figure 4.25a, i.e.

$$X = L\cos\alpha + D\sin\alpha, \tag{4.39}$$

$$Y = L\sin\alpha - D\cos\alpha. \tag{4.40}$$

The student should note, in particular, that the force Y *acts in the direction of blade motion*, giving positive work production.

For a symmetrical aerofoil, the direction of the tangential force Y is the same for both positive and negative values of α, as indicated in Figure 4.25b. If the aerofoils are secured to a rotor drum to form a turbine row, as in Figure 4.24, they will *always* rotate in the direction of the positive tangential force regardless of whether the air is approaching from above or below. With a time-varying, bi-directional air flow the torque produced will fluctuate cyclically but can be smoothed to a large extent by means of a high inertia rotor/generator.

It will be observed from the velocity diagrams that a residual swirl velocity is present for both directions of flow. It was suggested by Raghunathan *et al.* (1995) that the swirl losses at turbine exit can be reduced by the use of guide vanes.

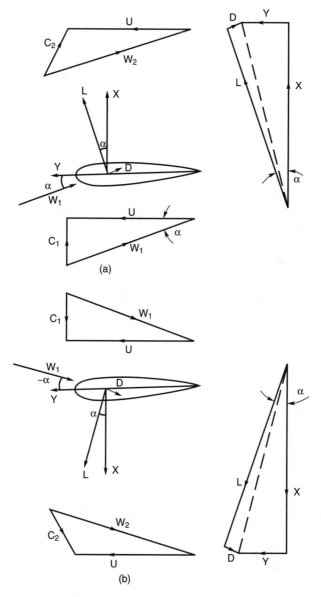

FIG. 4.25. Velocity and force vectors acting on a blade of a Wells turbine in motion: (a) upward absolute flow onto blade moving at speed U; (b) downward absolute flow onto blade moving at speed U.

Two-dimensional flow analysis

The performance of the Wells turbine can be predicted by means of blade element theory. In this analysis the turbine annulus is considered to be made up of a series of concentric elementary rings, each ring being treated separately as a two-dimensional cascade.

The power output from an elementary ring of area $2\pi r \, dr$ is given by

$$dW = ZU \, dy,$$

where Z is the number of blades and the tangential force on each blade element is

$$dY = C_y(\tfrac{1}{2}\rho w_1^2 l) dr.$$

The axial force acting on the blade elements at radius r is $Z \, dX$, where

$$dX = C_X(\tfrac{1}{2}\rho w_1^2 l) dr,$$

and where C_x, C_y are the axial and tangential force coefficients. Now the axial force on all the blade elements at radius r can be equated to the pressure force acting on the elementary ring:

$$2\pi r(p_1 - p_2) dr = ZC_x(\tfrac{1}{2}\rho w_1^2 l) dr,$$

$$\therefore \frac{(p_1 - p_2)}{\tfrac{1}{2}\rho c_x^2} = \frac{ZC_x l}{2\pi r \sin^2 \alpha_1},$$

where $w_1 = c_x / \sin \alpha_1$.

An expression for the efficiency can now be derived from a consideration of all the power losses and the power output. The power lost due to the drag forces is $dW_f = w_1 \, dD$, where

$$dD = ZC_D(\tfrac{1}{2}\rho w_1^2 l) dr$$

and the power lost due to exit kinetic energy is given by

$$dW_k = (\tfrac{1}{2}c_2^2) d\dot{m},$$

where $d\dot{m} = 2\pi r \rho c_x \, dr$ and c_2 is the absolute velocity at exit. Thus, the aerodynamic efficiency, defined as power output/power input, can now be written as

$$\eta = \frac{\int_h^t dW}{\int_h^t (dW + dW_f + dW_k)}.$$

The predictions for non-dimensional pressure drop p^* and aerodynamic efficiency η determined by Raghunathan *et al.* (1995) are shown in Figure 4.26a and b, respectively, together with experimental results for comparison.

Design and performance variables

The primary input for the design of a Wells turbine is the air power based upon the pressure amplitude $(p_1 - p_2)$ and the volume flow rate Q at turbine inlet. The performance indicators are the pressure drop, power and efficiency and their variation with the flow rate. The aerodynamic design and consequent performance is a function of several variables which have been listed by Raghunathan. In non-dimensional form

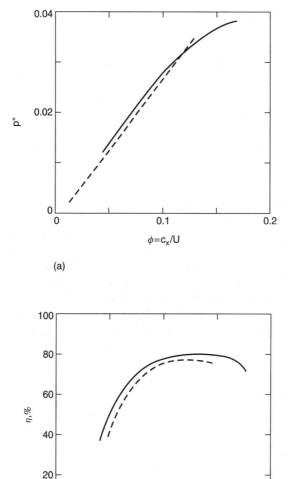

(a)

(b)

FIG. 4.26. Comparison of theory with experiment for the Wells turbine: ———— theory - - - - - experiment (adapted from Raghunathan 1995). (a) Non-dimensional pressure drop vs flow coefficient; (b) Efficiency vs flow coefficient.

these are:

flow coefficient $\quad\quad\quad\quad\phi = c_x/U$

solidity at mean radius $\quad\quad\sigma = \dfrac{2lZ}{\pi D_t(1 + \nu)}$.

hub/tip ratio $\quad\quad\quad\quad\quad\nu = D_h/D_t$

blade aspect ratio $\quad\quad\quad AR = \text{blade length/chord}$

blade tip clearance ratio $\quad\quad = t_c/D_t$

and also blade thickness ratio, turbulence level at inlet to turbine, frequency of waves and the relative Mach number. It was observed by Raghunathan *et al.* (1987) that the Wells turbine has a characteristic feature which makes it significantly different from most turbomachines: the absolute velocity of the flow is only a (small) fraction of the relative velocity. It is theoretically possible for transonic flow conditions to occur in the relative flow resulting in additional losses due to shock waves and an interaction with the boundary layers leading to flow separation. The effects of the variables listed above on the performance of the Wells turbine have been considered by Raghunathan (1995) and a summary of some of the main findings is given below.

Effect of flow coefficient

The flow coefficient ϕ is a measure of the angle of incidence of the flow and the aerodynamic forces developed are critically dependent upon this parameter. Typical results based on predictions and experiments of the non-dimensional pressure drop $p^* = \Delta p/(\rho \omega^2 D_t^2)$ and efficiency are shown in Figure 4.26. For a Wells turbine a linear relationship exists between pressure drop and the flowrate (Figure 4.26a) and this fact can be employed when making a match between a turbine and an oscillating water column which also has a similar characteristic.

The aerodynamic efficiency η (Figure 4.26b) is shown to increase up to a certain value, after which it decreases, because of boundary layer separation.

Effect of blade solidity

The solidity is a measure of the blockage offered by the blades to the flow of air and is an important design variable. The pressure drop across the turbine is, clearly, proportional to the axial force acting on the blades. An increase of solidity increases the axial force and likewise the pressure drop. Figure 4.27 shows how the variations of peak efficiency and pressure drop are related to the amount of the solidity.

Raghunathan gives correlations between pressure drop and efficiency with solidity:

$$p^*/p_0^* = 1 - \sigma^2 \text{ and } \eta/\eta_0 = \tfrac{1}{2}(1 - \sigma^2),$$

where the subscript 0 refers to values for a two-dimensional isolated aerofoil ($\sigma = 0$). A correlation between pressure drop and solidity (for $\sigma > 0$) was also expressed as

$$p^* = A\sigma^{1.6}$$

where A is a constant.

Effect of hub to tip ratio

The hub/tip ratio ν is an important parameter as it controls the volume flow rate through the turbine but also influences the stall conditions, the tip leakage and, most importantly, the ability of the turbine to run up to operating speed. Values of $\nu < 0.6$ are recommended for design.

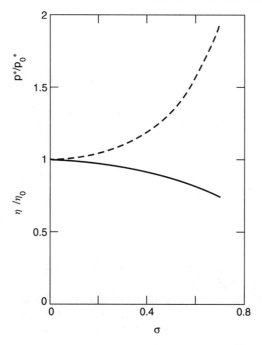

FIG. 4.27. Variation of peak efficiency and non-dimensional pressure drop (in comparison to the values for an isolated aerofoil) vs solidity: - - - pressure _____ efficiency (adapted from Raghunathan *et al.* 1995).

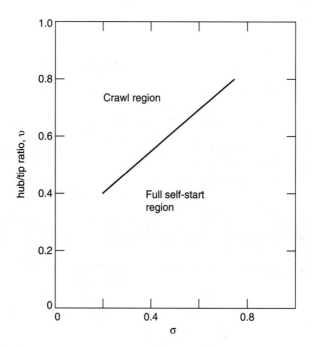

FIG. 4.28. Self-starting capability of the Wells turbine (adapted from Raghunathan *et al.* 1995).

The starting behaviour of the Wells turbine

When a Wells turbine is started from rest the incoming relative flow will be at 90 degrees to the rotor blades. According to the choice of the design parameters the blades could be severely stalled and, consequentially, the tangential force *Y* will be small and the acceleration negligible. In fact, if and when this situation occurs the turbine may only accelerate up to a speed much lower than the design operational speed, a phenomenon called *crawling*. The problem can be avoided either by choosing a suitable combination of hub/tip ratio and solidity values at the design stage or, by some other means such as incorporating a starter drive. Values of hub/tip ratio and solidity which have been found to allow self-starting of the Wells turbine are indicated in Figure 4.28.

References

Ainley, D. G. and Mathieson, G. C. R. (1951). A method of performance estimation for axial flow turbines. *ARC R&M 2974.*

Cooke, D. H. (1985). On prediction of off-design multistage turbine pressures by Stodola's ellipse. *J. Eng. Gas Turbines Power, Trans. Am. Soc. Mech. Engrs.*, **107**, 596–606.

Craig, H. R. M. and Cox, H. J. A. (1971). Performance estimation of axial flow turbines. *Proc. Instn. Mech. Engrs.*, **185**, 407–424.

Den Hartog, J. P. (1952). *Advanced Strength of Materials.* McGraw-Hill.

Dunham, J. and Came, P. M. (1970). Improvements to the Ainley–Mathieson method of turbine performance prediction. *Trans Am. Soc. Mech. Engrs., J. Eng. Power*, **92**, 252–6.

Dunham, J. and Panton, J. (1973). Experiments on the design of a small axial turbine. *Conference Publication 3*, Instn. Mech. Engrs.

Egli, A. (1936). The flow characteristics of variable-speed reaction steam turbines. *Trans. Am. Soc. Mech. Engrs.*, **58**.

Emmert, H. D. (1950). Current design practices for gas turbine power elements. *Trans. Am. Soc. Mech. Engrs.*, **72**, Pt. 2.

Freeman, JAW. (1955). High temperature materials. *Gas Turbines and Free Piston Engines*, Lecture 5, University of Michigan, Summer Session.

Gato, L. C. and Falcào, A. F. de O. (1984). On the theory of the Wells turbine. *J. Eng. Power, Trans. Am. Soc. Mech. Engrs.*, **106** (also as 84-GT-5).

Horlock, J. H. (1958). A rapid method for calculating the "off-design" performance of compressors and turbines. *Aeronaut. Quart.*, **9**.

Horlock, J. H. (1960). Losses and efficiencies in axial-flow turbines. *Int. J. Mech. Sci.* **2**.

Horlock, J. H. (1966). *Axial Flow Turbines.* Butterworths. (1973 reprint with corrections, Huntington, New York: Krieger.)

Japikse, D. (1986). Life evaluation of high temperature turbomachinery. In *Advanced Topics in Turbomachine Technology.* Principal Lecture Series, No. 2. (David Japikse, ed.) pp. 5–1 to 5–47, Concepts ETI.

Kacker, S. C. and Okapuu, U. (1982). A mean line prediction method for axial flow turbine efficiency. *J. Eng. Power. Trans. Am. Soc. Mech. Engrs.*, **104**, 111–9.

Kearton, W. J. (1958). *Steam Turbine Theory and Practice.* (7th edn). Pitman.

Le Grivès, E. (1986). Cooling techniques for modern gas turbines. In *Advanced Topics in Turbomachinery Technology* (David Japikse, ed.) pp. 4–1 to 4–51, Concepts ETI.

Mallinson, D. H. and Lewis, W. G. E. (1948). The part-load performance of various gas-turbine engine schemes. *Proc. Instn. Mech. Engrs.*, **159**.

Raghunathan, S. Setoguchi, T. and Kaneko, K. (1991). The Wells air turbine subjected to inlet flow distortion and high levels of turbulence. *Heat and Fluid Flow*, **8**, No. 2.

Raghunathan, S., Setoguchi, T. and Kaneko, K. (1991). Aerodynamics of monoplane Wells turbine – a review. *Proc. Conf. on Offshore Mechanics and Polar Engineering.*, Edinburgh.

Raghunathan, S., Curran, R. and Whittaker, T. J. T. (1995). Performance of the Islay Wells air turbine. *Proc. Instn Mech. Engrs.*, **209**, 55–62.

Raghunathan, S. (1995). A methodology for Wells turbine design for wave energy conversion. *Proc. Instn Mech. Engrs.*, **209**, 221–32.

Shapiro, A. H., Soderberg, C. R., Stenning, A. H., Taylor, E. S. and Horlock, J. H. (1957). Notes on Turbomachinery. Department of Mechanical Engineering, Massachusetts Institute of Technology. Unpublished.

Smith, G. E. (1986). Vibratory stress problems in turbomachinery. *Advanced Topics in Turbo-machine Technology*. Principal Lecture Series, No. 2. (David Japikse, ed.) pp. 8–1 to 8–23, Concepts ETI.

Soderberg, C. R. (1949). Unpublished note. Gas Turbine Laboratory, Massachusetts Institute of Technology.

Stenning, A. H. (1953). Design of turbines for high energy fuel, low power output applications. D.A.C.L. Report 79, Massachusetts Institute of Technology.

Stodola, A. (1945). *Steam and Gas Turbines*, (6th edn). Peter Smith, New York.

Wells, A. A. (1976). Fluid driven rotary transducer. British Patent 1595700.

Wilde, G. L. (1977). The design and performance of high temperature turbines in turbofan engines. *1977 Tokyo Joint Gas Turbine Congress*, co-sponsored by Gas Turbine Soc. of Japan, the Japan Soc. of Mech. Engrs and the Am. Soc. of Mech. Engrs., pp. 194–205.

Wilson, D. G. (1987). New guidelines for the preliminary design and performance prediction of axial-flow turbines. *Proc. Instn. Mech. Engrs.*, **201**, 279–290.

Problems

1. Show, for an axial flow turbine stage, that the *relative* stagnation enthalpy across the rotor row does not change. Draw an enthalpy–entropy diagram for the stage labelling all salient points.

Stage reaction for a turbine is defined as the ratio of the static enthalpy drop in the rotor to that in the stage. Derive expressions for the reaction in terms of the flow angles and draw velocity triangles for reactions of zero, 0.5 and 1.0.

2. (a) An axial flow turbine operating with an overall stagnation pressure of 8 to 1 has a polytropic efficiency of 0.85. Determine the total-to-total efficiency of the turbine.

(b) If the exhaust Mach number of the turbine is 0.3, determine the total-to-static efficiency.

(c) If, in addition, the exhaust velocity of the turbine is 160 m/s, determine the inlet total temperature.

Assume for the gas that $C_P = 1.175$ kJ/(kg K) and $R = 0.287$ kJ/(kg K).

3. The mean blade radii of the rotor of a mixed flow turbine are 0.3 m at inlet and 0.1 m at outlet. The rotor rotates at 20,000 rev/min and the turbine is required to produce 430 kW. The flow velocity at nozzle exit is 700 m/s and the flow direction is at 70° to the meridional plane.

Determine the absolute and relative flow angles and the absolute exit velocity if the gas flow is 1 kg/s and the velocity of the through-flow is constant through the rotor.

4. In a Parson's reaction turbine the rotor blades are similar to the stator blades but with the angles measured in the opposite direction. The efflux angle relative to each row of blades is 70 deg from the axial direction, the exit velocity of steam from the stator blades is 160 m/s, the blade speed is 152.5 m/s and the axial velocity is constant. Determine the specific work done by the steam per stage.

A turbine of 80% internal efficiency consists of ten such stages as described above and receives steam from the stop valve at 1.5 MPa and 300°C. Determine, with the aid of a Mollier chart, the condition of the steam at outlet from the last stage.

5. Values of pressure (kPa) measured at various stations of a zero-reaction gas turbine stage, all at the mean blade height, are shown in the table given below.

Stagnation pressure	Static pressure
Nozzle entry 414	Nozzle exit 207
Nozzle exit 400	Rotor exit 200

The mean blade speed is 291 m/s, inlet stagnation temperature 1100 K, and the flow angle at nozzle exit is 70 deg measured from the axial direction. Assuming the magnitude and direction of the velocities at entry and exit of the stage are the same, determine the total-to-total efficiency of the stage. Assume a perfect gas with $C_p = 1.148\,\text{kJ/(kg°C)}$ and $\gamma = 1.333$.

6. In a certain axial flow turbine stage the axial velocity c_x is constant. The absolute velocities entering and leaving the stage are in the axial direction. If the flow coefficient c_x/U is 0.6 and the gas leaves the stator blades at 68.2 deg from the axial direction, calculate:

(i) the stage loading factor, $\Delta W/U^2$;
(ii) the flow angles relative to the rotor blades;
(iii) the degree of reaction;
(iv) the total-to-total and total-to-static efficiencies.

The Soderberg loss correlation, eqn. (4.12) should be used.

7. An axial flow gas turbine stage develops 3.36 MW at a mass flow rate of 27.2 kg/s. At the stage entry the stagnation pressure and temperature are 772 kPa and 727°C, respectively. The static pressure at exit from the nozzle is 482 kPa and the corresponding absolute flow direction is 72° to the axial direction. Assuming the axial velocity is constant across the stage and the gas enters and leaves the stage without any absolute swirl velocity, determine:

(1) the nozzle exit velocity;
(2) the blade speed;
(3) the total-to-static efficiency;
(4) the stage reaction.

The Soderberg correlation for estimating blade row losses should be used. For the gas assume that $C_P = 1.148\,\text{kJ/(kg K)}$ and $R = 0.287\,\text{kJ/(kg K)}$.

8. Derive an approximate expression for the total-to-total efficiency of a turbine stage in terms of the enthalpy loss coefficients for the stator and rotor when the absolute velocities at inlet and outlet are **not** equal.

A steam turbine stage of high hub/tip ratio is to receive steam at a stagnation pressure and temperature of 1.5 MPa and 325°C respectively. It is designed for a blade speed of 200 m/s and the following *blade* geometry was selected:

	Nozzles	Rotor
Inlet angle, deg	0	48
Outlet angle, deg	70.0	56.25
Space/chord ratio, s/l	0.42	–
Blade length/axial chord ratio, H/b	2.0	2.1
Max. thickness/axial chord	0.2	0.2

The deviation angle of the flow from the rotor row is known to be 3 deg on the evidence of cascade tests at the design condition. In the absence of cascade data for the nozzle row, the designer estimated the deviation angle from the approximation $0.19\,\theta s/l$ where θ is the blade camber in degrees. Assuming the incidence onto the nozzles is zero, the incidence onto the rotor 1.04 deg and the axial velocity across the stage is constant, determine:

(i) the axial velocity;
(ii) the stage reaction and loading factor;
(iii) the approximate total-to-total stage efficiency on the basic of Soderberg's loss correlation, assuming Reynolds number effects can be ignored;
(iv) by means of a large steam chart (Mollier diagram) the stagnation temperature and pressure at stage exit.

9. (a) A single-stage axial flow turbine is to be designed for zero reaction without any absolute swirl at rotor exit. At nozzle inlet the stagnation pressure and temperature of the gas are 424 kPa and 1100 K. The static pressure at the mean radius between the nozzle row and rotor entry is 217 kPa and the nozzle exit flow angle is 70°.

Sketch an appropriate Mollier diagram (or a $T-s$ diagram) for *this* stage allowing for the effects of losses and sketch the corresponding velocity diagram. Hence, using Soderberg's correlation to calculate blade row losses, determine for the mean radius,

(1) the nozzle exit velocity,
(2) the blade speed,
(3) the total-to-static efficiency.

(b) Verify for this turbine stage that the total-to-total efficiency is given by

$$\frac{1}{\eta_{tt}} = \frac{1}{\eta_{ts}} - \left(\frac{\phi}{2}\right)^2$$

where $\phi = c_x/U$. Hence, determine the value of the total-to-total efficiency.
Assume for the gas that $C_p = 1.15\,\text{kJ/(kg K)}$ and $\gamma = 1.333$.

10. (a) Prove that the centrifugal stress at the root of an untapered blade attached to the drum of an axial flow turbomachine is given by

$$\sigma_c = \pi \rho_m N^2 A_n/1800,$$

where ρ_m = density of blade material, N = rotational speed of drum and A_n = area of the flow annulus.

(b) The preliminary design of an axial-flow gas turbine stage with stagnation conditions at stage entry of $p_{01} = 400\,\text{kPa}$, $T_{01} = 850\,\text{K}$, is to be based upon the following data *applicable to the mean radius*:

Flow angle at nozzle exit, $\alpha_2 = 63.8\,\text{deg}$
Reaction, $R = 0.5$
Flow coefficient, $c_x/U_m = 0.6$

Static pressure at stage exit, $p_3 = 200\,\text{kPa}$
Estimated total-to-static efficiency, $\eta_{ts} = 0.85$.

Assuming that the axial velocity is unchanged across the stage, determine:

(1) the specific work done by the gas;
(2) the blade speed;
(3) the static temperature at stage exit.

(c) The blade material has a density of $7850\,\text{kg/m}^3$ and the maximum allowable stress in the rotor blade is 120 MPa. Taking into account only the centrifugal stress, assuming untapered blades and constant axial velocity at all radii, determine for a mean flow rate of 15 kg/s:

(1) the rotor speed (rev/min);
(2) the mean diameter;
(3) the hub/tip radius ratio.

For the gas assume that $C_P = 1050\,\text{J/(kg K)}$ and $R = 287\,\text{J/(kg K)}$.

11. The design of a single-stage axial-flow turbine is to be based on constant axial velocity with axial discharge from the rotor blades directly to the atmosphere.

The following design values have been specified:

Mass flow rate	16.0 kg/s
Initial stagnation temperature, T_{01}	1100 K
Initial stagnation pressure, p_{01}	230 kN/m^2
Density of blading material, ρ_m	7850 kg/m^3
Maximum allowable centrifugal stress at blade root,	1.7×10^8 N/m^2
Nozzle profile loss coefficient, $Y_P = (p_{01} - p_{02})/(p_{02} - p_2)$	0.06
Taper factor for blade stressing, K	0.75

In addition the following may be assumed:

Atmospheric pressure, p_3	102 kPa
Ratio of specific heats, γ	1.333
Specific heat at constant pressure, C_P	1150 J/(kg K)

In the design calculations values of the parameters at the mean radius are as follows:

Stage loading coefficient, $\psi = \Delta W/U^2$	1.2
Flow coefficient, $\phi = c_x/U$	0.35
Isentropic velocity ratio, U/c_0	0.61
where $c_0 = \sqrt{[2(h_{01} - h_{3SS})]}$	

Determine:

(1) the velocity triangles at the mean radius;
(2) the required annulus area (based on the density at the mean radius);
(3) the maximum allowable rotational speed;
(4) the blade tip speed and the hub/tip radius ratio.

CHAPTER 5

Axial-flow Compressors and Fans

A solemn, strange and mingled air, 't was sad by fits, by starts was wild.
(W. COLLINS, *The Passions.*)

Introduction

The idea of using a form of *reversed turbine* as an axial compressor is as old as the reaction turbine itself. It is recorded by Stoney (1937) that Sir Charles Parsons obtained a patent for such an arrangement as early as 1884. However, simply reversing a turbine for use as a compressor gives efficiencies which are, according to Howell (1945), less than 40% for machines of high pressure ratio. Parsons actually built a number of these machines (*circa* 1900), with blading based upon improved propeller sections. The machines were used for blast furnace work, operating with delivery pressures between 10 and 100 kPa. The efficiency attained by these early, low pressure compressors was about 55%; the reason for this low efficiency is now attributed to blade stall. A high pressure ratio compressor (550 kPa delivery pressure) was also built by Parsons but is reported by Stoney to have "run into difficulties". The design, comprising two axial compressors in series, was abandoned after many trials, the flow having proved to be unstable (presumably due to *compressor surge*). As a result of low efficiency, axial compressors were generally abandoned in favour of multistage centrifugal compressors with their higher efficiency of 70–80%.

It was not until 1926 that any further development on axial compressors was undertaken when A. A. Griffith outlined the basic principles of his aerofoil theory of compressor and turbine design. The subsequent history of the axial compressor is closely linked with that of the aircraft gas turbine and has been recorded by Cox (1946) and Constant (1945). The work of the team under Griffith at the Royal Aircraft Establishment, Farnborough, led to the conclusion (confirmed later by rig tests) that efficiencies of at least 90% could be achieved for 'small' stages, i.e. low pressure ratio stages.

The early difficulties associated with the development of axial-flow compressors stemmed mainly from the fundamentally different nature of the flow process compared with that in axial-flow turbines. Whereas in the axial turbine the flow relative to each blade row is *accelerated*, in axial compressors it is *decelerated*. It is now widely known that although a fluid can be rapidly accelerated through a passage and sustain a small or moderate loss in total pressure the same is not true for a rapid deceleration. In the latter case large losses would arise as a result of severe

137

stall caused by a large adverse pressure gradient. So as to limit the total pressure losses during flow diffusion it is necessary for the rate of deceleration (and turning) in the blade passages to be severely restricted. (Details of these restrictions are outlined in Chapter 3 in connection with the correlations of Lieblein and Howell.) It is mainly because of these restrictions that axial compressors need to have many stages for a given pressure ratio compared with an axial turbine which needs only a few. Thus, the reversed turbine experiment tried by Parsons was doomed to a low operating efficiency.

The performance of axial compressors depends upon their usage category. Carchedi and Wood (1982) described the design and development of a single-shaft 15-stage axial-flow compressor which provided a 12 to 1 pressure ratio at a mass flow of 27.3 kg/s for a 6 MW industrial gas turbine. The design was based on subsonic flow and the compressor was fitted with variable stagger stator blades to control the position of the low-speed surge line. In the field of aircraft gas turbines, however, the engine designer is more concerned with *maximising* the work done per stage while retaining an acceptable level of overall performance. Increased stage loading almost inevitably leads to some aerodynamic constraint. This constraint will be increased Mach number, possibly giving rise to shock-induced boundary layer separation or increased losses arising from poor diffusion of the flow. Wennerstrom (1990) has outlined the history of highly loaded axial-flow compressors with special emphasis on the importance of reducing the number of stages and the ways that improved performance can be achieved. Since about 1970 a significant and special change occurred with respect to one design feature of the axial compressor and that was the introduction of low aspect ratio blading. It was not at all obvious why blading of large chord would produce any performance advantage, especially as the trend was to try to make engines more compact and lighter by using high aspect ratio blading. Wennerstrom (1989) has reviewed the increased usage of low aspect ratio blading in aircraft axial-flow compressors and reported on the high loading capability, high efficiency and good range obtained with this type of blading. One early application was an axial-flow compressor that achieved a pressure ratio of 12.1 in only five stages, with an isentropic efficiency of 81.9% and an 11% stall margin. The blade tip speed was 457 m/s and the flow rate per unit frontal area was 192.5 kg/s/m^2. It was reported that the mean aspect ratio ranged from a "high" of 1.2 in the first stage to less than 1.0 in the last three stages. A related later development pursued by the US Air Force was an alternative inlet stage with a rotor mean aspect ratio of 1.32 which produced, at design, a pressure ratio of 1.912 with an isentropic efficiency of 85.4% and an 11% stall margin. A maximum efficiency of 90.9% was obtained at a pressure ratio of 1.804 and lower rotational speed.

The flow within an axial-flow compressor is exceedingly complex which is one reason why research and development on compressors has proliferated over the years. In the following pages a very simplified and basic study is made of this compressor so that the student can grasp the essentials.

Two-dimensional analysis of the compressor stage

The analysis in this chapter is simplified (as it was for axial turbines) by assuming the flow is two-dimensional. This approach can be justified if the blade height is small compared with the mean radius. Again, as for axial turbines, the flow is assumed to be invariant in the circumferential direction and that no spanwise (radial) velocities occur. Some of the three-dimensional effects of axial turbomachines are considered in Chapter 6.

To illustrate the layout of an axial compressor, Figure 5.1(a) shows a sectional drawing of the three-shaft compressor system of the Rolls-Royce RB211 gas-turbine engine. The very large blade on the left is part of the fan rotor which is on one shaft; this is followed by two, six-stage compressors of the "core" engine, each on its own shaft. A *compressor stage* is defined as a rotor blade row followed by a stator blade row. Figure 5.1b shows some of the blades of the first stage of the low-pressure compressor opened out into a plane array. The rotor blades (black) are fixed to the

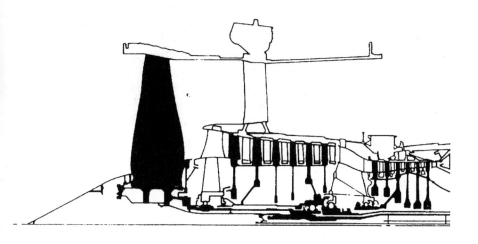

(a)

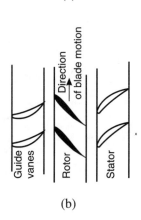

(b)

FIG. 5.1. Axial-flow compressor and blading arrays. (a) Section of the compression system of the RB211-535E4 gas-turbine engine (courtesy of Rolls-Royce plc). (b) Development of the first stage-blade rows and inlet guide vanes.

rotor drum and the stator blades are fixed to the outer casing. The blades upstream of the first rotor row are inlet guide vanes. These are not considered to be a part of the first stage and are treated separately. Their function is quite different from the other blade rows since, by directing the flow away from the axial direction, they act to *accelerate* the flow rather than diffuse it. Functionally, inlet guide vanes are the same as turbine nozzles; they increase the kinetic energy of the flow at the expense of the pressure energy.

Velocity diagrams of the compressor stage

The velocity diagrams for the stage are given in Figure 5.2 and the convention is adopted throughout this chapter of accepting all angles and swirl velocities in this figure as positive. As for axial turbine stages, a *normal* compressor stage is one where the absolute velocities and flow directions at stage outlet are the same as at stage inlet. The flow from a previous stage (or from the guide vanes) has a velocity c_1 and direction α_1; substracting vectorially the blades speed U gives the inlet relative velocity w_1 at angle β_1 (the axial direction is the datum for all angles). Relative to the blades of the rotor, the flow is turned to the direction β_2 at outlet with a relative velocity w_2. Clearly, by adding vectorially the blade speed U on to w_2 gives the absolute velocity from the rotor, c_2 at angle α_2. The stator blades deflect the flow towards the axis and the exit velocity is c_3 at angle α_3. For the normal stage $c_3 = c_1$ and $\alpha_3 = \alpha_1$. It will be noticed that as drawn in Figure 5.2, both the

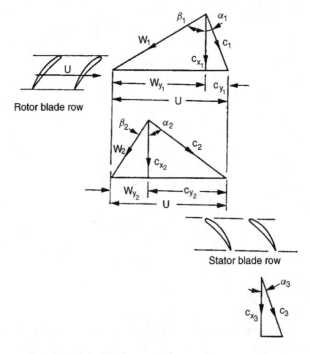

FIG. 5.2. Velocity diagrams for a compressor stage.

relative velocity in the rotor and the absolute velocity in the stator are diffused. It will be shown later in this chapter, that the relative amount of diffusion of kinetic energy in the rotor and stator rows, significantly influences the stage efficiency.

Thermodynamics of the compressor stage

The specific work done by the rotor on the fluid, from the steady flow energy equation (assuming adiabatic flow) and momentum equation is

$$\Delta W = \dot{W}_p / \dot{m} = h_{02} - h_{01} = U(c_{y2} - c_{y1}). \tag{5.1}$$

In Chapter 4 it was proved for all axial turbomachines that $h_{0\mathrm{rel}} = h + \frac{1}{2}w^2$ is constant in the rotor. Thus,

$$h_1 + \tfrac{1}{2}w_1^2 = h_2 + \tfrac{1}{2}w_2^2. \tag{5.2}$$

This is a valid result as long as there is no radial shift of the streamlines across the rotor (i.e. $U_1 = U_2$).

Across the stator, h_0 is constant, and

$$h_2 + \tfrac{1}{2}c_2^2 = h_3 + \tfrac{1}{2}c_3^2. \tag{5.3}$$

The compression process for the complete stage is represented on a Mollier diagram in Figure 5.3, which is generalised to include the effects of irreversibility.

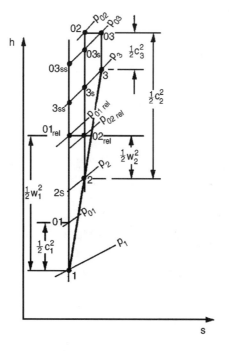

FIG. 5.3. Mollier diagram for an axial compressor stage.

Stage loss relationships and efficiency

From eqns. (5.1) and (5.3) the actual work performed by the rotor on unit mass of fluid is $\Delta W = h_{03} - h_{01}$. The reversible or *minimum* work required to attain the same final stagnation pressure as the real process is,

$$\Delta W_{min} = h_{03ss} - h_{01}$$

$$= (h_{03} - h_{01}) - (h_{03s} - h_{03ss}) - (h_{03} - h_{03s})$$

$$\doteq \Delta W - (T_{03}/T_2)(h_2 - h_{2s}) - (T_{03}/T_3)(h_3 - h_{3s}),$$

using the approximation that $\Delta h = T\Delta s$.

The temperature rise in a compressor stage is only a small fraction of the absolute temperature level and therefore, to a *close* approximation.

$$\Delta W_{min} = \Delta W - (h_2 - h_{2s}) - (h_3 - h_{3s}). \tag{5.4}$$

Again, because of the small stage temperature rise, the density change is also small and it is reasonable to assume incompressibility for the fluid. This approximation is applied *only* to the stage and a *mean* stage density is implied; across a multistage compressor an appreciable density change can be expected.

The enthalpy losses in eqn. (5.4) can be expressed as stagnation pressure losses as follows. As $h_{02} = h_{03}$ then,

$$h_3 - h_2 = \tfrac{1}{2}(c_2^2 - c_3^2)$$

$$= [(p_{02} - p_2) - (p_{03} - p_3)]/\rho, \tag{5.5}$$

since $p_0 - p = \tfrac{1}{2}\rho c^2$ for an incompressible fluid.

Along the isentrope $2 - 3_s$ in Figure 5.3, $Tds = 0 = dh - (1/\rho)dp$, and so,

$$h_{3s} - h_2 = (p_3 - p_2)/\rho. \tag{5.6}$$

Thus, subtracting eqn. (5.6) from eqn. (5.5)

$$h_3 - h_{3s} = (p_{02} - p_{03})/\rho = (1/\rho)\Delta p_{0stator}. \tag{5.7}$$

Similarly, for the rotor,

$$h_2 - h_{2s} = (p_{01rel} - p_{02rel})/\rho = (1/\rho)\Delta p_{0rotor}. \tag{5.8}$$

The total-to-total stage efficiency is,

$$\eta_{tt} = \frac{\dot{W}_{p\,min}}{\dot{W}_p} \doteq 1 - \frac{(h_2 - h_{2s}) + (h_3 - h_{3s})}{(h_{03} - h_{01})}$$

$$\doteq 1 - \frac{\Delta p_{0stator} + \Delta p_{0rotor}}{\rho(h_{03} - h_{01})} \tag{5.9}$$

It is to be observed that eqn. (5.9) also has direct application to pumps and fans.

Reaction ratio

For the case of *incompressible and reversible* flow it is permissible to define the reaction R, as the ratio of static pressure rise in the rotor to the static pressure rise in the stage

$$R = (p_2 - p_1)/(p_3 - p_1). \tag{5.10a}$$

If the flow is both *compressible and irreversible* a more general definition of R is the ratio of the rotor static enthalpy rise to the stage static enthalpy rise,

$$R = (h_2 - h_1)/(h_3 - h_1). \tag{5.10b}$$

From eqn. (5.2), $h_2 - h_1 = \frac{1}{2}(w_1^2 - w_2^2)$. For normal stages ($c_1 = c_3$), $h_3 - h_1 = h_{03} - h_{01} = U(c_{y2} - c_{y1})$. Substituting into eqn. (5.10b)

$$R = \frac{w_1^2 - w_2^2}{2U(c_{y2} - c_{y1})} \tag{5.10c}$$

$$= \frac{(w_{y1} + w_{y2})(w_{y1} - w_{y2})}{2U(c_{y2} - c_{y1})},$$

where it is assumed that c_x is constant across the stage. From Figure 5.2, $c_{y2} = U - w_{y2}$ and $c_{y1} = U - w_{y1}$ so that $c_{y2} - c_{y1} = w_{y1} - w_{y2}$. Thus,

$$R = (w_{y1} + w_{y2})/(2U) = (c_x/U)\tan\beta_m, \tag{5.11}$$

where

$$\tan\beta_m = \frac{1}{2}(\tan\beta_1 + \tan\beta_2). \tag{5.12}$$

An alternative useful expression for reaction can be found in terms of the fluid outlet angles from each blade row in a stage. With $w_{y1} = U - c_{y1}$, eqn. (5.11) gives,

$$R = \frac{1}{2} + (\tan\beta_2 - \tan\alpha_1)c_x/(2U). \tag{5.13}$$

Both expressions for reaction given above may be derived on a basis of incompressible, reversible flow, together with the definition of reaction in eqn. (5.10a).

Choice of reaction

The reaction ratio is a design parameter which has an important influence on stage efficiency. Stages having 50% reaction are widely used as the adverse (retarding) pressure gradient through the rotor and stator rows is equally shared. This choice of reaction minimises the tendency of the blade boundary layers to separate from the solid surfaces, thus avoiding large stagnation pressure losses.

If $R = 0.5$, then $\alpha_1 = \beta_2$ from eqn. (5.13), and the velocity diagram is symmetrical. The stage enthalpy rise is equally distributed between the rotor and stator rows.

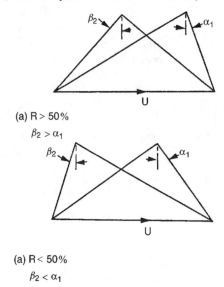

(a) R > 50%

$\beta_2 > \alpha_1$

(a) R < 50%

$\beta_2 < \alpha_1$

FIG. 5.4. Asymmetry of velocity diagrams for reactions greater or less than 50%.

If $R > 0.5$ then $\beta_2 > \alpha_1$ and the velocity diagram is skewed to the *right* as shown in Figure 5.4a. The static enthalpy rise in the rotor exceeds that in the stator (this is also true for the static pressure rise).

If $R < 0.5$ then $\beta_2 < \alpha_1$ and the velocity diagram is skewed to the *left* as indicated in Figure 5.4b. Clearly, the stator enthalpy (and pressure) rise exceeds that in the rotor.

In axial turbines the limitation on stage work output is imposed by rotor blade stresses but, in axial compressors, stage performance is limited by Mach number considerations. If Mach number effects could be ignored, the permissible temperature rise, based on incompressible flow cascasde limits, increases with the amount of reaction. With a limit of 0.7 on the allowable Mach number, the temperature rise and efficiency are at a maximum with a reaction of 50%, according to Horlock (1958).

Stage loading

The stage loading factor ψ is another important design parameter of a compressor stage and is one which strongly affects the off-design performance characteristics. It is defined by

$$\psi = \frac{h_{03} - h_{01}}{U^2} = \frac{c_{y2} - c_{y1}}{U}. \tag{5.14a}$$

With $c_{y2} = U - w_{y2}$ this becomes,

$$\psi = 1 - \phi(\tan \alpha_1 + \tan \beta_2), \tag{5.14b}$$

where $\phi = c_x/U$ is called the *flow coefficient*.

The stage loading factor may also be expressed in terms of the lift and drag coefficients for the *rotor*. From Figure 3.5, replacing α_m with β_m, the tangential blade force on the *moving* blades per unit span is,

$$Y = L \cos \beta_m + D \sin \beta_m$$

$$= L \cos \beta_m \left(1 + \frac{C_D}{C_L} \tan \beta_m \right),$$

where $\tan \beta_m = \frac{1}{2}(\tan \beta_1 + \tan \beta_2)$.

Now $C_L = L/(\frac{1}{2}\rho w_m^2 l)$ hence substituting for L above,

$$Y = \frac{1}{2}\rho c_x^2 l C_L \sec \beta_m (1 + \tan \beta_m C_D/C_L). \qquad (5.15)$$

The work done by *each* moving blade per second is YU and is transferred to the fluid through *one* blade passage during that period. Thus, $YU = \rho s c_x (h_{03} - h_{01})$.

Therefore, the stage loading factor may now be written

$$\psi = \frac{h_{03} - h_{01}}{U^2} = \frac{Y}{\rho s c_x U}. \qquad (5.16)$$

Substituting eqn. (5.15) in eqn. (5.16) the final result is

$$\psi = (\phi/2) \sec \beta_m (l/s)(C_L + C_D \tan \beta_m). \qquad (5.17)$$

In Chapter 3, the approximate analysis indicated that maximum efficiency is obtained when the mean flow angle is 45 deg. The corresponding optimum stage loading factor at $\beta_m = 45$ deg is,

$$\psi_{\text{opt}} = (\phi/\sqrt{2})(l/s)(C_L + C_D). \qquad (5.18)$$

Since $C_D \ll C_L$ in the normal low loss operating range, it is permissible to drop C_D from eqn. (5.18).

Simplified off-design performance

Horlock (1958) has considered how the stage loading behaves with varying flow coefficient, ϕ and how this off-design performance is influenced by the choice of design conditions. Now cascade data suggests that fluid *outlet angles* β_2 (for the rotor) and $\alpha_1 (= \alpha_3)$ for the stator, *do not change appreciably* for a range of incidence up to the stall point. The simplification may therefore be made that, for a given stage,

$$\tan \alpha_1 + \tan \beta_2 = t = \text{constant}. \qquad (5.19)$$

Inserting this expression into eqn. (5.14b) gives

$$\psi = 1 - \phi t. \qquad (5.20a)$$

An inspection of eqns. (5.20a) and (5.14a) indicates that the stagnation enthalpy rise of the stage increases as the mass flow is reduced, when running at constant

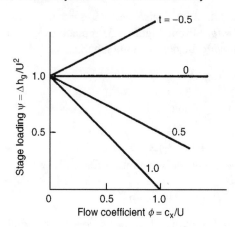

FIG. 5.5. Simplified off-design performance of a compressor stage (adapted from Horlock 1958).

rotational speed, provided t is positive. The effect is shown in Figure 5.5, where ψ is plotted against ϕ for several values of t.

Writing $\psi = \psi_d$ and $\phi = \phi_d$ for conditions at the design point, then

$$\psi_d = 1 - \phi_d t. \tag{5.20b}$$

The values of ψ_d and ϕ_d chosen for a particular stage design, determines the value of t. Thus t is fixed without regard to the degree of reaction and, therefore, the variation of stage loading at off-design conditions is not dependent on the choice of design reaction. However, from eqn. (5.13) it is apparent that, except for the case of 50% reaction when $\alpha_1 = \beta_2$, the reaction *does* change away from the design point. For design reactions exceeding 50% ($\beta_2 > \alpha_1$), the reaction decreases towards 50% as ϕ decreases; conversely, for design reactions less than 50% the reaction approaches 50% with diminishing flow coefficient.

If t is eliminated between eqns. (5.20a) and (5.20b) the following expression results,

$$\frac{\psi}{\psi_d} = \frac{1}{\psi_d} - \frac{\phi}{\phi_d}\left(\frac{1 - \psi_d}{\psi_d}\right). \tag{5.21}$$

This equation shows that, for a given design stage loading ψ_d, the fractional change in stage loading corresponding to a fractional change in flow coefficient is always the same, independent of the stage reaction. In Figure 5.6 it is seen that heavily loaded stages ($\psi_d \to 1$) are the most flexible producing little variation of ψ with change of ϕ. Lightly loaded stages ($\psi_d \to 0$) produce large changes in ψ with changing ϕ. Data from cascade tests show that ψ_d is limited to the range 0.3 to 0.4 for the most efficient operation and so substantial variations of ψ can be expected away from the design point.

In order to calculate the pressure rise at off-design conditions the variation of stage efficiency with flow coefficient is required. For an ideal stage (no losses) the pressure rise in incompressible flow is given by

$$\psi = \frac{\Delta h}{U^2} = \frac{\Delta p}{\rho U^2}. \tag{5.22}$$

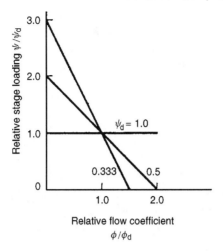

FIG. 5.6. Effect of design stage loading (ψ_d) on simplified off-design performance characteristics (adapted from Horlock 1958).

Stage pressure rise

Consider first the *ideal* compressor stage which has no stagnation pressure losses. Across the rotor row $p_{0\mathrm{rel}}$ is constant and so

$$p_2 - p_1 = \tfrac{1}{2}\rho(w_1^2 - w_2^2). \tag{5.23a}$$

Across the stator row p_0 is constant and so

$$p_3 - p_2 = \tfrac{1}{2}\rho(c_2^2 - c_3^2). \tag{5.23b}$$

Adding together the pressure rise for each row and considering a normal stage ($c_3 = c_1$), gives

$$(p_3 - p_1)2/\rho = (c_2^2 - w_2^2) + (w_1^2 - c_1^2). \tag{5.24}$$

For either velocity triangle (Figure 5.2), the *cosine rule* gives $c^2 - U^2 + w^2 = 2Uw\cos(\pi/2 - \beta)$ or

$$c^2 - w^2 = U^2 - 2Uw_y. \tag{5.25}$$

Substituting eqn. (5.25) into the stage pressure rise,

$$2(p_3 - p_1)/\rho = (U^2 - 2Uw_{y2}) - (U^2 - 2Uw_{y1})$$
$$= 2U(w_{y1} - w_{y2}).$$

Again, referring to the velocity diagram, $w_{y1} - w_{y2} = c_{y2} - c_{y1}$ and

$$(p_3 - p_1)/\rho = U(c_{y2} - c_{y1}) = h_3 - h_1. \tag{5.26}$$

It is noted that, for an isentropic process, $T\mathrm{d}s = 0 = \mathrm{d}h - (1/\rho)\mathrm{d}p$ and therefore, $\Delta h = (1/\rho)\Delta p$.

The pressure rise in a real stage (involving irreversible processes) can be determined if the stage efficiency is known. Defining the stage efficiency η_s as the ratio of the isentropic enthalpy rise to the actual enthalpy rise corresponding to the same *finite* pressure change, (cf. Figure 2.7), this can be written as

$$\eta_s = (\Delta h_{is})/(\Delta h) = (1/\rho)\Delta p/\Delta h.$$

Thus,

$$(1/\rho)\Delta p = \eta_s \Delta h = \eta_s U \Delta c_y. \tag{5.27}$$

If $c_1 = c_3$, then η_s is a very close approximation of the total-to-total efficiency η_{tt}. Although the above expressions are derived for incompressible flow they are, nevertheless, a valid approximation for compressible flow if the stage temperature (and pressure) rise is small.

Pressure ratio of a multistage compressor

It is possible to apply the preceding analysis to the determination of multistage compressor pressure ratios. The procedure requires the calculation of pressure and temperature changes for a single stage, the stage exit conditions enabling the density at entry to the following stage to be found. This calculation is repeated for each stage in turn until the required final conditions are satisfied. However, for compressors having identical stages it is more convenient to resort to a simple compressible flow analysis. An illustrative example is given below.

EXAMPLE 5.1. A multistage axial compressor is required for compressing air at 293 K, through a pressure ratio of 5 to 1. Each stage is to be 50% reaction and the mean blade speed 275 m/s, flow coefficient 0.5, and stage loading factor 0.3, are taken, for simplicity, as constant for all stages. Determine the flow angles and, the number of stages required if the stage efficiency is 88.8%. Take $C_p = 1.005$ kJ/(kg°C) and $\gamma = 1.4$ for air.

Solution. From eqn. (5.14a) the stage load factor can be written as,

$$\psi = \phi(\tan \beta_1 - \tan \beta_2).$$

From eqn. (5.11) the reaction is

$$R = \frac{\phi}{2}(\tan \beta_1 + \tan \beta_2).$$

Solving for $\tan \beta_1$ and $\tan \beta_2$ gives

$$\tan \beta_1 = (R + \psi/2)/\phi \text{ and } \tan \beta_2 = (R - \psi/2)/\phi.$$

Calculating β_1 and β_2 and observing for $R = 0.5$ that the velocity diagram is symmetrical,

$$\beta_1 = \alpha_2 = 52.45 \text{ deg and } \beta_2 = \alpha_1 = 35 \text{ deg}.$$

Writing the stage load factor as $\psi = C_p \Delta T_0 / U^2$, then the stage stagnation temperature rise is,

$$\Delta T_0 = \psi U^2 / C_p = 0.3 \times 275^2 / 1005 = 22.5°C.$$

It is reasonable to take the stage efficiency as equal to the polytropic efficiency since the stage temperature rise of an axial compressor is small. Denoting compressor inlet and outlet conditions by subscripts I and II respectively then, from eqn. (2.33),

$$\frac{T_{0II}}{T_{0I}} = 1 + \frac{N \Delta T_0}{T_{0I}} = \left(\frac{p_{0II}}{p_{0I}}\right)^{(\gamma-1)/\eta_p \gamma},$$

where N is the required number of stages. Thus

$$N = \frac{T_{01}}{\Delta T_0}\left[\left(\frac{p_{0II}}{p_{0I}}\right)^{(\gamma-1)/\eta_p \gamma} - 1\right] = \frac{293}{22.5}[5^{1/3.11} - 1] = 8.86.$$

A suitable number of stages is therefore 9.

The overall efficiency is found from eqn. (2.36).

$$\eta_{tt} = \left[\left(\frac{p_{0II}}{p_{0I}}\right)^{(\gamma-1)/\gamma} - 1\right] \bigg/ \left[\left(\frac{p_{0II}}{p_{0I}}\right)^{(\gamma-1)/\eta_p \gamma} - 1\right]$$

$$= [5^{1/3.5} - 1]/[5^{1/3.11} - 1] = 86.3\%.$$

Estimation of compressor stage efficiency

In eqn. (5.9) the amount of the actual stage work $(h_{03} - h_{01})$ can be found from the velocity diagram. The losses in total pressure may be estimated from cascade data. This data is incomplete however, as it only takes account of the blade profile loss. Howell (1945) has subdivided the total losses into three categories as shown in Figure 3.11.

 (i) Profile losses on the blade surfaces.
 (ii) Skin friction losses on the annulus walls.
(iii) "Secondary" losses by which he meant all losses not included in (i) and (ii) above.

In performance estimates of axial compressor and fan stages the *overall* drag coefficient for the blades of each row is obtained from

$$C_D = C_{Dp} + C_{Da} + C_{Ds}$$

$$= C_{Dp} + 0.02 \, s/H + 0.018 C_L^2 \qquad (5.28)$$

using the empirical values given in Chapter 3.

Although the subject matter of this chapter is primarily concerned with two-dimensional flows, there is an interesting three-dimensional aspect which cannot be ignored. In multistage axial compressors the annulus wall boundary layers

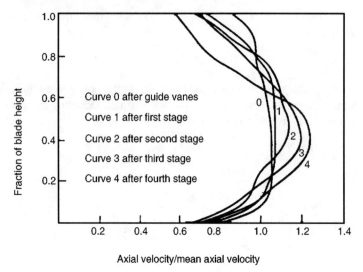

Fraction of blade height

Curve 0 after guide vanes

Curve 1 after first stage

Curve 2 after second stage

Curve 3 after third stage

Curve 4 after fourth stage

Axial velocity/mean axial velocity

FIG. 5.7. Axial velocity profiles in a compressor (Howell 1945). (Courtesy of the Institution of Mechanical Engineers).

rapidly thicken through the first few stages and the axial velocity profile becomes increasingly peaked. This effect is illustrated in Figure 5.7, from the experimental results of Howell (1945), which shows axial velocity traverses through a four-stage compressor. Over the central region of the blade, the axial velocity is higher than the mean value based on the through flow. The mean blade section (and most of the span) will, therefore, do less work than is estimated from the velocity triangles based on the mean axial velocity. In theory it would be expected that the tip and root sections would provide a compensatory effect because of the low axial velocity in these regions. Due to stalling of these sections (and tip leakage) no such work increase actually occurs, and the net result is that the work done by the whole blade is below the design figure. Howell (1945) suggested that the stagnation enthalpy rise across a stage could be expressed as

$$h_{03} - h_{01} = \lambda U(c_{y2} - c_{y1}),\tag{5.29}$$

where λ is a "work done". For multistage compressors Howell recommended for λ a mean value of 0.86. Using a similar argument for axial turbines, the increase in axial velocity at the pitch-line gives an *increase* in the amount of work done, which is then roughly cancelled out by the loss in work done at the blade ends. Thus, for turbines, no "work done" factor is required as a correction in performance calculations.

Other workers have suggested that λ should be high at entry (0.96) where the annulus wall boundary layers are thin, reducing progressively in the later stages of the compressor (0.85). Howell (1950) has given mean "work done" factors for compressors with varying numbers of stages, as in Figure 5.9. For a four-stage compressor the value of λ would be 0.9 which would be applied to all four stages.

Smith (1970) commented upon the rather pronounced deterioration of compressor performance implied by the example given in Figure 5.7 and suggested that things are not so bad as suggested. As an example of modern practice he gave the axial

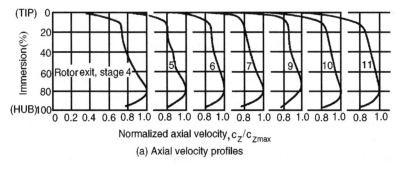

(a) Axial velocity profiles

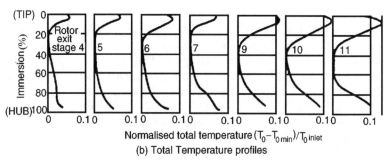

(b) Total Temperature profiles

FIG. 5.8. Traverse measurements obtained from a 12-stage compressor (Smith 1970). (Courtesy of the Elsevier Publishing Co).

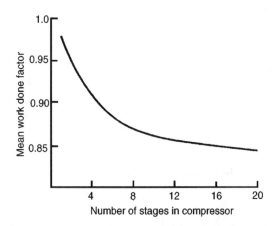

FIG. 5.9. Mean work done factor in compressors (Howell and Bonham 1950). (Courtesy of the Institution of Mechanical Engineers).

velocity distributions through a twelve-stage axial compressor, Figure 5.8(a). This does illustrate that rapid changes in velocity distribution still occur in the first few stages, but that the profile settles down to a fairly constant shape thereafter. This phenomenon has been referred to as *ultimate steady flow*.

Smith also provided curves of the spanwise variation in total temperature, Figure 5.8(b), which shows the way losses increase from midpassage towards the annulus walls. An inspection of this figure shows also that the excess total temperature near the end walls increases in magnitude and extent as the flow passes through the compressor. Work on methods of predicting annulus wall boundary layers in turbomachines and their effects on performance are being actively pursued in many countries. Although rather beyond the scope of this textbook, it may be worth mentioning two papers for students wishing to advance their studies further. Mellor and Balsa (1972) offer a mathematical flow model based on the pitchwise-averaged turbulent equations of motion for predicting axial flow compressor performance whilst Daneshyar *et al.* (1972) review and compare different existing methods for predicting the growth of annulus wall boundary layers in turbomachines.

EXAMPLE 5.2. The last stage of an axial flow compressor has a reaction of 50% at the design operating point. The cascade characteristics, which correspond to each row at the mean radius of the stage, are shown in Figure 3.12. These apply to a cascade of circular arc camber line blades at a space–chord ratio 0.9, a blade inlet angle of 44.5 deg and a blade outlet angle of −0.5 deg. The blade height–chord ratio is 2.0 and the work done factor can be taken as 0.86 when the mean radius relative incidence $(i - i^*)/\varepsilon^*$ is 0.4 (the operating point).

For this operating condition, determine

(i) the nominal incidence i^* and nominal deflection ε^*;
(ii) the inlet and outlet flow angles for the rotor;
(iii) the flow coefficient and stage loading factor;
(iv) the rotor lift coefficient;
(v) the overall drag coefficient of each row;
(vi) the stage efficiency.

The density at entrance to the stage is $3.5\,\text{kg/m}^3$ and the mean radius blade speed is 242 m/s. Assuming the density across the stage is constant and ignoring compressibility effects, estimate the stage pressure rise.

In the solution given below the *relative flow* onto the rotor is considered. The notation used for flow angles is the same as for Figure 5.2. For blade angles, β' is therefore used instead of α' for the sake of consistency.

Solution. (i) The nominal deviation is found using eqns. (3.39) and (3.40). With the camber $\theta = \beta'_1 - \beta'_2 = 44.5° - (-0.5°) = 45°$ and the space/chord ratio, $s/l = 0.9$, then

$$\delta^* = [0.23 + \beta_2^*/500]\theta(s/l)^{1/2}$$

But $\quad \beta_2^* = \delta^* + \beta'_2 = \delta^* - 0.5$

$$\therefore \delta^* = [0.23 + (\delta^* + \beta'_2)/500] \times 45 \times (0.9)^{1/2}$$

$$= [0.229 + \delta^*/500] \times 42.69 = 9.776 + 0.0854\,\delta^*$$

$$\therefore \delta^* = 10.69°$$

$$\therefore \beta_2^* = \delta^* + \beta'_2 = 10.69 - 0.5$$

$$\simeq 10.2°$$

The nominal deflection $\varepsilon^* = 0.8\epsilon_{max}$ and, from Figure 3.12, $\varepsilon_{max} = 37.5°$. Thus, $\varepsilon^* = 30°$ and the nominal incidence is

$$i^* = \beta_2^* + \varepsilon^* - \beta_1'$$
$$= 10.2 + 30 - 44.5 = -4.3°.$$

(ii) At the operating point $i = 0.4\varepsilon^* + i^* = 7.7°$. Thus, the actual inlet flow angle is

$$\beta_1 = \beta_1' + i = 52.2°.$$

From Figure 3.12 at $i = 7.7°$, the deflection $\varepsilon = 37.5°$ and the flow outlet angle is

$$\beta_2 = \beta_1 - \varepsilon = 14.7°.$$

(iii) From Figure 5.2, $U = c_{x1}(\tan\alpha_1 + \tan\beta_1) = c_{x2}(\tan\alpha_2 + \tan\beta_2)$. For $c_x = $ constant across the stage and $R = 0.5$

$$\beta_1 = \alpha_2 = 52.2° \text{ and } \beta_2 = \alpha_1 = 14.7°$$

and the flow coefficient is

$$\phi = \frac{c_x}{U} = \frac{1}{\tan\alpha_1 + \tan\beta_1} = 0.644.$$

The stage loading factor, $\psi = \Delta h_0/U^2 = \lambda\phi(\tan\alpha_2 - \tan\alpha_1)$ using eqn. (5.29). Thus, with $\lambda = 0.86$,

$$\psi = 0.568.$$

(iv) The lift coefficient can be obtained using eqn. (3.18)

$$C_L = 2(s/l)\cos\beta_m(\tan\beta_1 - \tan\beta_2)$$

ignoring the small effect of the drag coefficient. Now $\tan\beta_m = (\tan\beta_1 + \tan\beta_2)/2$. Hence $\beta_m = 37.8°$ and so

$$C_L = 2 \times 0.9 \times 0.7902 \times 1.027 = 1.46.$$

(v) Combining eqns. (3.7) and (3.17) the drag coefficient is

$$C_D = \frac{s}{l}\left(\frac{\Delta p_0}{\frac{1}{2}\rho w_1^2}\right)\frac{\cos^3\beta_m}{\cos^2\beta_1}.$$

Again using Figure 3.12 at $i = 7.7°$, the profile total pressure loss coefficient $\Delta p_0/(\frac{1}{2}\rho w_1^2) = 0.032$, hence the profile drag coefficient for the blades of either row is

$$C_{D_p} = 0.9 \times 0.032 \times 0.7902^3/0.6129^2 = 0.038.$$

Taking account of the annulus wall drag coefficient C_{Da} and the secondary loss drag coefficient C_{Ds}

$$C_{Da} = 0.02(s/l)(l/H) = 0.02 \times 0.9 \times 0.5 = 0.009$$
$$C_{Ds} = 0.018C_L^2 = 0.018 \times 1.46^2 = 0.038.$$

Thus, the *overall* drag coefficient, $C_D = C_{D_p} + C_{D_a} + C_{D_s} = 0.084$ and this applies to each row of blades. If the reaction had been other than 0.5 the drag coefficients for each blade row would have been computed separately.

(vi) The total-to-total stage efficiency, using eqn. (5.9) can be written as

$$\eta_{tt} = 1 - \frac{\Sigma \Delta p_0 / \rho}{\psi U^2} = 1 - \frac{\Sigma \Delta p_0 / (\frac{1}{2}\rho c_x^2)}{2\psi / \phi^2} = 1 - \frac{(\zeta_R + \zeta_S)\phi^2}{2\psi}$$

where ζ_R and ζ_S are the overall total pressure loss coefficients for the rotor and stator rows respectively. From eqn. (3.17)

$$\zeta_s = (l/s)C_D \sec^3 \alpha_m.$$

Thus, with $\zeta_S = \zeta_R$

$$\eta_{tt} = 1 - \frac{\phi^2 C_D (l/s)}{\psi \cos^3 \alpha_m}$$

$$= 1 - \frac{0.644^2 \times 0.084}{0.568 \times 0.7903^3 \times 0.9} = 0.862.$$

From eqn. (5.27), the pressure rise across the stage is

$$\Delta p = \eta_{tt} \psi \rho U^2 = 0.862 \times 0.568 \times 3.5 \times 242^2$$

$$= 100 \, \text{kPa}.$$

Stall and surge phenomena in compressors

Casing treatment

It was discovered in the late 1960s that the stall of a compressor could be delayed to a lower mass flow by a suitable treatment of the compressor casing. Given the right conditions this can be of great benefit in extending the range of stall-free operation. Numerous investigations have since been carried out on different types of casing configurations under widely varying flow conditions to demonstrate the range of usefulness of the treatment.

Greitzer *et al.* (1979) observed that two types of stall could be found in a compressor blade row, namely, "blade stall" or "wall stall". Blade stall is, roughly speaking, a two-dimensional type of stall where a significant part of the blade has a large wake leaving the blade suction surface. Wall stall is a stall connected with the boundary layer on the outer casing. Figure 5.10 illustrates the two types of stall. Greitzer *et al.* found that the response to casing treatment depended conspicuously upon the type of stall encountered.

The influence of a grooved casing treatment on the stall margin of a model axial compressor rotor was investigated experimentally. Two rotor builds of different blade solidities, σ, (chord/space ratio) but with all the other parameters identical, were tested. Greitzer emphasised that the motive behind the use of different solidities was simply a convenient way to change the type of stall from a blade stall to a wall stall and that the benefit of casing treatment was unrelated to the amount of solidity

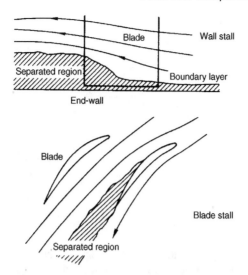

FIG. 5.10. Compressor stall inception (adapted from Greitzer *et al.* 1979).

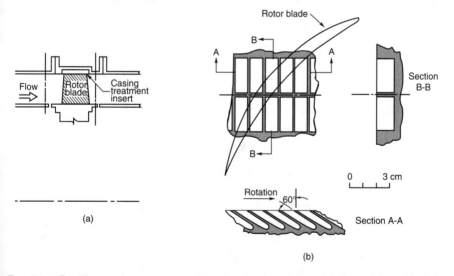

FIG. 5.11. Position and appearance of casing treatment insert (adapted from Greitzer *et al.* 1979).

of the blade row. The position of the casing treatment insert in relation to the rotor blade row is shown in Figure 5.11a and the appearance of the grooved surface used is illustrated in Figure 5.11b. The grooves, described as "axial skewed" and extending over the middle 44% of the blade, have been used in a wide variety of compressors.

As predicted from their design study, the high solidity blading ($\sigma = 2$) resulted in the production of a wall stall, while the low solidity ($\sigma = 1$) blading gave a blade stall. Figure 5.12 shows the results obtained for non-dimensionalised wall static

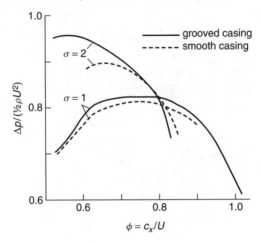

FIG. 5.12. Effects of casing treatment and solidity on compressor characteristics (adapted from Greitzer *et al.* 1979 and data points removed for clarity).

pressure rise, $\Delta p/(\frac{1}{2}\rho U^2)$, across the rotor plotted against the mean radius flow coefficient, $\phi = c_x/U$, for the four conditions tested. The extreme left end of each curve represents the *onset* of stall. It can be seen that there is a marked difference in the curves for the two solidities. For the high solidity configuration there is a higher static peak pressure rise and the decline does not occur until ϕ is much lower than the low solidity configuration. However, the most important difference in performance is the change in the stall point with and without the casing treatment. It can be seen that with the grooved casing a substantial change in the range of ϕ occurred with the high solidity blading. However, for the low solidity blading there is only a marginal difference in range. The shape of the performance curve is significantly affected for the high solidity rotor blades, with a substantial increase in the peak pressure rise brought about by the grooved casing treatment.

The conclusion reached by Greitzer *et al.* (1979) is that casing treatment is highly effective in delaying the onset of stall when the compressor row is more prone to *wall stall* than *blade stall*. However, despite this advantage casing treatment has not been generally adopted in industry. The major reason for this ostensible rejection of the method appears to be that a performance penalty is attached to it. The more effective the casing treatment, the more the stage efficiency is reduced.

Smith and Cumsty (1984) made an extensive series of experimental investigations to try to discover the reasons for the effectiveness of casing treatment and the underlying causes for the loss in compressor efficiency. At the simplest level it was realised that the slots provide a route for fluid to pass from the pressure surface to the suction surface allowing a small proportion of the flow to be recirculated. The approaching boundary layer fluid tends to have a high absolute swirl and is, therefore, suitably orientated to enter the slots. Normally, with a smooth wall the high swirl would cause energy to be wasted but, with the casing treatment, the flow entering the slot is turned and reintroduced back into the main flow near the blade's leading edge with its absolute swirl direction reversed. The re-entrant flow has, in effect, flowed upstream along the slot to a lower pressure region.

Rotating stall and surge

A salient feature of a compressor performance map, such as Figure 1.10, is the limit to stable operation known as the *surge line*. This limit can be reached by reducing the mass flow (with a throttle valve) whilst the rotational speed is maintained constant.

When a compressor goes into surge the effects are usually quite dramatic. Generally, an increase in noise level is experienced, indicative of a pulsation of the air flow and of mechanical vibration. Commonly, there are a small number of predominant frequencies superimposed on a high background noise. The lowest frequencies are usually associated with a *Helmholtz-type of resonance* of the flow through the machine, with the inlet and/or outlet volumes. The higher frequencies are known to be due to *rotating stall* and are of the same order as the rotational speed of the impeller.

Rotating stall is a phenomenon of axial-compressor flow which has been the subject of many detailed experimental and theoretical investigations and the matter is still not fully resolved. An early survey of the subject was given by Emmons *et al.* (1959). Briefly, when a blade row (usually the rotor of a compressor reaches the "stall point", the blades instead of all stalling together as might be expected, stall in separate patches and these stall patches, moreover, travel around the compressor annulus (i.e. they rotate).

That stall patches *must* propagate from blade to blade has a simple physical explanation. Consider a portion of a blade row, as illustrated in Figure 5.13 to be affected by a stall patch. This patch must cause a partial obstruction to the flow which is deflected on both sides of it. Thus, the incidence of the flow on to the blades on the right of the stall cell is reduced but, the incidence to the left is increased. As these blades are already close to stalling, the net effect is for the stall patch to move to the left; the motion is then self-sustaining.

There is a strong practical reason for the wide interest in rotating stall. Stall patches travelling around blade rows load and unload each blade at some frequency related to the speed and number of the patches. This frequency may be close to a natural frequency of blade vibration and there is clearly a need for accurate prediction of the conditions producing such a vibration. Several cases of blade failure due to resonance induced by rotating stall have been reported, usually with serious consequences to the whole compressor.

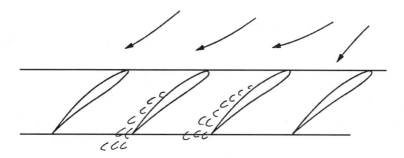

Fig. 5.13. Model illustrating mechanism of stall cell propagation: partial blockage due to stall patch deflects flow, increasing incidence to the left and decreasing incidence to the right.

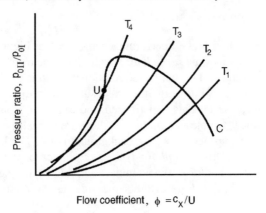

FIG. 5.14. Stability of operation of a compressor (adapted from Horlock 1958).

It is possible to distinguish between surge and propagating stall by the unsteadiness, or otherwise, of the total mass flow. The characteristic of stall propagation is that the flow passing through the annulus, summed over the whole area, is steady with time; the stall cells merely redistribute the flow over the annulus. Surge, on the other hand, involves an axial oscillation of the total mass flow, a condition highly detrimental to efficient compressor operation.

The conditions determining the point of surge of a compressor have not yet been completely determined satisfactorily. One physical explanation of this breakdown of the flow is given by Horlock (1958).

Figure 5.14 shows a constant rotor speed compressor characteristic (C) of pressure ratio plotted against flow coefficient. A second set of curves (T_1, T_2, etc.) are superimposed on this figure showing the pressure loss characteristics of the throttle for various fixed throttle positions. The intersection of curves T with compressor curve C denotes the various operating points of the combination. A state of *flow stability* exists if the throttle curves at the point of intersection have a greater (positive) slope than the compressor curve. That this is so may be illustrated as follows. Consider the operating point at the intersection of T_2 with C. If a small reduction of flow should momentarily occur, the compressor will produce a greater pressure rise and the throttle resistance will fall. The flow rate must, of necessity, increase so that the original operating point is restored. A similar argument holds if the flow is temporarily augmented, so that the flow is completely stable at this operating condition.

If, now, the operating point is at point U, unstable operation is possible. A small reduction in flow will cause a greater reduction in compressor pressure ratio than the corresponding pressure ratio across the throttle. As a consequence of the increased resistance of the throttle, the flow will decrease even further and the operating point U is clearly unstable. By inference, neutral stability exists when the slopes of the throttle pressure loss curves equal the compressor pressure rise curve.

Tests on low pressure ratio compressors appear to substantiate this explanation of instability. However, for high rotational speed multistage compressors the above argument does not seem sufficient to describe surging. With high speeds no stable operation appears possible on constant speed curves of positive slope and surge appears to occur when this slope is zero or even a little negative. A more complete understanding of surge in multistage compressors is only possible from a detailed study of the individual stages performance and their interaction with one another.

Control of flow instabilities

Important and dramatic advances have been made in recent years in the understanding and controlling of surge and rotating stall. Both phenomena are now regarded as the mature forms of the natural oscillatory modes of the compression system (see Moore and Greizer 1986). The flow model they considered predicts that an initial disturbance starts with a very small amplitude but quickly grows into a large amplitude form. Thus, the stability of the compressor is equivalent to the stability of these small amplitude waves that exist just prior to stall or surge (Haynes *et al.* 1994). Only a very brief outline can be given of the advances in the understanding of these unstable flows and the means now available for controlling them. Likewise only a few of the many papers written on these topics are cited.

Epstein *et al.* (1989) first suggested that surge and rotating stall could be prevented by using active feedback control to damp the hydrodynamic disturbances while they were still of small amplitude. Active suppression of surge was subsequently demonstrated on a centrifugal compressor by Ffowcs Williams and Huang (1989), also by Pinsley *et al.* (1991) and on an axial compressor by Day (1993). Shortly after this Paduano *et al.* (1993) demonstrated active suppression of rotating stall in a single-stage low-speed axial compressor. By damping the small amplitude waves rotating about the annulus prior to stall, they increased the stable flow range of the compressor by 25%. The control scheme adopted comprised a circumferential array of hot wires just upstream of the compressor and a set of 12 individually actuated vanes upstream of the rotor used to generate the rotating disturbance structure required for control. Haynes *et al.* (1994), using the same control scheme as Paduano *et al.*, actively stabilised a three-stage, low-speed axial compressor and obtained an 8% increase in the operating flow range.

Gysling and Greitzer (1995) employed a different strategy using aeromechanical feedback to suppress the onset of rotating stall in a low-speed axial compressor. Figure 5.15 shows a schematic of the aeromechanical feedback system they used. An auxiliary injection plenum chamber is fed by a high pressure source so that high momentum air is injected upsteam towards the compressor rotor. The amount of air injected at a given circumferential position is governed by an array of locally reacting reed valves able to respond to perturbations in the static pressure upstream of the compressor. The reeds valves, which were modelled as mass-spring-dampers, regulated the amount of high-pressure air injected into the face of the compressor. The cantilevered reeds were designed to deflect upward to allow an increase of the injected flow, whereas a downward deflection decreases the injection.

A qualitative explanation of the stabilising mechanism has been given by Gysling and Greitzer (1995):

Consider a disturbance to an initally steady, axisymmetric flow, which causes a small decrease in axial velocity in one region of the compressor annulus. In this

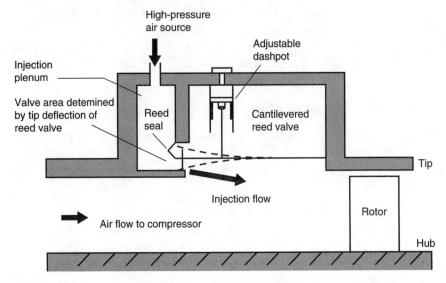

FIG. 5.15. Schematic of the aeromechanical feedback system used to suppress the onset of rotating stall (adapted from Gysling and Greitzer 1995).

region, the static pressure in the potential flow field upstream of the compressor will increase. The increase in static pressure deflects the reed valves in that region, increasing the amount of high momentum fluid injected and, hence, the local mass flow and pressure rise across the compressor. The net result is an increase in pressure rise across the compressor in the region of decreased axial velocity. The feedback thus serves to add a negative component to the real part of the compressor pressure rise versus mass flow transfer function.

Only a small amount (4%) of the overall mass flow through the compressor was used for aeromechanical feedback, enabling the stall flow coefficient of the compression system to be reduced by 10% compared to the stalling flow coefficient with the same amount of steady-state injection.

It is claimed that the research appears to be the first demonstration of dynamic control of rotating stall in an axial compressor using aeromechanical feedback.

Axial-flow ducted fans

In essence, an axial-flow fan is simply a single-stage compressor of low pressure (and temperature) rise, so that much of the foregoing theory of this chapter is valid for this class of machine. However, because of the high space–chord ratio used in many axial fans, a simplified theoretical approach based on *isolated aerofoil theory* is often used. This method can be of use in the design of ventilating fans (usually of high space–chord) in which aerodynamic interference between adjacent blades can be assumed negligible. Attempts have been made to extend the scope of isolated aerofoil theory to less widely spaced blades by the introduction of an *interference factor*; for instance, the ratio k of the lift force of a single blade in a cascade to the

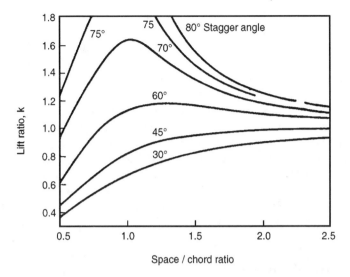

FIG. 5.16. Weinig's results for lift ratio of a cascade of thin flat plates, showing dependence on stagger angle and space/chord ratio (adapted from Wislicenus (1947).

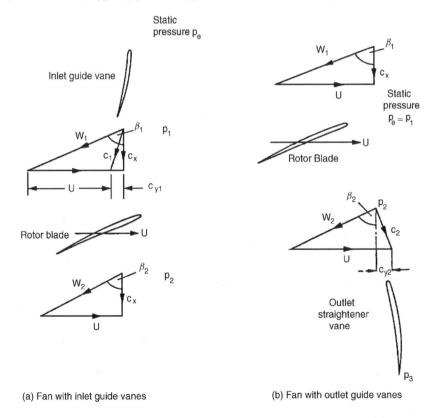

(a) Fan with inlet guide vanes

(b) Fan with outlet guide vanes

FIG. 5.17. Two simple types of axial-flow fan and their associated velocity diagrams (after Van Niekerk 1958).

lift force of a single isolated blade. As a guide to the degree of this interference, the exact solution obtained by Weinig (1935) and used by Wislicenus (1947) for a row of thin flat plates is of value and is shown in Figure 5.16. This illustrates the dependence of k on space–chord ratio for several stagger angles. The rather pronounced effect of stagger for moderate space–chord ratios should be noted as well as the asymptotic convergence of k towards unity for higher space–chord ratios.

Two simple types of axial-flow fan are shown in Figure 5.17 in which the inlet and outlet flows are entirely axial. In the first type (a), a set of guide vanes provide a contra-swirl and the flow is restored to the axial direction by the rotor. In the second type (b), the rotor imparts swirl in the direction of blade motion and the flow is restored to the axial direction by the action of outlet *straighteners* (or outlet guide vanes). The theory and design of both the above types of fan have been investigated by Van Niekerk (1958) who was able to formulate expressions for calculating the optimum sizes and fan speeds using blade element theory.

Blade element theory

A blade element at a given radius can be defined as an aerofoil of vanishingly small span. In fan-design theory it is commonly assumed that each such element operates as a two-dimensional aerofoil, behaving completely independently of conditions at any other radius. Now the forces impressed upon the fluid by unit span of a single stationary blade have been considered in some detail already, in Chapter 3. Considering an *element of a rotor* blade dr, at radius r, the elementary axial and tangential forces, dX and dY respectively, exerted on the fluid are, referring to Figure 3.5,

$$dX = (L \sin \beta_m - D \cos \beta_m)dr, \tag{5.30}$$

$$dY = (L \cos \beta_m + D \sin \beta_m)dr, \tag{5.31}$$

where $\tan \beta_m = \frac{1}{2}\{\tan \beta_1 + \tan \beta_2\}$ and L, D are the lift and drag on unit span of a blade.

Writing $\tan \gamma = D/L = C_D/C_L$ then,

$$dX = L(\sin \beta_m - \tan \gamma \cos \beta_m)dr.$$

Introducing the lift coefficient $C_L = L/(\frac{1}{2}\rho w_m^2 l)$ for the rotor blade (cf. eqn. (3.16a)) into the above expression and rearranging,

$$dX = \frac{\rho c_x^2 l C_L dr}{2 \cos^2 \beta_m} \cdot \frac{\sin(\beta_m - \gamma)}{\cos \gamma}, \tag{5.32}$$

where $c_x = w_m \cos \beta_m$.

The torque exerted by *one* blade element at radius r is rdY. If there are Z blades the elementary torque is

$$d\tau = rZdY$$

$$= rZL(\cos \beta_m + \tan \gamma \sin \beta_m)dr,$$

after using eqn. (5.31). Substituting for L and rearranging,

$$d\tau = \frac{\rho c_x^2 l Z C_L r dr}{2 \cos^2 \beta_m} \cdot \frac{\cos(\beta_m - \gamma)}{\cos \gamma}. \tag{5.33}$$

Now the work done by the rotor in unit time equals the product of the stagnation enthalpy rise and the mass flow rate; for the elementary ring of area $2\pi r dr$,

$$\Omega d\tau = (C_p \Delta T_0) d\dot{m}, \tag{5.34}$$

where Ω is the rotor angular velocity and the element of mass flow, $d\dot{m} = \rho c_x 2\pi r dr$. Substituting eqn. (5.33) into eqn. (5.34), then

$$C_p \Delta T_0 = C_p \Delta T = C_L \frac{U c_x l \cos(\beta_m - \gamma)}{2 s \cos^2 \beta_m \cos \gamma}. \tag{5.35}$$

where $s = 2\pi r/Z$. Now the static temperature rise equals the stagnation temperature rise when the velocity is unchanged across the fan; this, in fact, is the case for both types of fan shown in Figure 5.17.

The increase in static pressure of the *whole* of the fluid crossing the rotor row may be found by equating the total axial force on all the blade elements at radius r with the product of static pressure rise and elementary area $2\pi r dr$, or

$$Z dX = (p_2 - p_1) 2\pi r dr.$$

Using eqn. (5.32) and rearranging,

$$p_2 - p_1 = C_L \frac{\rho c_x^2 l \sin(\beta_m - \gamma)}{2 s \cos^2 \beta_m \cos \gamma} \tag{5.36}$$

Note that, so far, all the above expressions are applicable to both types of fan shown in Figure 5.17.

Blade element efficiency

Consider the fan type shown in Figure 5.17a fitted with guide vanes at inlet. The pressure rise across this fan is equal to the rotor pressure rise $(p_2 - p_1)$ *minus* the drop in pressure across the guide vanes $(p_e - p_1)$. The ideal pressure rise across the fan is given by the product of density and $C_p \Delta T_0$. Fan designers define a blade element efficiency

$$\eta_b = \{(p_2 - p_1) - (p_e - p_1)\}/(\rho C_p \Delta T_0). \tag{5.37}$$

The drop in static pressure across the guide vanes, assuming *frictionless* flow for simplicity, is

$$p_e - p_1 = \tfrac{1}{2}\rho(c_1^2 - c_x^2) = \tfrac{1}{2}\rho c_{y1}^2. \tag{5.38}$$

Now since the change in swirl velocity across the rotor is equal and opposite to the swirl produced by the guide vanes, the work done per unit mass flow, $C_p \Delta T_0$ is equal to $U c_{y1}$. Thus the second term in eqn. (5.37) is

$$(p_e - p_1)/(\rho C_p \Delta T_0) = c_{y1}/(2U). \tag{5.39}$$

Combining eqns. (5.35), (5.36) and (5.39) in eqn. (5.37), then

$$\eta_b = (c_x/U)\tan(\beta_m - \gamma) - c_{y1}/(2U). \tag{5.40a}$$

The foregoing exercise can be repeated for the second type of fan having outlet straightening vanes and, assuming frictionless flow through the "straighteners", the rotor blade element efficiency becomes,

$$\eta_b = (c_x/U)\tan(\beta_m - \gamma) + c_{y2}/(2U). \tag{5.40b}$$

Some justification for ignoring the losses occurring in the guide vanes is found by observing that the ratio of guide vane pressure change to rotor pressure rise is normally small in ventilating fans. For example, in the first type of fan

$$(p_e - p_1)/(p_2 - p_1) \doteq (\tfrac{1}{2}\rho c_{y1}^2)/(\rho U c_{y1}) = c_{y1}/2(U),$$

the tangential velocity c_{y1} being rather small compared with the blade speed U.

Lift coefficient of a fan aerofoil

For a specified blade element geometry, blade speed and lift/drag ratio the temperature and pressure rises can be determined if the lift coefficient is known. An estimate of lift coefficient is most easily obtained from two-dimensional aerofoil potential flow theory. Glauert (1959) shows for isolated aerofoils of small camber and thickness, that

$$C_L = 2\pi \sin\chi, \tag{5.41}$$

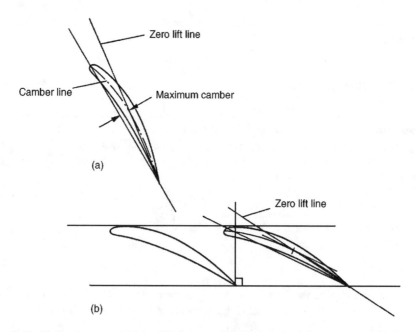

FIG. 5.18. Method suggested by Wislicenus (1947) for obtaining the zero lift line of cambered aerofoils.

where χ is the angle between the flow direction and *line of zero lift* of the aerofoil. For an isolated, cambered aerofoil Wislicenus (1947) suggested that the zero lift line may be found by joining the trailing edge point with the point of maximum camber as depicted in Figure 5.18a. For fan blades experiencing some interference effects from adjacent blades, the modified lift coefficient of a blade may be estimated by assuming that Weinig's results for flat plates (Figure 5.15) are valid for the slightly cambered, finite thickness blades, and

$$C_L = 2\pi k \sin \chi. \tag{5.41a}$$

When the vanes overlap (as they may do at sections close to the hub), Wislicenus suggested that the zero lift line may be obtained by the line connecting the trailing edge point with the maximum camber of that portion of blade which is not overlapped (Figure 5.18b).

The extension of both blade element theory and cascade data to the design of complete fans is discussed in considerable detail by Wallis (1961).

References

Carchedi, F. (1982). Design and development of a 12:1 pressure ratio compressor for the Ruston 6-MW gas turbine. *J. Eng. Power, Trans. Am. Soc. Mech. Engrs.*, **104**, 823–31.

Constant, H. (1945). The early history of the axial type of gas turbine engine. *Proc. Instn. Mech. Engrs.*, **153**.

Cox, H. Roxbee. (1946). British aircraft gas turbines. *J. Aero. Sci.*, **13**.

Daneshyar, M., Horlock, J. H. and Marsh, H. (1972). Prediction of annulus wall boundary layers in axial flow turbomachines. *AGARDograph No. 164*. Advisory Group for Aerospace Research and Development.

Day, I. J. (1993). Stall inception in axial flow compressors. *J. Turbomachinery, Trans. Am. Soc. Mech. Engrs.*, **115**, 1–9.

Emmons, H. W., Kronauer, R. E. and Rocket, J. A. (1959). A survey of stall propagation – experiment and theory. *Trans. Am. Soc. Mech. Engrs.*, Series D, **81**.

Epstein, A. H., Ffowcs Williams, J. E. and Greitzer, E. M. (1989). Active suppression of aerodynamic instabilities in turbomachines. *J. of Propulsion and Power*, **5**, 204–11.

Ffowcs Williams, J. E. and Huang, X. Y. (1989). Active stabilization of compressor surge. *J. Fluid Mech.*, **204**, 204–262.

Glauert, H. (1959). *The Elements of Aerofoil and Airscrew Theory*. Cambridge UP.

Greitzer, E. M., Nikkanen, J. P., Haddad, D. E., Mazzawy, R. S. and Joslyn, H. D. (1979). A fundamental criterion for the application of rotor casing treatment. *J. Fluid Eng., Trans. Am. Soc. Mech. Engrs.*, **101**, 237–43.

Gysling, D. L. and Greitzer, E. M. (1995). Dynamic control of rotating stall in axial flow compressors using aeromechanical feedback. *J. Turbomachinery, Trans. Am. Soc. Mech. Engrs.*, **117**, 307–19.

Haynes, J. M., Hendricks, G. J. and Epstein, A. H. (1994). Active stabilization of rotating stall in a three-stage axial compressor. *J. Turbomachinery, Trans. Am. Soc. Mech. Engrs.*, **116**, 226–37.

Horlock, J. H. (1958). *Axial Flow Compressors*. Butterworths (1973). Reprint with supplemental material, Huntington, New York: Kreiger.

Howell, A. R. (1945). Fluid dynamics of axial compressors. *Proc. Instn. Mech. Engrs.*, **153**.

Howell, A. R. and Bonham, R. P. (1950). Overall and stage characteristics of axial flow compressors. *Proc. Instn. Mech. Engrs.*, **163**.

Mellor, G. L. and Balsa, T. F. (1972). The prediction of axial compressor performance with emphasis on the effect of annulus wall boundary layers. *Agardograph No. 164*. Advisory Group for Aerospace Research and Development.

Moore, F. K. and Greitzer, E. M. (1986). A theory of post stall transients in axial compression systems: Parts I & II. *J. Eng. Gas Turbines Power, Trans. Am. Soc. Mech. Engrs.*, **108**, 68–76.

Paduano, J. P., *et al.*, (1993). Active control of rotating stall in a low speed compressor. *J. Turbomachinery, Trans. Am. Soc. Mech. Engrs.*, **115**, 48–56.

Pinsley, J. E., Guenette, G. R., Epstein, A. H. and Greitzer, E. M. (1991). Active stabilization of centrifugal compressor surge. *J. Turbomachinery, Trans. Am. Soc. Mech. Engrs.*, **113**, 723–32.

Smith, G. D. J. and Cumpsty, N. A. (1984). Flow phenomena in compressor casing treatment. *J. Eng. Gas Turbines and Power, Trans. Am. Soc. Mech. Engrs.*, **106**, 532–41.

Smith, L. H., Jr. (1970). Casing boundary layers in multistage compressors. *Proceedings of Symposium on Flow Research on Blading*. L. S. Dzung (ed.), Elsevier.

Stoney, G. (1937). Scientific activities of the late Hon. Sir Charles Parsons, F.R.S., *Engineering*, **144**.

Van Niekerk, C. G. (1958). Ducted fan design theory. *J. Appl. Mech.*, **25**.

Wallis, R. A. (1961). *Axial Flow Fans, Design and Practice*. Newnes.

Weinig, F. (1935). *Die Stroemung um die Schaufeln von Turbomaschinen*, Joh. Ambr. Barth, Leipzig.

Wennerstrom, A. J. (1989). Low aspect ratio axial flow compressors: Why and what it means. *J. Turbomachinery, Trans Am. Soc. Mech. Engrs.*, **111**, 357–65.

Wennerstrom, A. J. (1990). Highly loaded axial flow compressors: History and current development. *J. Turbomachinery, Trans. Am. Soc. Mech. Engrs.*, **112**, 567–78.

Wislicenus, G. F. (1947). *Fluid Mechanics of Turbomachinery*. McGraw-Hill.

Problems

(*Note*: In questions 1 to 4 and 8 take $R = 287 \, \text{J/(kg°C)}$ and $\gamma = 1.4$.)

1. An axial flow compressor is required to deliver 50 kg/s of air at a stagnation pressure of 500 kPa. At inlet to the first stage the stagnation pressure is 100 kPa and the stagnation temperature is 23°C. The hub and tip diameters at this location are 0.436 m and 0.728 m. At the mean radius, which is constant through all stages of the compressor, the reaction is 0.50 and the absolute air angle at stator exit is 28.8 deg for all stages. The speed of the rotor is 8000 rev/min. Determine the number of similar stages needed assuming that the polytropic efficiency is 0.89 and that the axial velocity at the mean radius is constant through the stages and equal to 1.05 times the average axial velocity.

2. Derive an expression for the degree of reaction of an axial compressor stage in terms of the flow angles relative to the rotor and the flow coefficient.

Data obtained from early cascade tests suggested that the limit of efficient working of an axial-flow compressor stage occurred when

(i) a *relative* Mach number of 0.7 on the rotor is reached;
(ii) the flow coefficient is 0.5;
(iii) the relative flow angle at rotor outlet is 30 deg measured from the axial direction;
(iv) the stage reaction is 50%.

Find the limiting stagnation temperature rise which would be obtained in the first stage of an axial compressor working under the above conditions and compressing air at an inlet *stagnation* temperature of 289 K. Assume the axial velocity is constant across the stage.

3. Each stage of an axial flow compressor is of 0.5 reaction, has the same mean blade speed and the same flow outlet angle of 30 deg relative to the blades. The mean flow coefficient is constant for all stages at 0.5. At entry to the first stage the stagnation temperature is 278 K, the stagnation pressure 101.3 kPa, the static pressure is 87.3 kPa and the flow area $0.372\,m^2$. Using compressible flow analysis determine the axial velocity and the mass flow rate.

Determine also the shaft power needed to drive the compressor when there are 6 stages and the mechanical efficiency is 0.99.

4. A sixteen-stage axial flow compressor is to have a pressure ratio of 6.3. Tests have shown that a stage total-to-total efficiency of 0.9 can be obtained for each of the first six stages and 0.89 for each of the remaining ten stages. Assuming constant work done in each stage and similar stages find the compressor overall total-to-total efficiency. For a mass flow rate of 40 kg/s determine the power required by the compressor. Assume an inlet total temperature of 288 K.

5. At a particular operating condition an axial flow compressor has a reaction of 0.6, a flow coefficient of 0.5 and a stage loading, defined as $\Delta h_0/U^2$ of 0.35. If the flow exit angles for each blade row may be assumed to remain unchanged when the mass flow is throttled, determine the reaction of the stage and the stage loading when the air flow is reduced by 10% at constant blade speed. Sketch the velocity triangles for the two conditions.

Comment upon the likely behaviour of the flow when further reductions in air mass flow are made.

6. The proposed design of a compressor rotor blade row is for 59 blades with a circular arc camber line. At the mean radius of 0.254 m the blades are specified with a camber of 30 deg, a stagger of 40 deg and a chord length of 30 mm. Determine, using Howell's correlation method, the nominal outlet angle, the nominal deviation and the nominal inlet angle. The tangent difference approximation, proposed by Howell for nominal conditions $(0 \leqslant \alpha_2^* \leqslant 40°)$, can be used:

$$\tan \alpha_1^* - \tan \alpha_2^* = 1.55/(1 + 1.5\,s/l).$$

Determine the nominal lift coefficient given that the blade drag coefficient $C_D = 0.017$.

Using the data for relative deflection given in Figure 3.17, determine the flow outlet angle and lift coefficient when the incidence $i = 1.8$ deg. Assume that the drag coefficient is unchanged from the previous value.

7. The preliminary design of an axial flow compressor is to be based upon a simplified consideration of the mean diameter conditions. Suppose that the stage characteristics of a repeating stage of such a design are as follows:

Stagnation temperature rise	25°C
Reaction ratio	0.6
Flow coefficient	0.5
Blade speed	275 m/s

The gas compressed is air with a specific heat at constant pressure of 1.005 kJ/(kg°C). Assuming constant axial velocity across the stage and equal absolute velocities at inlet and outlet, determine the relative flow angles for the rotor.

Physical limitations for this compressor dictate that the space/chord ratio is unity at the mean diameter. Using Howell's correlation method, determine a suitable camber at the mid-height of the rotor blades given that the incidence angle is zero. Use the tangent difference approximation:

$$\tan \beta_1^* - \tan \beta_2^* = 1.55/(1 + 1.5\,s/l)$$

for nominal conditions and the data of Figure 3.17 for finding the design deflection. (*Hint.* Use several trial values of θ to complete the solution.)

8. Air enters an axial flow compressor with a stagnation pressure and temperature of 100 kPa and 293 K, leaving at a stagnation pressure of 600 kPa. The hub and tip diameters at entry to the first stage are 0.3 m and 0.5 m. The flow Mach number *after* the inlet guide vanes is 0.7 at the mean diameter. At this diameter, which can be assumed constant for all the compressor stages, the reaction is 50%, the axial velocity to mean blade speed ratio is 0.6 and the absolute flow angle is 30 deg at the exit from all stators. The type of blading used for this compressor is designated "free-vortex" the axial velocity is constant for each stage.

Assuming isentropic flow through the inlet guide vanes and a small stage efficiency of 0.88, determine:

(1) the air velocity at exit from the IGVs at the mean radius;
(2) the air mass flow and rotational speed of the compressor;
(3) the specific work done in each stage;
(4) the overall efficiency of the compressor;
(5) the number of compressor stages required and the power needed to drive the compressor;
(6) consider the implications of rounding the number of stages to an integer value if the pressure ratio *must* be maintained at 6 for the same values of blade speed and flow coefficient.

NB. In the following problems on axial-flow fans the medium is air for which the density is taken to be $1.2 \, \text{kg/m}^3$.

9. (a) The volume flow rate through an axial-flow fan fitted with inlet guide vanes is $2.5 \, \text{m}^3/\text{s}$ and the rotational speed of the rotor is 2604 rev/min. The rotor blade tip radius is 23 cm and the root radius is 10 cm. Given that the stage static pressure increase is 325 Pa and the blade element efficiency is 0.80, determine the angle of the flow leaving the guide vanes at the tip, mean and root radii.

(b) A diffuser is fitted at exit to the fan with an area ratio of 2.5 and an effectiveness of 0.82. Determine the overall increase in static pressure and the air velocity at diffuser exit.

10. The rotational speed of a four-bladed axial-flow fan is 2900 rev/min. At the mean radius of 16.5 cm the rotor blades operate at $C_L = 0.8$ with $C_D = 0.045$. The inlet guide vanes produce a flow angle of 20° to the axial direction and the axial velocity through the stage is constant at 20 m/s.

For the mean radius, determine:

(1) the rotor relative flow angles;
(2) the stage efficiency;
(3) the rotor static pressure increase;
(4) the size of the blade chord needed for this duty.

11. A diffuser is fitted to the axial fan in the previous problem which has an efficiency of 70% and an area ratio of 2.4. Assuming that the flow at entry to the diffuser is uniform and axial in direction, and the losses in the entry section and the guide vanes are negligible, determine:

(1) the static pressure rise and the pressure recovery factor of the diffuser;
(2) the loss in total pressure in the diffuser;
(3) the overall efficiency of the fan and diffuser.

CHAPTER 6

Three-dimensional Flows in Axial Turbomachines

It cost much labour and many days before all these things were brought to perfection. (DEFOE, *Robinson Crusoe*.)

Introduction

IN CHAPTERS 4 and 5 the fluid motion through the blade rows of axial turbomachines was assumed to be two-dimensional in the sense that radial (i.e. spanwise) velocities did not exist. This is a not unreasonable assumption for axial turbomachines of high hub–tip ratio. However, with hub–tip ratios less than about 4/5, radial velocities through a blade row may become appreciable, the consequent redistribution of mass flow (with respect to radius) seriously affecting the outlet velocity profile (and flow angle distribution). It is the temporary imbalance between the strong centrifugal forces exerted on the fluid and radial pressures restoring equilibrium which is responsible for these radial flows. Thus, to an observer travelling with a fluid particle, radial motion will continue until sufficient fluid is transported (radially) to change the pressure distribution to that necessary for equilibrium. The flow in an annular passage in which there is no radial component of velocity, whose streamlines lie in circular, cylindrical surfaces and which is axisymmetric, is commonly known as *radial equilibrium* flow.

An analysis called *the radial equilibrium method*, widely used for three-dimensional design calculations in axial compressors and turbines, is based upon the assumption that any radial flow which may occur, is completed *within* a blade row, the flow *outside* the row then being in radial equilibrium. Figure 6.1 illustrates the nature of this assumption. The other assumption that the flow is axisymmetric implies that the effect of the discrete blades is not transmitted to the flow.

Theory of radial equilibrium

Consider a small element of fluid of mass dm, shown in Figure 6.2, of unit depth and subtending an angle $d\theta$ at the axis, rotating about the axis with tangential velocity, c_θ at radius r. The element is in radial equilibrium so that the pressure forces balance the centrifugal forces;

$$(p + dp)(r + dr)d\theta - prd\theta - (p + \tfrac{1}{2}dp)drd\theta = dmc_\theta^2/r.$$

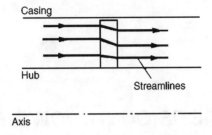

FIG. 6.1. Radial equilibrium flow through a rotor blade row.

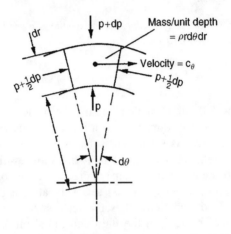

FIG. 6.2. A fluid element in radial equilibrium ($c_r = 0$).

Writing $dm = \rho r d\theta dr$ and ignoring terms of the second order of smallness the above equation reduces to,

$$\frac{1}{\rho}\frac{dp}{dr} = \frac{c_\theta^2}{r}. \tag{6.1}$$

If the swirl velocity c_θ and density are known functions of radius, the radial pressure variation along the blade length can be determined,

$$p_{\text{tip}} - p_{\text{root}} = \int_{\text{root}}^{\text{tip}} \rho c_\theta^2 \frac{dr}{r}. \tag{6.2a}$$

For an incompressible fluid

$$p_{\text{tip}} - p_{\text{root}} = \rho \int_{\text{root}}^{\text{tip}} c_\theta^2 \frac{dr}{r}. \tag{6.2b}$$

The stagnation enthalpy is written (with $c_r = 0$)

$$h_0 = h + \tfrac{1}{2}(c_x^2 + c_\theta^2) \tag{6.3}$$

therefore,

$$\frac{dh_0}{dr} = \frac{dh}{dr} + c_x\frac{dc_x}{dr} + c_\theta\frac{dc_\theta}{dr}. \tag{6.4}$$

The thermodynamic relation $T ds = dh - (1/\rho)dp$ can be similarly written

$$T\frac{ds}{dr} = \frac{dh}{dr} - \frac{1}{\rho}\frac{dp}{dr}. \tag{6.5}$$

Combining eqns. (6.1), (6.4) and (6.5), eliminating dp/dr and dh/dr, the *radial equilibrium equation* may be obtained,

$$\frac{dh_0}{dr} - T\frac{ds}{dr} = c_x\frac{dc_x}{dr} + \frac{c_\theta}{r}\frac{d}{dr}(rc_\theta). \tag{6.6}$$

If the stagnation enthalpy h_0 and entropy s remain the same at all radii, $dh_0/dr = ds/dr = 0$, eqn. (6.6) becomes,

$$c_x\frac{dc_x}{dr} + \frac{c_\theta}{r}\frac{d}{dr}(rc_\theta) = 0. \tag{6.6a}$$

Equation (6.6a) will hold for the flow between the rows of an adiabatic, reversible (ideal) turbomachine in which rotor rows either deliver or receive equal work at all radii. Now if the flow is incompressible, instead of eqn. (6.3) use $p_0 = p + \frac{1}{2}\rho(c_x^2 + c_\theta^2)$ to obtain

$$\frac{1}{\rho}\frac{dp_0}{dr} = \frac{1}{\rho}\frac{dp}{dr} + c_x\frac{dc_x}{dr} + c_\theta\frac{dc_\theta}{dr}. \tag{6.7}$$

Combining eqns. (6.1) and (6.7), then

$$\frac{1}{\rho}\frac{dp_0}{dr} = c_x\frac{dc_x}{dr} + \frac{c_\theta}{r}\frac{d}{dr}(rc_\theta). \tag{6.8}$$

Equation (6.8) clearly reduces to eqn. (6.6a) in a turbomachine in which equal work is delivered at all radii and the total pressure losses across a row are uniform with radius.

Equation (6.6a) may be applied to two sorts of problem as follows: (i) the design (or indirect) problem – in which the tangential velocity distribution is specified and the axial velocity variation is found, or (ii) the direct problem – in which the swirl angle distribution is specified, the axial and tangential velocities being determined.

The indirect problem

1. Free-vortex flow

This is a flow where the product of radius and tangential velocity remains constant (i.e. $rc_\theta = K$, a constant). The term "vortex-free" might be more appropriate as the vorticity (to be precise we mean *axial* vorticity component) is then zero.

Consider an element of an ideal inviscid fluid rotating about some fixed axis, as indicated in Figure. 6.3. The *circulation* Γ, is defined as the line integral of velocity around a curve enclosing an area A, or $\Gamma = \oint c ds$. The *vorticity* at a point is defined as, the limiting value of circulation $\delta\Gamma$ divided by area δA, as δA becomes vanishingly small. Thus vorticity, $\omega = d\Gamma/dA$.

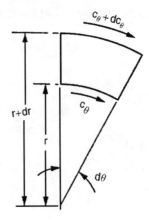

FIG. 6.3. Circulation about an element of fluid.

For the element shown in Figure 6.3, $c_r = 0$ and

$$d\Gamma = (c_\theta + dc_\theta)(r + dr)d\theta - c_\theta r d\theta$$

$$= \left(\frac{dc_\theta}{dr} + \frac{c_\theta}{r} \right) r d\theta dr$$

ignoring the product of small terms. Thus, $\omega = d\Gamma/dA = (1/r)d(rc_\theta)/dr$. If the vorticity is zero, $d(rc_\theta)/dr$ is also zero and, therefore, rc_θ is constant with radius.

Putting $rc_\theta = $ constant in eqn. (6.6a), then $dc_x/dr = 0$ and so $c_x = a$ constant. This information can be applied to the incompressible flow through a free-vortex compressor or turbine stage, enabling the radial variation in flow angles, reaction and work to be found.

Compressor stage. Consider the case of a compressor stage in which $rc_{\theta 1} = K_1$ before the rotor and $rc_{\theta 2} = K_2$ after the rotor, where K_1, K_2 are constants. The work done by the rotor on unit mass of fluid is

$$\Delta W = U(c_{\theta 2} - c_{\theta 1}) = \Omega r(K_2/r - K_1/r)$$

$$= \text{constant.}$$

Thus, the work done is equal at all radii.

The relative flow angles (see Figure 5.2) entering and leaving the rotor are

$$\tan \beta_1 = \frac{U}{c_x} - \tan a_1 = \frac{\Omega r - K_1/r}{c_x},$$

$$\tan \beta_2 = \frac{U}{c_x} - \tan a_2 = \frac{\Omega r - K_2/r}{c_x}.$$

in which $c_{x1} = c_{x2} = c_x$ for incompressible flow.

In Chapter 5, reaction in an axial compressor is defined by

$$R = \frac{\text{static enthalpy rise in the rotor}}{\text{static enthalpy rise in the stage}}.$$

For a normal stage ($\alpha_1 = \alpha_3$) with c_x constant across the stage, the reaction was shown to be

$$R = \frac{c_x}{2U}(\tan \beta_1 + \tan \beta_2). \qquad (5.11)$$

Substituting values of $\tan \beta_1$ and $\tan \beta_2$ into eqn. (5.11), the reaction becomes

$$R = 1 - \frac{k}{r^2}, \qquad (6.9)$$

where

$$k = (K_1 + K_2)/(2\Omega).$$

It will be clear that as k is positive, the reaction increases from root to tip. Likewise, from eqn. (6.1) we observe that as c_θ^2/r is always positive (excepting $c_\theta = 0$), so static pressure increases from root to tip. For the free-vortex flow $rc_\theta = K$, the static pressure variation is obviously $p/\rho = \text{constant} - K/(2r^2)$ upon integrating eqn. (6.1).

EXAMPLE 6.1. An axial flow compressor stage is designed to give free-vortex tangential velocity distributions for all radii before and after the rotor blade row. The tip diameter is constant and 1.0 m; the hub diameter is 0.9 m and constant for the stage. At the rotor tip the flow angles are as follows

Absolute inlet angle, α_1	$= 30\,\text{deg.}$
Relative inlet angle, β_1	$= 60\,\text{deg.}$
Absolute outlet angle, α_2	$= 60\,\text{deg.}$
Relative outlet angle, β_2	$= 30\,\text{deg.}$

Determine,

(i) the axial velocity;
(ii) the mass flow rate;
(iii) the power absorbed by the stage;
(iv) the flow angles at the hub;
(v) the reaction ratio of the stage at the hub;

given that the rotational speed of the rotor is 6000 rev/min and the gas density is 1.5 kg/m^3 which can be assumed constant for the stage. It can be further assumed that stagnation enthalpy and entropy are constant before and after the rotor row for the purpose of simplifying the calculations.

Solution. (i) The rotational speed, $\Omega = 2\pi N/60 = 628.4\,\text{rad/s}$.
Therefore blade tip speed, $U_t = \Omega r_t = 314.2\,\text{m/s}$ and blade speed at hub, $U_h = \Omega r_h = 282.5\,\text{m/s}$.
From the velocity diagram for the stage (e.g. Figure 5.2), the blade tip speed is

$$U_t = c_x(\tan 60° + \tan 30°) = c_x(\sqrt{3} + 1/\sqrt{3}).$$

Therefore $c_x = 136\,\text{m/s}$, constant at all radii by eqn. (6.6a).

(ii) The rate of mass flow, $\dot{m} = \pi(r_t^2 - r_h^2)\rho c_x$

$$= \pi(0.5^2 - 0.45^2)1.5 \times 136 = 30.4 \,\text{kg/s}.$$

(iii) The power absorbed by the stage,

$$\dot{W}_c = \dot{m}U_t(c_{\theta 2t} - c_{\theta 1t})$$

$$= \dot{m}U_t c_x(\tan \alpha_{2t} - \tan \alpha_{1t})$$

$$= 30.4 \times 314.2 \times 136(\sqrt{3} - 1/\sqrt{3})$$

$$= 1.5 \,\text{MW}.$$

(iv) At inlet to the rotor tip,

$$c_{\theta 1t} = c_x \tan \alpha_1 = 136/\sqrt{3} = 78.6 \,\text{m/s}.$$

The absolute flow is a free-vortex, $rc_\theta = \text{constant}$.
  Therefore $c_{\theta 1h} = c_{\theta 1t}(r_t/r_h) = 78.6 \times 0.5/0.45 = 87.3 \,\text{m/s}.$
  At outlet to the rotor tip,

$$c_{\theta 2t} = c_x \tan \alpha_2 = 136 \times \sqrt{3} = 235.6 \,\text{m/s}.$$

Therefore $c_{\theta 2h} = c_{\theta 2t}(r_t/r_h) = 235.6 \times 0.5/0.45 = 262 \,\text{m/s}.$
  The flow angles at the hub are,

$$\tan \alpha_1 = c_{\theta 1h}/c_x = 87.3/136 = 0.642,$$

$$\tan \beta_1 = U_h/c_x - \tan \alpha_1 = 1.436,$$

$$\tan \alpha_2 = c_{\theta 2h}/c_x = 262/136 = 1.928,$$

$$\tan \beta_2 = U_h/c_x - \tan \alpha_2 = 0.152.$$

Thus $\alpha_1 = 32.75°, \beta_1 = 55.15°, \alpha_2 = 62.6°, \beta_2 = 8.64°$ at the hub.
  (v) The reaction at the hub can be found by several methods. With eqn. (6.9)

$$R = 1 - k/r^2$$

and noticing that, from symmetry of the velocity triangles,

$$R = 0.5 \text{ at } r = r_t, \text{ then } k = 0.5r_t^2.$$

Therefore $R_h = 1 - 0.5(0.5/0.45)^2 = 0.382.$

The velocity triangles will be asymmetric and similar to those in Figure 5.4(b).
  The simplicity of the flow under free-vortex conditions is, superficially, very attractive to the designer and many compressors have been designed to conform to this flow. (Constant (1945, 1953) may be consulted for an account of early British compressor design methods.) Figure 6.4 illustrates the variation of fluid angles and Mach numbers of a typical compressor stage designed for free-vortex flow. Characteristic of this flow are the large fluid deflections near the inner wall and high Mach numbers near the outer wall, both effects being deleterious to efficient performance.

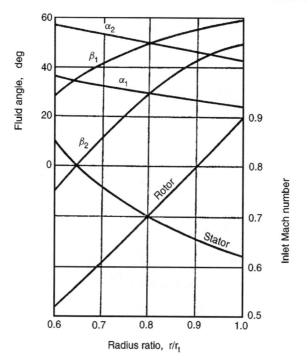

FIG. 6.4. Variation of fluid angles and Mach numbers of a free-vortex compressor stage with radius (adapted from Howell 1945).

A further serious disadvantage is the large amount of rotor twist from root to tip which adds to the expense of blade manufacture.

Many types of vortex design have been proposed to overcome some of the disadvantages set by free-vortex design and several of these are compared by Horlock (1958). Radial equilibrium solutions for the work and axial velocity distributions of some of these vortex flows in an axial compressor stage are given below.

2. Forced vortex

This is sometimes called "solid-body" rotation because c_θ varies directly with r. At entry to the rotor assume h_{01} is constant and $c_{\theta 1} = K_1 r$.

With eqn. (6.6a)

$$\frac{d}{dr}\left(\frac{c_{x1}^2}{2}\right) = -K_1\frac{d}{dr}(K_1 r^2)$$

and, after integrating,

$$c_{x1}^2 = \text{constant} - 2K_1^2 r^2. \tag{6.10}$$

After the rotor $c_{\theta 2} = K_2 r$ and $h_{02} - h_{01} = U(c_{\theta 2} - c_{\theta 1}) = \Omega(K_2 - K_1)r^2$. Thus, as the work distribution is non-uniform, the radial equilibrium equation in the form eqn. (6.6) is required for the flow after the rotor.

$$\frac{dh_{02}}{dr} = 2\Omega(K_2 - K_1)r = \frac{d}{dr}\left(\frac{c_{x2}^2}{2}\right) + K_2\frac{d}{dr}(K_2 r^2).$$

After rearranging and integrating

$$c_{x2}^2 = \text{ constant } - 2[K_2^2 - \Omega(K_2 - K_1)]r^2. \tag{6.11}$$

The constants of integration in eqns. (6.10) and (6.11) can be found from the continuity of mass flow, i.e.

$$\frac{\dot{m}}{2\pi\rho} = \int_{r_h}^{r_t} c_{x1}r\,dr = \int_{r_h}^{r_t} c_{x2}r\,dr, \tag{6.12}$$

which applies to the assumed incompressible flow.

3. General whirl distribution

The tangential velocity distribution is given by

$$c_{\theta 1} = ar^n - b/r \text{ (before rotor)}, \tag{6.13a}$$

$$c_{\theta 2} = ar^n + b/r \text{ (after rotor)}. \tag{6.13b}$$

The distribution of work for all values of the index n is constant with radius so that if h_{01} is uniform, h_{02} is also uniform with radius. From eqns. (6.13)

$$\Delta W = h_{02} - h_{01} = U(c_{\theta 2} - c_{\theta 1}) = 2b\Omega. \tag{6.14}$$

Selecting different values of n gives several of the tangential velocity distributions commonly used in compressor design. With $n = 0$, or zero power blading, it leads to the so-called "exponential" type of stage design (included as an exercise at the end of this chapter). With $n = 1$, or *first power blading*, the stage design is called (incorrectly, as it transpires later) "constant reaction".

First power stage design. For a given stage temperature rise the discussion in Chapter 5 would suggest the choice of 50% reaction at all radii for the highest stage efficiency. With swirl velocity distributions

$$c_{\theta 1} = ar - b/r, \quad c_{\theta 2} = ar + b/r \tag{6.15}$$

before and after the rotor respectively, and rewriting the expression for reaction, eqn. (5.11), as

$$R = 1 - \frac{c_x}{2U}(\tan\alpha_1 + \tan\alpha_2), \tag{6.16}$$

then, using eqn. (6.15),

$$R = 1 - a/\Omega = \text{ constant}. \tag{6.17}$$

Implicit in eqn. (6.16) is the assumption that the axial velocity across the rotor remains constant which, of course, is tantamount to ignoring radial equilibrium.

The axial velocity *must* change in crossing the rotor row so that eqn. (6.17) is only a crude approximation at the best. Just how crude is this approximation will be indicated below.

Assuming constant stagnation enthalpy at entry to the stage, integrating eqn. (6.6a), the axial velocity distributions before and after the rotor are

$$c_{x1}^2 = \text{constant} - 4a(\tfrac{1}{2}ar^2 - b\ln r),$$ (6.18a)

$$c_{x2}^2 = \text{constant} - 4a(\tfrac{1}{2}ar^2 + b\ln r),$$ (6.18b)

More conveniently, these expressions can be written non-dimensionally as,

$$\left(\frac{c_{x1}}{U_t}\right)^2 = A_1 - \left(\frac{2a}{\Omega}\right)^2 \left[\frac{1}{2}\left(\frac{r}{r_t}\right)^2 - \frac{b}{ar_t^2}\ln\left(\frac{r}{r_t}\right)\right],$$ (6.19a)

$$\left(\frac{c_{x2}}{U_t}\right)^2 = A_2 - \left(\frac{2a}{\Omega}\right)^2 \left[\frac{1}{2}\left(\frac{r}{r_t}\right)^2 + \frac{b}{ar_t^2}\ln\left(\frac{r}{r_t}\right)\right],$$ (6.19b)

in which $U_t = \Omega r_t$ is the tip blade speed. The constants A_1, A_2 are not entirely arbitrary as the continuity equation, eqn. (6.12), must be satisfied.

EXAMPLE 6.2. As an illustration consider a single stage of an axial-flow air compressor of hub-tip ratio 0.4 with a nominally constant reaction (i.e. according to eqn. (6.17)) of 50%. Assuming incompressible, inviscid flow, a blade tip speed of 300 m/s, a blade tip diameter of 0.6 m, and a stagnation temperature rise of 16.1°C, determine the radial equilibrium values of axial velocity before and after the rotor. The axial velocity far upstream of the rotor at the casing is 120 m/s. Take C_p for air as 1.005 kJ/(kg°C).

Solution: The constants in eqn. (6.19) can be easily determined. From eqn. (6.17)

$$2a/\Omega = 2(1 - R) = 1.0.$$

Combining eqns. (6.14) and (6.17)

$$\frac{b}{ar_t^2} = \frac{\Delta W}{2\Omega^2(1 - R)r_t^2} = \frac{C_p \cdot \Delta T_0}{2U_t^2(1 - R)}$$

$$= \frac{1005 \times 16.1}{300^2} = 0.18.$$

The inlet axial velocity distribution is completely specified and the constant A_1 solved. From eqn. (6.19a)

$$\left(\frac{c_{x1}}{U_t}\right)^2 = A_1 - [\tfrac{1}{2}(r/r_t)^2 - 0.18\ln(r/r_t)].$$

At $r = r_t$, $c_{x1}/U_t = 0.4$ and hence $A_1 = 0.66$.

Although an explicit solution for A_2 can be worked out from eqn. (6.19b) and eqn. (6.12), it is far speedier to use a semigraphical procedure. For an arbitrarily selected value of A_2, the distribution of c_{x2}/U_t is known. Values of $(r/r_t) \cdot (c_{x2}/U_t)$

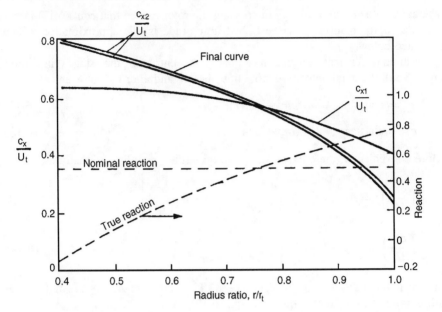

FIG. 6.5. Solution of exit axial-velocity profile for a first power stage.

and $(r/r_t) \cdot (c_{x1}/U_t)$ are plotted against r/r_t and the areas under these curves compared. New values of A_2 are then chosen until eqn. (6.12) is satisfied. This procedure is quite rapid and normally requires only two or three attempts to give a satisfactory solution. Figure 6.5 shows the final solution of c_{x2}/U_t obtained after three attempts. The solution is,

$$\left(\frac{c_{x2}}{U_t}\right)^2 = 0.56 - \left[\frac{1}{2}\left(\frac{r}{r_t}\right)^2 + 0.18 \ln\left(\frac{r}{r_t}\right)\right].$$

It is illuminating to calculate the actual variation in reaction taking account of the change in axial velocity. From eqn. (5.10c) the true reaction across a normal stage is,

$$R' = \frac{w_1^2 - w_2^2}{2U(c_{\theta 2} - c_{\theta 1})}.$$

From the velocity triangles, Figure 5.2,

$$w_1^2 - w_2^2 = (w_{\theta 1} + w_{\theta 2})(w_{\theta 1} - w_{\theta 2}) + (c_{x1}^2 - c_{x2}^2).$$

As $w_{\theta 1} + w_{\theta 2} = 2U - (c_{\theta 1} + c_{\theta 2})$ and $w_{\theta 1} - w_{\theta 2} = c_{\theta 2} - c_{\theta 1}$,

$$R' = 1 - \frac{c_{\theta 1} + c_{\theta 2}}{2U} + \frac{c_{x1}^2 - c_{x2}^2}{2U(c_{\theta 2} - c_{\theta 1})}.$$

For the first power swirl distribution, eqn. (6.15),

$$R' = 1 - \frac{a}{\Omega} + \frac{c_{x1}^2 - c_{x2}^2}{4\Omega b}.$$

From the radial equilibrium solution in eqns. (6.19), after some rearrangement,

$$\frac{c_{x1}^2 - c_{x2}^2}{4\Omega b} = \frac{A_1 - A_2}{2\psi_t} + \left(\frac{2a}{\Omega}\right) \ln\left(\frac{r}{r_t}\right),$$

where

$$\psi_t = \frac{\Delta W}{U_t^2} = \frac{C_p \Delta T_0}{\Omega^2 r_t^2}.$$

In the above example, $1 - a/\Omega = \frac{1}{2}$, $\psi t = 0.18$

$$R' = 0.778 + \ln(r/r_t).$$

The true reaction variation is shown in Figure 6.5 and it is evident that eqn. (6.17) is *invalid* as a result of axial velocity changes.

The direct problem

The flow angle variation is specified in the direct problem and the radial equilibrium equation enables the solution of c_x and c_θ to be found. The general radial equilibrium equation can be written in the form

$$\frac{dh_0}{dr} - T\frac{ds}{dr} = \frac{c_\theta^2}{r} + c\frac{dc}{dr}$$

$$= \frac{c^2 \sin^2 \alpha}{r} + c\frac{dc}{dr}, \tag{6.20}$$

as $c_\theta = c \sin \alpha$.

If both dh_0/dr and ds/dr are zero, eqn. (6.20) integrated gives

$$\log c = -\int \sin^2 \alpha \frac{dr}{r} + \text{constant}$$

or, if $c = c_m$ at $r = r_m$, then

$$\frac{c}{c_m} = \exp\left(-\int_{r_m}^{r} \sin^2 \alpha \frac{dr}{r}\right). \tag{6.21}$$

If the flow angle α is held constant, eqn. (6.21) simplifies still further,

$$\frac{c}{c_m} = \frac{c_x}{c_{xm}} = \frac{c_\theta}{c_{\theta m}} = \left(\frac{r}{r_m}\right)^{-\sin^2 \alpha} \tag{6.22}$$

The vortex distribution represented by eqn. (6.22) is frequently employed in practice as untwisted blades are relatively simple to manufacture.

The general solution of eqn. (6.20) can be found by introducing a suitable *integrating factor* into the equation. Multiplying throughout by $\exp[2\int \sin^2 \alpha dr/r]$ it follows that

$$\frac{d}{dr}\left\{c^2 \exp\left[2\int \sin^2 \alpha dr/r\right]\right\} = 2\left(\frac{dh_0}{dr} - T\frac{ds}{dr}\right)\exp\left[2\int \sin^2 \alpha dr/r\right].$$

After integrating and inserting the limit $c = c_m$ at $r = r_m$, then

$$c^2 \exp\left[2\int^r \sin^2\alpha\,dr/r\right] - c_m^2 \exp\left[2\int^{r_m} \sin^2\alpha\,dr/r\right]$$

$$= 2\int_{r_m}^r \left(\frac{dh_0}{dr} - T\frac{ds}{dr}\right)\exp\left[2\int \sin^2\alpha\,dr/r\right]dr. \tag{6.23}$$

Particular solutions of eqn. (6.23) can be readily obtained for simple radial distributions of α, h_0 and s. Two solutions are considered here in which both $2dh_0/dr = kc_m^2/r_m$ and $ds/dr = 0$, k being an arbitrary constant

(i) Let $a = 2\sin^2\alpha$. Then $\exp[2\int \sin^2\alpha\,dr/r] = r^a$ and, hence

$$\left(\frac{c}{c_m}\right)^2 \left(\frac{r}{r_m}\right)^a = 1 + \frac{k}{1+a}\left[\left(\frac{r}{r_m}\right)^{1+a} - 1\right]. \tag{6.23a}$$

Equation (6.22) is obtained immediately from this result with $k = 0$.

(ii) Let $br/r_m = 2\sin^2\alpha$. Then,

$$c^2 \exp(br/r_m) - c_m^2 \exp(b) = (kc_m^2/r_m)\int_{r_m}^r \exp(br/r_m)dr$$

and eventually,

$$\left(\frac{c}{c_m}\right)^2 = \frac{k}{b} + \left(1 - \frac{k}{b}\right)\exp\left[b\left(1 - \frac{r}{r_m}\right)\right]. \tag{6.23b}$$

Compressible flow through a fixed blade row

In the blade rows of high-performance gas turbines, fluid velocities approaching, or even exceeding, the speed of sound are quite normal and compressibility effects may no longer be ignored. A simple analysis is outlined below for the inviscid flow of a perfect gas through a *fixed* row of blades which, nevertheless, can be extended to the flow through moving blade rows.

The radial equilibrium equation, eqn. (6.6), applies to *compressible* flow as well as incompressible flow. With constant stagnation enthalpy and constant entropy, a free-vortex flow therefore implies uniform axial velocity downstream of a blade row, regardless of any *density* changes incurred in passing through the blade row. In fact, for high-speed flows there *must* be a density change in the blade row which implies a streamline shift as shown in Figure 6.1. This may be illustrated by considering the free-vortex flow of a perfect gas as follows. In radial equilibrium,

$$\frac{1}{\rho}\frac{dp}{dr} = \frac{c_\theta^2}{r} = \frac{K^2}{r^3} \text{ with } c_\theta = K/r.$$

For reversible adiabatic flow of a perfect gas, $\rho = Ep^{1/\gamma}$, where E is constant. Thus

$$\int p^{-1/\gamma}dp = EK^2\int r^{-3}dr + \text{constant},$$

therefore

$$p = \left[\text{constant} - \left(\frac{\gamma - 1}{2\gamma} \right) \frac{EK^2}{r^2} \right]^{\gamma/(\gamma - 1)} \tag{6.24}$$

For this free-vortex flow the pressure, and therefore the density also, must be larger at the casing than at the hub. The density difference from hub to tip may be appreciable in a high-velocity, high-swirl angle flow. If the fluid is without swirl at entry to the blades the density will be uniform. Therefore, from continuity of mass flow there must be a redistribution of fluid in its passage across the blade row to compensate for the changes in density. Thus, for this blade row, the continuity equation is,

$$\dot{m} = \rho_1 A_1 c_{x1} = 2\pi c_{x2} \int_{r_h}^{r_t} \rho_2 r \, dr, \tag{6.25}$$

where ρ_2 is the density of the swirling flow, obtainable from eqn. (6.24).

Constant specific mass flow

Although there appears to be no evidence that the redistribution of the flow across blade rows is a source of inefficiency, it has been suggested by Horlock (1966) that the radial distribution of c_θ for each blade row is chosen so that the product of axial velocity and density is constant with radius, i.e.

$$d\dot{m}/dA = \rho c_x = \rho c \cos \alpha = \rho_m c_m \cos \alpha_m = \text{constant} \tag{6.26}$$

where subscript m denotes conditions at $r = r_m$. This *constant specific mass flow design* is the logical choice when radial equilibrium theory is applied to compressible flows as the assumption that $c_r = 0$ is then likely to be realised.

Solutions may be determined by means of a simple numerical procedure and, as an illustration of one method, a turbine stage is considered here. It is convenient to assume that the stagnation enthalpy is uniform at nozzle entry, the entropy is constant throughout the stage and the fluid is a perfect gas. At nozzle exit under these conditions the equation of radial equilibrium, eqn. (6.20), can be written as

$$dc/c = -\sin^2 \alpha \, dr/r. \tag{6.27}$$

From eqn. (6.1), nothing that at constant entropy the acoustic velocity $a = \sqrt{(dp/d\rho)}$,

$$\frac{1}{\rho} \frac{dp}{dr} = \frac{1}{\rho} \left(\frac{dp}{d\rho} \right) \left(\frac{d\rho}{dr} \right) = \frac{a^2}{\rho} \frac{d\rho}{dr} = \frac{c^2}{r} \sin^2 \alpha,$$

$$\therefore d\rho/\rho = M^2 \sin^2 \alpha \, dr/r \tag{6.28}$$

where the flow Mach number

$$M = c/a = c/\sqrt{(\gamma R T)}. \tag{6.28a}$$

The isentropic relation between temperature and density for a perfect gas is

$$T/T_m = (\rho/\rho_m)^{\gamma - 1}$$

which after logarithmic differentiation gives

$$dT/T = (\gamma - 1)d\rho/\rho. \tag{6.29}$$

Using the above set of equations the procedure for determining the nozzle exit flow is as follows. Starting at $r = r_m$, values of c_m, α_m, T_m and ρ_m are assumed to be known. For a small finite interval Δr, the changes in velocity Δc, density $\Delta \rho$, and temperature ΔT can be computed using eqns. (6.27), (6.28) and (6.29) respectively. Hence, at the new radius $r = r_m + \Delta r$ the velocity $c = c_m + \Delta c$, the density $\rho = \rho_m + \Delta \rho$ and temperature $T = T_m + \Delta T$ are obtained. The corresponding flow angle α and Mach number M can now be determined from eqns. (6.26) and (6.28a) respectively. Thus, all parameters of the problem are known at radius $r = r_m + \Delta r$. This procedure is repeated for further increments in radius to the casing and again from the mean radius to the hub.

Figure 6.6 shows the distributions of flow angle and Mach number computed with this procedure for a turbine nozzle blade row of 0.6 hub/tip radius ratio. The input data used was $\alpha_m = 70.4$ deg and $M = 0.907$ at the mean radius. Air was assumed at a stagnation pressure of 859 kPa and a stagnation temperature of 465 K. A remarkable feature of these results is the almost uniform swirl angle which is obtained.

With the nozzle exit flow fully determined the flow at rotor outlet can now be computed by a similar procedure. The procedure is a little more complicated than that for the nozzle row because the specific work done by the rotor is not uniform with radius. Across the rotor, using the notation of Chapter 4,

$$h_{o2} - h_{o3} = U(c_{\theta 2} + c_{\theta 3}) \tag{6.30}$$

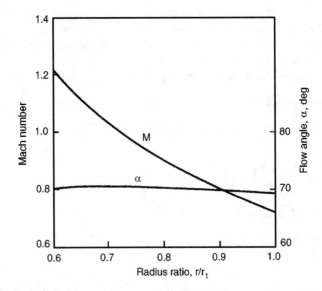

FIG. 6.6. Flow angle and Mach number distributions with radius of a nozzle blade row designed for constant specific mass flow.

and hence the gradient in stagnation enthalpy after the rotor is

$$dh_{o3}/dr = -d[U(c_{\theta 2} + c_{\theta 3})]/dr = -d(Uc_{\theta 2})/dr - d(Uc_3 \sin \alpha_3)/dr.$$

After differentiating the last term,

$$-dh_o = d(Uc_{\theta 2}) + U(c \sin \alpha dr/r + \sin \alpha dc + c \cos \alpha d\alpha) \tag{6.30a}$$

the subscript 3 having now been dropped.
From eqn. (6.20) the radial equilibrium equation applied to the rotor exit flow is

$$dh_o = c^2 \sin^2 \alpha dr/r + cdc. \tag{6.30b}$$

After logarithmic differentiation of $\rho c \cos \alpha =$ constant,

$$d\rho/\rho + dc/c = \tan \alpha \, d\alpha. \tag{6.31}$$

Eliminating successively dh_o between eqns. (6.30a) and (6.30b), $d\rho/\rho$ between eqns. (6.28) and (6.31) and finally $d\alpha$ from the resulting equations gives

$$\frac{dc}{c}\left(1 + \frac{c_\theta}{U}\right) = -\sin^2 \alpha \left\{\frac{d(rc_\theta)}{rc_\theta} + \left(1 + \frac{c_\theta}{U} + M_x^2\right)\frac{dr}{r}\right\} \tag{6.32}$$

where $M_x = M \cos \alpha = c \cos \alpha / \sqrt{(\gamma RT)}$ and the static temperature

$$T = T_3 = T_{o3} - c_3^2/(2C_p)$$

$$= T_{o2} - [U(c_{\theta 2} + c_{\theta 3}) + \tfrac{1}{2}c_3^2]/C_p. \tag{6.33}$$

The verification of eqn. (6.32) is left as an exercise for the diligent student.

Provided that the exit flow angle α_3 at $r = r_m$ and the mean rotor blade speeds are specified, the velocity distribution, etc., at rotor exit can be readily computed from these equations.

Off-design performance of a stage

A turbine stage is considered here although, with some minor modifications, the analysis can be made applicable to a compressor stage.

Assuming the flow is at constant entropy, apply the radial equilibrium equation, eqn. (6.6), to the flow on both sides of the rotor, then

$$\frac{dh_{03}}{dr} = \frac{dh_{02}}{dr} - \Omega\frac{d}{dr}(rc_{\theta 2} + rc_{\theta 3}) = c_{x3}\frac{dc_{x3}}{dr} + \frac{c_{\theta 3}}{r}\frac{d}{dr}(rc_{\theta 3}).$$

Therefore

$$c_{x2}\frac{dc_{x2}}{dr} + \left(\frac{c_{\theta 2}}{r} - \Omega\right)\frac{d}{dr}(rc_{\theta 2}) = c_{x3}\frac{dc_{x3}}{dr} + \left(\frac{c_{\theta 3}}{r} + \Omega\right)\frac{d}{dr}(rc_{\theta 3}).$$

Substituting $c_{\theta 3} = c_{x3} \tan \beta_3 - \Omega r$ into the above equation, then, after some simplification,

$$c_{x2}\frac{dc_{x2}}{dr} + \left(\frac{c_{\theta 2}}{r} - \Omega\right)\frac{d}{dr}(rc_{\theta 2}) = c_{x3}\frac{dc_{x3}}{dr} + \frac{c_{x3}}{r}\tan \beta_3\frac{d}{dr}(rc_{x3} \tan \beta_3)$$

$$- 2\Omega c_{x3} \tan \beta_3. \tag{6.34}$$

In a particular problem the quantities c_{x2}, $c_{\theta2}$, β_3 are known functions of radius and Ω can be specified. Equation (6.34) is thus a first order differential equation in which c_{x3} is unknown and may best be solved, in the general case, by numerical iteration. This procedure requires a guessed value of c_{x3} at the hub and, by applying eqn. (6.34) to a small interval of radius Δr, a new value of c_{x3} at radius $r_h + \Delta r$ is found. By repeating this calculation for successive increments of radius a complete velocity profile c_{x3} can be determined. Using the continuity relation

$$\int_{r_h}^{r_t} c_{x3} r \mathrm{d}r = \int_{r_h}^{r_t} c_{x2} r \mathrm{d}r,$$

this initial velocity distribution can be integrated and a new, more accurate, estimate of c_{x3} at the hub then found. Using this value of c_{x3} the step-by-step procedure is repeated as described and again checked by continuity. This iterative process is normally rapidly convergent and, in most cases, three cycles of the calculation enables a sufficiently accurate exit velocity profile to be found.

The off-design performance may be obtained by making the approximation that the rotor relative exit angle β_3 and the nozzle exit angle α_2 remain constant at a particular radius with a change in mass flow. This approximation is not unrealistic as cascade data (see Chapter 3) suggest that fluid angles at outlet from a blade row alter very little with change in incidence up to the stall point.

Although any type of flow through a stage may be successfully treated using this method, rather more elegant solutions in closed form can be obtained for a few special cases. One such case is outlined below for a free-vortex turbine stage whilst other cases are already covered by eqns. (6.21)–(6.23).

Free-vortex turbine stage

Suppose, for simplicity, a free-vortex stage is considered where, at the design point, the flow at rotor exit is completely axial (i.e. without swirl). At stage entry the flow is again supposed completely axial and of constant stagnation enthalpy h_{01}. Free-vortex conditions prevail at entry to the rotor, $rc_{\theta2} = rc_{x2} \tan a_2 = $ constant. The problem is to find how the axial velocity distribution at rotor exit varies as the mass flow is altered away from the design value.

At off-design conditions the relative rotor exit angle β_3 is assumed to remain equal to the value β^* at the design mass flow (* denotes design conditions). Thus, referring to the velocity triangles in Figure 6.7, at off-design conditions the swirl velocity $c_{\theta3}$ is evidently non-zero,

$$c_{\theta3} = c_{x3} \tan \beta_3 - U$$

$$= c_{x3} \tan \beta_3^* - \Omega r. \tag{6.35}$$

At the design condition, $c_{\theta3}^* = 0$ and so

$$c_{x3}^* \tan \beta_3^* = \Omega r. \tag{6.36}$$

Combining eqns. (6.35) and (6.36)

$$c_{\theta3} = \Omega r \left(\frac{c_{x3}}{c_{x3}^*} - 1 \right). \tag{6.37}$$

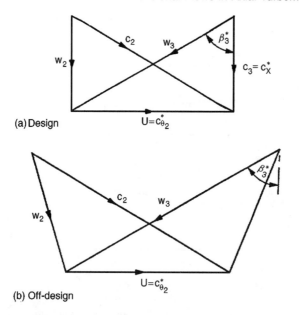

(a) Design

(b) Off-design

FIG. 6.7. Design and off-design velocity triangles for a free-vortex turbine stage.

The radial equilibrium equation at rotor outlet gives

$$\frac{dh_{03}}{dr} = c_{x3}\frac{dc_{x3}}{dr} + \frac{c_{\theta 3}}{r}\frac{d}{dr}(rc_{\theta 3}) = -\Omega\frac{d}{dr}(rc_{\theta 3}),$$ (6.38)

after combining with eqn. (6.33), nothing that $dh_{02}/dr = 0$ and that $(d/dr)(rc_{\theta 2}) = 0$ at all mass flows. From eqn. (6.37),

$$\Omega + \frac{c_{\theta 3}}{r} = \Omega\frac{c_{x3}}{c_{x3}^*}, \; rc_{\theta 3} = \Omega r^2\left(\frac{c_{x3}}{c_{x3}^*} - 1\right),$$

which when substituted into eqn. (6.38) gives,

$$-\frac{dc_{x3}}{dr} = \frac{\Omega^2}{c_{x3}^*}\left[2r\left(\frac{c_{x3}}{c_{x3}^*} - 1\right) + \frac{r^2}{c_{x3}^*}\frac{dc_{x3}}{dr}\right].$$

After rearranging,

$$\frac{dc_{x3}}{c_{x3} - c_{x3}^*} = \frac{-d(\Omega^2 r^2)}{(c_{x3}^{*2} + \Omega^2 r^2)}.$$ (6.39)

Equation (6.39) is immediately integrated in the form

$$\frac{c_{x3} - c_{x3}^*}{c_{x3m} - c_{x3}^*} = \frac{c_{x3}^{*2} + \Omega^2 r_m^2}{c_{x3}^{*2} + \Omega^2 r^2}$$ (6.40)

where $c_{x3} = c_{x3m}$ at $r = r_m$. Equation (6.40) is more conveniently expressed in a non-dimensional form by introducing flow coefficients $\phi = c_{x3}/U_m$, $\phi^* = c_{x3}^*/U_m$

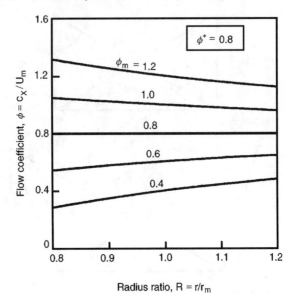

FIG. 6.8. Off-design rotor exit flow coefficients.

and $\phi_m = c_{x3m}/U_m$. Thus,

$$\frac{\phi/\phi^* - 1}{\phi_m/\phi^* - 1} = \frac{\phi^{*2} + 1}{\phi^{*2} + (r/r_m)^2},$$ (6.40a)

If r_m is the mean radius then $c_{x3m} \doteq c_{x1}$ and, therefore, ϕ_m provides an approximate measure of the overall flow coefficient for the machine (N.B. c_{x1} is uniform).

The results of this analysis are shown in Figure 6.8 for a representative design flow coefficient $\phi^* = 0.8$ at several different off-design flow coefficients ϕ_m, with $r/r_m = 0.8$ at the hub and $r/r_m = 1.2$ at the tip. It is apparent for values of $\phi_m < \phi^*$, that c_{x3} increases from hub to tip; conversely for $\phi_m > \phi^*$, c_{x3} decreases towards the tip.

The foregoing analysis is only a special case of the more general analysis of free-vortex turbine and compressor flows (Horlock and Dixon 1966) in which rotor exit swirl, $rc_{\theta3}^*$ is constant (at design conditions), is included. However, from Horlock and Dixon, it is quite clear that even for fairly large values of α_{3m}^*, the value of ϕ is little different from the value found when $\alpha_3^* = 0$, all other factors being equal. In Figure 6.8 values of ϕ are shown when $\alpha_{3m}^* = 31.4°$ at $\phi_m = 0.4(\phi^* = 0.8)$ for comparison with the results obtained when $\alpha_3^* = 0$.

It should be noted that the rotor efflux flow at off-design conditions is *not* a free vortex.

Actuator disc approach

In the radial equilibrium design method it was assumed that all radial motion took place within the blade row. However, in most turbomachines of low hub–tip ratio, appreciable radial velocities can be measured outside the blade row. Figure 6.9,

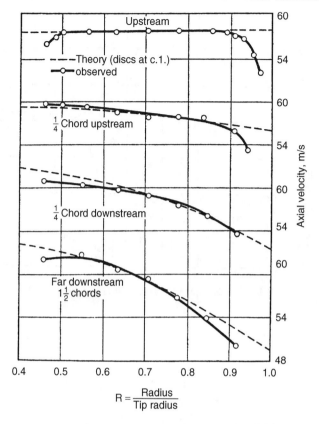

FIG. 6.9. Variation of the distribution in axial velocity through a row of guide vanes (adapted from Hawthorne and Horlock 1962).

taken from a review paper by Hawthorne and Horlock (1962), shows the distribution of the axial velocity component at various axial distances upstream and downstream of an isolated row of stationary inlet guide vanes. This figure clearly illustrates the appreciable redistribution of flow in regions outside of the blade row and that radial velocities must exist in these regions. For the flow through a single row of rotor blades, the variation in pressure (near the hub and tip) and variation in axial velocity (near the hub) both as functions of axial position, are shown in Figure 6.10, also taken from Hawthorne and Horlock. Clearly, radial equilibrium is not established entirely within the blade row.

A more accurate form of three-dimensional flow analysis than radial equilibrium theory is obtained with the *actuator disc* concept. The idea of an actuator disc is quite old and appears to have been first used in the theory of propellers; it has since evolved into a fairly sophisticated method of analysing flow problems in turbomachinery. To appreciate the idea of an actuator disc, imagine that the axial width of each blade row is shrunk while, at the same time, the space–chord ratio, the blade angles and overall length of machine are maintained constant. As the deflection through each blade row for a given incidence is, apart from Reynolds number and Mach number effects (cf. Chapter 3 on cascades), fixed by the cascade

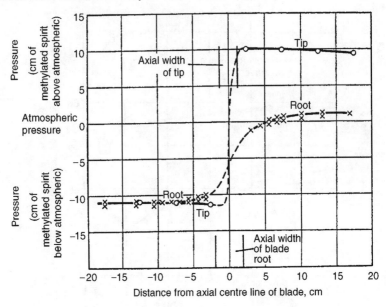

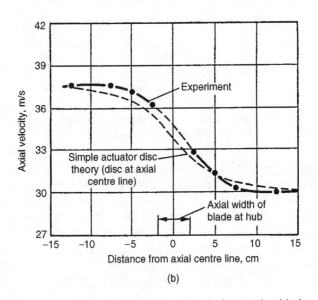

(b)

FIG. 6.10. (a) Pressure variation in the neighbourhood of a rotating blade row. (b) Axial velocity at the hub in the neighbourhood of a rotating blade row (adapted from Hawthorne and Horlock 1962).

geometry, a blade row of reduced width may be considered to affect the flow in exactly the same way as the original row. In the limit as the axial width vanishes, the blade row becomes, conceptually, a *plane discontinuity* of tangential velocity – the actuator disc. Note that while the tangential velocity undergoes an abrupt change in direction, the axial and radial velocities are continuous across the disc.

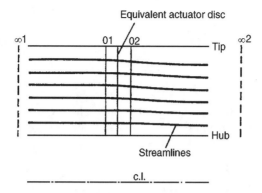

Fɪɢ. 6.11. The actuator disc assumption (after Horlock 1958).

An isolated actuator disc is depicted in Figure 6.11 with radial equilibrium established at fairly large axial distances from the disc. An approximate solution to the velocity fields upstream and downstream of the actuator can be found in terms of the axial velocity distributions, *far upstream* and *far downstream* of the disc. The detailed analysis exceeds the scope of this book, involving the solution of the equations of motion, the equation of continuity and the satisfaction of boundary conditions at the walls and disc. The form of the approximate solution is of considerable interest and is quoted below.

For convenience, conditions far upstream and far downstream of the disc are denoted by subscripts $\infty 1$ and $\infty 2$ respectively (Figure 6.11). Actuator disc theory proves that at the disc ($x = 0$), at any given radius, the axial velocity is equal to the *mean* of the axial velocities at $\infty 1$ and $\infty 2$ at the *same* radius, or

$$c_{x01} = c_{x02} = \tfrac{1}{2}(c_{x\infty 1} + c_{x\infty 2}). \tag{6.41}$$

Subscripts 01 and 02 denote positions immediately upstream and downstream respectively of the actuator disc. Equation (6.41) is known as the *mean-value rule*.

In the downstream flow field ($x \geq 0$), the *difference* in axial velocity at some position (x, r_A) to that at position $(x = \infty, r_A)$ is conceived as a velocity perturbation. Referring to Figure 6.12, the axial velocity perturbation at the disc $(x = 0, r_A)$ is

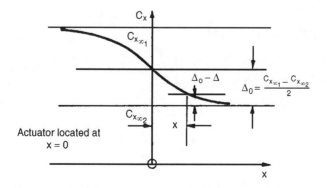

Fɪɢ. 6.12. Variation in axial velocity with axial distance from the actuator disc.

denoted by Δ_0 and at position (x, r_A) by Δ. The important result of actuator disc theory is that velocity perturbations *decay exponentially* away from the disc. This is also true for the upstream flow field ($x \leq 0$). The result obtained for the decay rate is

$$\Delta/\Delta_0 = 1 - \exp[\mp\pi x/(r_t - r_h)], \tag{6.42}$$

where the minus and plus signs above apply to the flow regions $x \geq 0$ and $x \leq 0$ respectively. Equation (6.42) is often called the *settling-rate rule*. Since $c_{x1} = c_{x01} + \Delta$, $c_{x2} = c_{x02} - \Delta$ and noting that $\Delta_0 = \frac{1}{2}(c_{x\infty 1} - c_{x\infty 2})$, eqns. (6.41) and (6.42) combine to give,

$$c_{x1} = c_{x\infty 1} - \frac{1}{2}(c_{x\infty 1} - c_{x\infty 2}) \exp[\pi x/(r_t - r_h)], \tag{6.43a}$$

$$c_{x2} = c_{x\infty 2} + \frac{1}{2}(c_{x\infty 1} - c_{x\infty 2}) \exp[-\pi x/(r_t - r_h)]. \tag{6.43b}$$

At the disc, $x = 0$, eqns. (6.43) reduce to eqn. (6.41). It is of particular interest to note, in Figures 6.9 and 6.10, how closely isolated actuator disc theory compares with experimentally derived results.

Blade row interaction effects

The spacing between consecutive blade rows in axial turbomachines is usually sufficiently small for mutual flow interactions to occur between the rows. This interference may be calculated by an extension of the results obtained from isolated actuator disc theory. As an illustration, the simplest case of two actuator discs situated a distance δ apart from one another is considered. The extension to the case of a large number of discs is given in Hawthorne and Harlock (1962).

Consider each disc in turn as though it were in isolation. Referring to Figure 6.13, disc A, located at $x = 0$, changes the far upstream velocity $c_{x\infty 1}$ to $c_{x\infty 2}$ far downstream. Let us suppose for simplicity that the effect of disc B, located at $x = \delta$, exactly cancels the effect of disc A (i.e. the velocity far upstream of disc B is $c_{x\infty 2}$ which changes to $c_{x\infty 1}$ far downstream). Thus, for disc A in isolation,

$$c_x = c_{x\infty 1} - \frac{1}{2}(c_{x\infty 1} - c_{x\infty 2}) \exp\left[\frac{-\pi |x|}{H}\right], \quad x \leq 0, \tag{6.44}$$

$$c_x = c_{x\infty 2} + \frac{1}{2}(c_{x\infty 1} - c_{x\infty 2}) \exp\left[\frac{-\pi |x|}{H}\right], \quad x \geq 0, \tag{6.45}$$

where $|x|$ denotes modulus of x and $H = r_t - r_h$.

For disc B in isolation,

$$c_x = c_{x\infty 2} - \frac{1}{2}(c_{x\infty 2} - c_{x\infty 1}) \exp\left[\frac{-\pi |x - \delta|}{H}\right], \quad x \leq \delta, \tag{6.46}$$

$$c_x = c_{x\infty 1} + \frac{1}{2}(c_{x\infty 2} - c_{x\infty 1}) \exp\left[\frac{-\pi |x - \delta|}{H}\right], \quad x \geq \delta. \tag{6.47}$$

Now the combined effect of the two discs is most easily obtained by extracting from the above four equations the velocity perturbations appropriate to a given

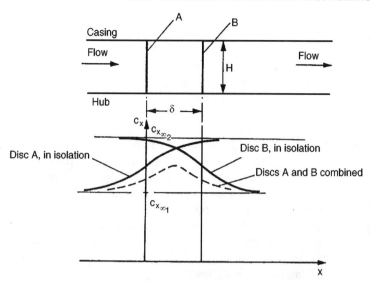

FIG. 6.13. Interaction between two closely spaced actuator discs.

region and adding these to the related radial equilibrium velocity. For $x \leq 0$, and to $c_{x\infty1}$ the perturbation velocities from eqns. (6.44) and (6.46).

$$c_x = c_{x\infty1} - \frac{1}{2}(c_{x\infty1} - c_{x\infty2})\left\{\exp\left[\frac{-\pi|x|}{H}\right] - \exp\left[\frac{-\pi|x - \delta|}{H}\right]\right\}. \quad (6.48)$$

For the region $0 \leq x \leq \delta$,

$$c_x = c_{x\infty2} + \frac{1}{2}(c_{x\infty1} - c_{x\infty2})\left\{\exp\left[\frac{-\pi|x|}{H}\right] + \exp\left[\frac{-\pi|x - \delta|}{H}\right]\right\}. \quad (6.49)$$

For the region $x \geq \delta$,

$$c_x = c_{x\infty1} + \frac{1}{2}(c_{x\infty1} - c_{x\infty2})\left\{\exp\left[\frac{-\pi|x|}{H}\right] - \exp\left[\frac{-\pi|x - \delta|}{H}\right]\right\}. \quad (6.50)$$

Figure 6.13 indicates the variation of axial velocity when the two discs are regarded as *isolated* and when they are *combined*. It can be seen from the above equations that as the gap between these two discs is increased, so the perturbations tend to vanish. Thus in turbomachines where δ/r, is fairly small (e.g. the front stages of aircraft axial compressors or the rear stages of condensing steam turbines), interference effects are strong and one can infer that the simpler radial equilibrium analysis is then inadequate.

Computer-aided methods of solving the through-flow problem

Although actuator disc theory has given a better understanding of the complicated meridional (the radial-axial plane) through-flow problem in turbomachines of simple geometry and flow conditions, its application to the design of axial-flow compressors has been rather limited. The extensions of actuator disc theory to the solution of the complex three-dimensional, compressible flows in compressors with varying hub and tip radii and non-uniform total pressure distributions were found to have become too unwieldy in practice. In recent years advanced computational methods have been successfully evolved for predicting the meridional compressible flow in turbomachines with flared annulus walls.

Reviews of numerical methods used to analyse the flow in turbomachines have been given by Gostelow *et al.* (1969), Japikse (1976), Macchi (1985) and Whitfield and Baines (1990) among many others. The literature on computer-aided methods of solving flow problems is now extremely extensive and no attempt is made here to summarise the progress. The real flow in a turbomachine is three-dimensional, unsteady, viscous and is usually compressible, if not transonic or even supersonic. According to Macchi the solution of the full equations of motion with the actual boundary conditions of the turbomachine is still beyond the capabilities of the most powerful modern computers. The best fully three-dimensional methods available are still only simplifications of the real flow.

Through-flow methods

In any of the so-called *through-flow* methods the equations of motion to be solved are simplified. First, the flow is taken to be steady in both the absolute and relative frames of reference. Secondly, outside of the blade rows the flow is assumed to be axisymmetric, which means that the effects of wakes from an upstream blade row are understood to have "mixed out" so as to give uniform circumferential conditions. Within the blade rows the effects of the blades themselves are modelled by using a passage averaging technique or an equivalent process. Clearly, with these major assumptions, solutions obtained with these through-flow methods can be only approximations to the real flow. As a step beyond this Stow (1985) has outlined the ways, supported by equations, of including the viscous flow effects into the flow calculations.

Three of the most widely used techniques for solving through-flow problems are:

(1) Streamline curvature, which is based on an iterative procedure, is described in some detail by Macchi (1985) and earlier by Smith (1966). It is the oldest and most widely used method for solving the through-flow problem in axial-flow turbomachines and has with the intrinsic capability of being able to handle variously shaped boundaries with ease. The method is widely used in the gas turbine industry.

(2) Matrix through-flow or finite difference solutions (Marsh 1968), where computations of the radial equilibrium flow field are made at a number of axial locations *within* each blade row as well as at the leading and trailing edges and outside of the blade row. An illustration of a typical computing mesh for a single blade row is shown in Figure 6.14.

(3) Time-marching (Denton 1985), where the computation starts from some assumed flow field and the governing equations are marched forward with time. The method, although slow because of the large number of iterations needed to reach a convergent solution, can be used to solve both subsonic and supersonic

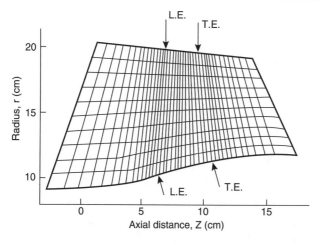

FIG. 6.14. Typical computational mesh for a single blade row (adapted from Macchi 1985).

flow. With the present design trend towards highly loaded blade rows, which can include patches of supersonic flow, this design method has considerable merit.

All three methods solve the same equations of fluid motion, energy and state for an axisymmetric flow through a turbomachine with varying hub and tip radii and therefore lead to the same solution. In the first method the equation for the meridional velocity $c_m = (c_r^2 + c_x^2)^{1/2}$ in a plane (at $x = x_a$) contain terms involving both the slope and curvature of the meridional streamlines which are estimated by using a polynominal curve-fitting procedure through points of equal stream function on neighbouring planes at $(x_a - dx)$ and $(x_a + dx)$. The major source of difficulty is in accurately estimating the curvature of the streamlines. In the second method a grid of calculating points is formed on which the stream function is expressed as a quasi-linear equation. A set of corresponding finite difference equations are formed which are then solved at all mesh points of the grid. A more detailed description of these methods is rather beyond the scope and intention of the present text.

Secondary flows

No account of three-dimensional motion in axial turbomachines would be complete without giving, at least, a brief description of secondary flow. When a fluid particle possessing *rotation* is turned (e.g. by a cascade) its axis of rotation is deflected in a manner analogous to the motion of a gyroscope, i.e. in a direction perpendicular to the direction of turning. The result of turning the rotation (or vorticity) vector is the formation of *secondary flows*. The phenomenon must occur to some degree in all turbomachines but is particularly in evidence in axial-flow compressors because of the thick boundary layers on the annulus walls. This case has been discussed in some detail by Horlock (1958), Preston (1953), Carter (1948) and many other writers.

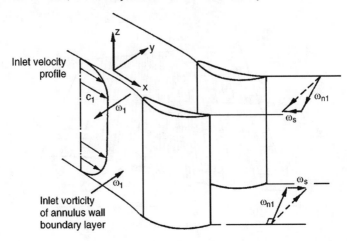

FIG. 6.15. Secondary vorticity produced by a row of guide vanes.

Consider the flow at inlet to the guide vanes of a compressor to be completely axial and with a velocity profile as illustrated in Figure 6.15. This velocity profile is non-uniform as a result of friction between the fluid and the wall; the vorticity of this boundary layer is normal to the approach velocity c_1 and of magnitude

$$\omega_1 = \frac{dc_1}{dz}, \tag{6.51}$$

where z is distance from the wall.

The direction of ω_1 follows from the right-hand screw rule and it will be observed that ω_1 is in opposite directions on the two annulus walls. This vector is turned by the cascade, thereby generating *secondary vorticity* parallel to the outlet stream direction. If the deflection angle ϵ is not large, the magnitude of the secondary vorticity ω_s is, approximately,

$$\omega_s = -2\epsilon \frac{dc_1}{dz}. \tag{6.52}$$

A swirling motion of the cascade exit flow is associated with the vorticity ω_s, as shown in Figure 6.16, which is in opposite directions for the two wall boundary layers. This secondary flow will be the *integrated* effect of the distribution of secondary vorticity along the blade length.

Now if the variation of c_1 with z is known or can be predicted, then the distribution of ω_s along the blade can be found using eqn. (6.52). By considering the secondary flow to be small perturbation of the two-dimensional flow from the vanes, the flow angle distribution can be calculated using a series solution developed by Hawthorne (1955). The actual analysis lies outside the scope (and purpose) of this book, however. Experiments on cascade show excellent agreement with these calculations provided there are but small viscous effects and no flow separations. Such a comparison has been given by Horlock (1963) and a typical result is shown in Figure 6.17. It is clear that the flow is *overturned* near the walls and *underturned*

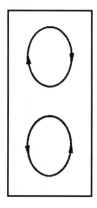

FIG. 6.16. Secondary flows at exit from a blade passage (viewed in upstream direction).

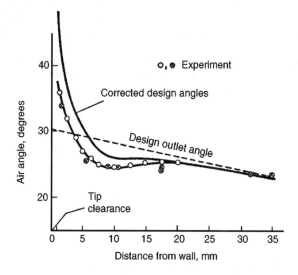

FIG. 6.17. Exit air angle from inlet guide vanes (adapted from Horlock 1963).

some distance away from the walls. It is known that this overturning is a source of inefficiency in compressors as it promotes stalling at the blade extremities.

References

Carter, A. D. S. (1948). Three-dimensional flow theories for axial compressors and turbines. *Proc. Instn. Mech. Engrs.*, **159**, 41.

Constant, H. (1945). The early history of the axial type of gas turbine engine. *Proc. Instn. Mech. Engrs.*, **153**.

Constant, H. (1953). *Gas Turbines and their Problems.* Todd.

Denton, J. D. (1985). Solution of the Euler equations for turbomachinery flows. Part 2. Three-dimensional flows. In *Thermodynamics and Fluid Mechanics of Turbomachinery*, Vol. 1 (A. S. Ücer, P. Stow and Ch. Hirsch, eds) pp. 313–47. Martinus Nijhoff.

Gostelow, J. P., Horlock, J. H. and Marsh, H. (1969). Recent developments in the aerodynamic design of axial flow compressors. *Proc. Instn. Mech. Engrs.*, **183**, Pt. 3N.

Hawthorne, W. R. (1955). Some formulae for the calculation of secondary flow in cascades. ARC Report 17,519.

Hawthorne, W. R. and Horlock, J. H. (1962). Actuator disc theory of the incompressible flow in axial compressors. *Proc. Instn. Mech. Engrs.*, **176**, 789.

Horlock, J. H. (1958). *Axial Flow Compressors*. Butterworths.

Horlock, J. H. (1963). Annulus wall boundary layers in axial compressor stages. *Trans. Am. Soc. Mech. Engrs.*, Series D, **85**.

Horlock, J. H. (1966). *Axial Flow Turbines*. Butterworths.

Horlock, J. H. and Dixon, S. L. (1966). The off-design performance of free vortex turbine and compressor stages. ARC. Report 27,612.

Howell, A. R. (1945). Fluid dynamics of axial compressors. *Proc. Instn. Mech. Engrs.*, **153**.

Japikse, D. (1976). Review – progress in numerical turbomachinery analysis. *J. Fluids Eng., Trans Am. Soc. Mech. Engrs.*, **98**, 592–606.

Macchi, E. (1985). The use of radial equilibrium and streamline curvature methods for turbomachinery design and prediction. In *Thermodynamics and Fluid Mechanics of Turbomachinery*, Vol. 1. (A. S. Ücer, P. Stow and Ch. Hirsch, eds) pp. 133–66. Martinus Nijhoff.

Marsh, H. (1968). A digital computer program for the through-flow fluid mechanics on an arbitrary turbomachine using a matrix method. *ARC, R&M 3509*.

Preston, J. H. (1953). A simple approach to the theory of secondary flows. *Aero. Quart.*, **5**, (3).

Smith, L. H., Jr. (1966) The radial-equilibrium equation of turbomachinery. *Trans. Am. Soc. Mech. Engrs.*, Series A, **88**.

Stow, P. (1985). Modelling viscous flows in turbomachinery. In *Thermodynamics and Fluid Mechanics of Turbomachinery*, Vol. 1. (A. S. Ücer, P. Stow and Ch. Hirsch, eds) pp. 37–71. Martinus Nijhoff.

Whitfield, A. and Baines, N. C. (1990). *Design of Radial Turbomachines*. Longman.

Problems

1. Derive the radial equilibrium equation for an incompressible fluid flowing with axisymmetric swirl through an annular duct.

Air leaves the inlet guide vanes of an axial flow compressor in radial equilibrium and with a free-vortex tangenital velocity distribution. The absolute static pressure and static temperature at the hub, radius 0.3 m, are 94.5 kPa and 293 K respectively. At the casing, radius 0.4 m, the absolute static pressure is 96.5 kPa. Calculate the flow angles at exit from the vanes at the hub and casing when the inlet absolute stagnation pressure is 101.3 kPa. Assume the fluid to be inviscid and incompressible. (Take $R = 0.287$ kJ/(kg°C) for air.).

2. A gas turbine stage has an initial absolute pressure of 350 kPa and a temperature of 565°C with negligible initial velocity. At the mean radius, 0.36 m, conditions are as follows:

Nozzle exit flow angle	68 deg
Nozzle exit absolute static pressure	207 kPa
Stage reaction	0.2

Determine the flow coefficient and stage loading factor at the mean radius and the reaction at the hub, radius 0.31 m, at the design speed of 8000 rev/min, given that stage is to have a free vortex swirl at this speed. You may assume that losses are absent. Comment upon the results you obtain.

(Take $C_p = 1.148$ kJ(kg°C) and $\gamma = 1.33$.)

3. Gas enters the nozzles of an axial flow turbine stage with uniform total pressure at a uniform velocity c_1 in the axial direction and leaves the nozzles at a constant flow angle α_2 to the axial direction. The absolute flow leaving the rotor c_3 is completely axial at all radii.

Using radial equilibrium theory and assuming no losses in total pressure show that

$$(c_3^2 - c_1^2)/2 = U_m c_{\theta m2} \left[1 - \left(\frac{r}{r_m}\right)^{\cos^2 \alpha_2}\right]$$

where U_m is the mean blade speed,

$c_{\theta m2}$ is the tangential velocity component at nozzle exit at the mean radius $r = r_m$.

(*Note:* The approximate $c_3 = c_1$ at $r = r_m$ is used to derive the above expression.)

4. Gas leaves an untwisted turbine nozzle at an angle α to the axial direction and in radial equilibrium. Show that the variation in axial velocity from root to tip, assuming total pressure is constant, is given by

$$c_x r^{\sin^2 \alpha} = \text{constant.}$$

Determine the axial velocity at a radius of 0.6 m when the axial velocity is 100 m/s at a radius of 0.3 m. The outlet angle α is 45 deg.

5. The flow at the entrance and exit of an axial-flow compressor rotor is in radial equilibrium. The distributions of the tangential components of absolute velocity with radius are:

$c_{\theta 1} = ar - b/r$, before the rotor,

$c_{\theta 2} = ar + b/r$, after the rotor,

where a and b are constants. What is the variation of work done with radius? Deduce expressions for the axial velocity distributions before and after the rotor, assuming incompressible flow theory and that the radial gradient of stagnation pressure is zero.

At the mean radius, $r = 0.3$ m, the stage loading coefficient, $\psi = \Delta W/U_t^2$ is 0.3, the reaction ratio is 0.5 and the mean axial velocity is 150 m/s. The rotor speed is 7640 rev/min. Determine the rotor flow inlet and outlet angles at a radius of 0.24 m given that the hub–tip ratio is 0.5. Assume that at the mean radius the axial velocity remained unchanged ($c_{x1} = c_{x2}$ at $r = 0.3$ m).

(*Note:* ΔW is the specific work and U_t the blade tip speed.)

6. An axial-flow turbine stage is to be designed for free-vortex conditions at exit from the nozzle row and for zero swirl at exit from the rotor. The gas entering the stage has a stagnation temperature of 1000 K, the mass flow rate is 32 kg/s, the root and tip diameters are 0.56 m and 0.76 m respectively, and the rotor speed is 8000 rev/min. At the rotor tip the stage reaction is 50% and the axial velocity is constant at 183 m/s. The velocity of the gas entering the stage is equal to that leaving.

Determine:

(i) the maximum velocity leaving the nozzles;
(ii) the maximum absolute Mach number in the stage;
(iii) the root section reaction;
(iv) the power output of the stage;
(v) the stagnation and static temperatures at stage exit.

(Take $R = 0.287$ kJ/(kg°C) and $C_p = 1.147$ kJ/(kg°C).)

7. The rotor blades of an axial-flow turbine stage are 100 mm long and are designed to receive gas at an incidence of 3 deg from a nozzle row. A free-vortex whirl distribution is to be maintained between nozzle exit and rotor entry. At rotor exit the absolute velocity is 150 m/s in the axial direction at all radii. The deviation is 5 deg for the rotor blades and zero for the nozzle blades at all radii. At the hub, radius 200 mm, the conditions are as follows:

Nozzle outlet angle	70 deg
Rotor blade speed	180 m/s
Gas speed at nozzle exit	450 m/s

Assuming that the axial velocity of the gas is constant across the stage, determine

 (i) the nozzle outlet angle at the tip;
 (ii) the rotor blade inlet angles at hub and tip;
(iii) the rotor blade outlet angles at hub and tip;
(iv) the degree of reaction at root and tip.

Why is it essential to have a *positive* reaction in a turbine stage?

8. The rotor and stator of an isolated stage in an axial-flow turbomachine are to be represented by two actuator discs located at axial positions $x = 0$ and $x = \delta$ respectively. The hub and tip diameters are constant and the hub–tip radius ratio r_h/r_t is 0.5. The rotor disc considered on its own has an axial velocity of 100 m/s far upstream and 150 m/s downstream at a constant radius $r = 0.75r_t$. The stator disc in isolation has an axial velocity of 150 m/s far upstream and 100 m/s far downstream at radius $r = 0.75r_t$. Calculate and plot the axial velocity variation between $-0.5 \leqslant x/r_t \leqslant 0.6$ at the given radius for each actuator disc in isolation and for the combined discs when

 (i) $\delta = 0.1r_t$, (ii) $\delta = 0.25r_t$, (iii) $\delta = r_t$.

CHAPTER 7

Centrifugal Pumps, Fans and Compressors

And to thy speed add wings. (MILTON, *Paradise Lost.*)

Introduction

This chapter is concerned with the elementary flow analysis and preliminary design of *radial-flow* work-absorbing turbomachines comprising pumps, fans and compressors. The major part of the discussion is centred around the compressor since the basic action of all these machines is, in most respects, the same.

Turbomachines employing centrifugal effects for increasing fluid pressure have been in use for more than a century. The earliest machines using this principle were, undoubtedly, hydraulic pumps followed later by ventilating fans and blowers. Cheshire (1945) recorded that a centrifugal compressor was incorporated in the build of the whittle turbojet engine.

> For the record, the first successful test flight of an aircraft powered by a turbojet engine was on August 27, 1939 at Marienebe Airfield, Waruemunde, Germany (Gas Turbine News (1989). The engine, designed by Hans von Ohain, incorporated an axial flow compressor. The Whittle turbojet engine, with the centrifugal compressor, was first flown on May 15, 1941 at Cranwell, England (see Hawthorne 1978).

Development of the centrifugal compressor continued into the mid-1950s but, long before this, it had become abundantly clear (Campbell and Talbert 1945, Moult and Pearson 1951 that for the increasingly larger engines required for aircraft propulsion the axial flow compressor was preferred. Not only was the frontal area (and drag) smaller with engines using axial compressors but also the efficiency for the same duty was better by as much as 3 or 4%. However, at very low air mass flow rates the efficiency of axial compressors drops sharply, blading is small and difficult to make accurately and the advantage lies with the centrifugal compressor.

In the mid-1960s the need for advanced military helicopters powered by small gas turbine engines provided the necessary impetus for further rapid development of the centrifugal compressor. The technological advances made in this sphere provided a spur to designers in a much wider field of existing centrifugal compressor applications, e.g. in small gas turbines for road vehicles and commercial helicopters as well as for diesel engine turbochargers, chemical plant processes, factory workshop air supplies and large-scale air-conditioning plant, etc.

199

Centrifugal compressors were the reasoned choice for refrigerating plants and compression-type heat pumps used in district heating schemes described by Hess (1985). These compressors with capacities ranging from below 1 MW up to nearly 30 MW were preferred because of their good economy, low maintenance and absolute reliability. Dean (1973) quoted total-to-static efficiencies of 80–84 per cent for small single-stage centrifugal compressors with pressure ratios of between 4 and 6. Higher pressure ratios than this have been achieved in single stages, but at reduced efficiency and a very limited airflow range (i.e. up to surge). For instance, Schorr *et al.* (1971) designed and tested a single-stage centrifugal compressor which gave a pressure ratio of 10 at an efficiency of 72 per cent but having an airflow range of only 10 per cent at design speed.

Came (1978) described a design procedure and the subsequent testing of a 6.5 pressure ratio centrifugal compressor incorporating 30 deg *back swept vanes*, giving an isentropic total-to-total efficiency for the impeller of over 85 per cent. The overall total-to-total efficiency for the stage was 76.5 per cent and, with a stage pressure ratio of 6.8 a surge margin of 15 per cent was realised. The use of back swept vanes and the avoidance of high vane loading were factors believed to have given a significant improvement in performance compared to an earlier unswept vane design.

Palmer and Waterman (1995) gave some details of an advanced two-stage centrifugal compressor used in a helicopter engine with a pressure ratio of 14, a mass flow rate of 3.3 kg/s and an overall total-to-total efficiency of 80 per cent. Both stages employed back swept vanes (approximately 47 deg) with a low aerodynamic loading achieved by having a relatively large number of vanes (19 full vanes and 19 splitter vanes).

An interesting and novel compressor is the "axi-fuge", a mixed flow design with a high efficiency potential, described by Wiggins (1986) and giving on test a pressure ratio of 6.5 at an isentropic efficiency (undefined) of 84 per cent. Essentially, the machine has a typical short centrifugal compressor annulus but actually contains six stages of rotor and stator blades similar to those of an axial compressor. The axi-fuge is claimed to have the efficiency and pressure ratio of an axial compressor of many stages but retains the compactness and structural simplicity of a centrifugal compressor.

Some definitions

Most of the pressure-increasing turbomachines in use are of the radial-flow type and vary from fans that produce pressure rises equivalent to a few millimetres of water to pumps producing heads of many hundreds of metres of water. The term *pump* is used when referring to machines that increase the pressure of a flowing liquid. The term *fan* is used for machines imparting only a small increase in pressure to a flowing gas. In this case the pressure rise is usually so small that the gas can be considered as being incompressible. A *compressor* gives a substantial rise in pressure to a flowing gas. For purposes of definition, the boundary between fans and compressors is often taken as that where the density ratio across the machine is 1.05. Sometimes, but more rarely nowadays, the term *blower* is used instead of fan.

A centrifugal compressor or pump consists essentially of a rotating *impeller* followed by *diffuser*. Figure 7.1 shows diagrammatically the various elements of a centrifugal compressor. Fluid is drawn in through the *inlet casing* into the *eye* of the impeller. The function of the impeller is to increase the energy level of the fluid by whirling it outwards, thereby increasing the angular momentum of the fluid. Both the static pressure and the velocity are increased within the impeller. The purpose of the diffuser is to convert the kinetic energy of the fluid leaving the impeller into pressure energy. This process can be accomplished by free diffusion in the annular space surrounding the impeller or, as indicated in Figure 7.1, by incorporating a row of fixed diffuser vanes which allows the diffuser to be made very much smaller. Outside the diffuser is a *scroll* or *volute* whose function is to collect the flow from the diffuser and deliver it to the outlet pipe. Often, in low-speed compressors and pump applications where simplicity and low cost count for more than efficiency, the volute follows immediately after the impeller.

The *hub* is the curved surface of revolution of the impeller $a - b$; the *shroud* is the curved surface $c - d$ forming the outer boundary to the flow of fluid. Impellers may be enclosed by having the shroud attached to the vane ends (called shrouded impellers) or unenclosed with a small clearance gap between the vane ends and the stationary wall. Whether or not the impeller is enclosed the surface, $c - d$ is generally called the shroud. Shrouding an impeller has the merit of eliminating

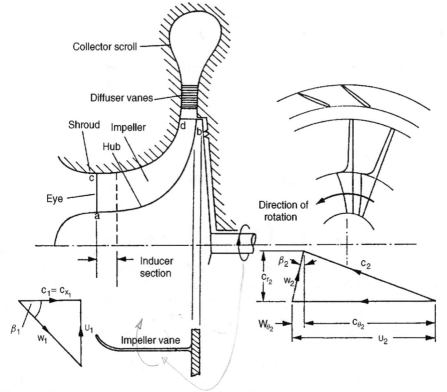

FIG. 7.1. Centrifugal compressor stage and velocity diagrams at impeller entry and exit.

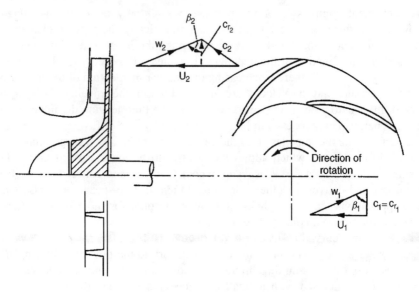

FIG. 7.2. Radial-flow pump and velocity triangles.

tip leakage losses but at the same time increases friction losses. NACA tests have demonstrated that shrouding of a single impeller appears to be detrimental at high speeds and beneficial at low speeds. At entry to the impeller the relative flow has a velocity w_1 at angle β_1 to the axis of rotation. This relative flow is turned into the axial direction by the *inducer section* or *rotating guide vanes* as they are sometimes called. The inducer starts at the eye and usually finishes in the region where the flow is beginning to turn into the radial direction. Some compressors of advanced design extend the inducer well into the radial flow region apparently to reduce the amount of relative diffusion.

To simplify manufacture and reduce cost, many fans and pumps are confined to a two-dimensional radial section as shown in Figure 7.2. With this arrangement some loss in efficiency can be expected. For the purpose of greatest utility, relations obtained in this chapter are generally in terms of the three-dimensional compressor configuration.

Theoretical analysis of a centrifugal compressor

The flow through a compressor stage is a highly complicated, three-dimensional motion and a full analysis presents many problems of the highest order of difficulty. However, we can obtain approximate solutions quite readily by simplifying the flow model. We adopt the so-called *one-dimensional* approach which assumes that the fluid conditions are uniform over certain flow cross-sections. These cross-sections are conveniently taken immediately before and after the impeller as well as at inlet and exit of the entire machine. Where inlet vanes are used to give prerotation to the fluid entering the impeller, the one-dimensional treatment is no longer valid and an extension of the analysis is then required (see Chapter 6).

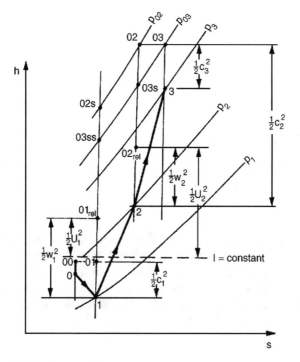

FIG. 7.3. Mollier diagram for the complete centrifugal compressor stage.

Inlet casing

The fluid is accelerated from velocity c_0 to velocity c_1 and the static pressure falls from p_0 to p_1 as indicated in Figure 7.3. Since the stagnation enthalpy is constant in steady, adiabatic flow without shaft work then $h_{00} = h_{01}$ or,

$$h_0 + \tfrac{1}{2}c_0^2 = h_1 + \tfrac{1}{2}c_1^2.$$

Some efficiency definitions appropriate to this process are stated in Chapter 2.

Impeller

The general three-dimensional motion has components of velocity c_r, c_θ, and c_x respectively in the radial, tangential and axial directions and $c^2 = c_r^2 + c_\theta^2 + c_x^2$.
Thus, from eqn. (2.12e), the rothalpy is

p 29

$$I = h + \tfrac{1}{2}(c_r^2 + c_\theta^2 + c_x^2 - 2Uc_\theta).$$

Adding and subtracting $\tfrac{1}{2}U^2$ this becomes

$$I = h + \tfrac{1}{2}\{(U - c_\theta)^2 + c_r^2 + c_x^2 - U^2\}. \tag{7.1}$$

From the velocity triangle, Figure 7.1, $U - c_\theta = w_\theta$ and together with $w^2 = c_r^2 + w_\theta^2 + c_x^2$, eqn. (7.1) becomes

$$I = h + \tfrac{1}{2}(w^2 - U^2)$$

or

$$I = h_{0\mathrm{rel}} - \tfrac{1}{2}U^2,$$

since

$$h_{0\mathrm{rel}} = h + \tfrac{1}{2}w^2.$$

Since $I_1 = I_2$ across the impeller, then

$$h_2 - h_1 = \tfrac{1}{2}(U_2^2 - U_1^2) + \tfrac{1}{2}(w_1^2 - w_2^2). \tag{7.2}$$

The above expression provides the reason why the static enthalpy rise in a centrifugal compressor is so large compared with a single-stage axial compressor. On the right-hand side of eqn. (7.2), the second term $\tfrac{1}{2}(w_2^2 - w_1^2)$, is the contribution from the diffusion of relative velocity and was obtained for axial compressors also. The first term, $\tfrac{1}{2}(U_2^2 - U_1^2)$, is the contribution due to the centrifugal action which is zero if the streamlines remain at the same radii before and after the impeller.

The relation between state points 1 and 2 in Figure 7.3 can be easily traced with the aid of eqn. (7.2)

Referring to Figure 7.1, and in particular the inlet velocity diagram, the absolute flow has no whirl component or angular momentum and $c_{\theta 1} = 0$. In centrifugal compressors and pumps this is the normal situation where the flow is free to enter axially. For such a flow the specific work done on the fluid, from eqn. (2.12c), is written as

$$\Delta W = U_2 c_{\theta 2} = h_{02} - h_{01} \tag{7.3a}$$

in the case of compressors, and

$$\Delta W = U_2 c_{\theta 2} = gH_i \tag{7.3b}$$

in the case of pumps, where H_i (the "ideal" head) is the total head rise across the pump excluding all internal losses. In high pressure ratio compressors it may be necessary to impart *prerotation* to the flow entering the impeller as a means of reducing a high relative inlet velocity. The effects of high relative velocity at the impeller inlet are experienced as Mach number effects in compressors and cavitation effects in pumps. The usual method of establishing prerotation requires the installation of a row of inlet guide vanes upstream of the impeller, the location depending upon the type of inlet. Unless contrary statements are made it will be assumed for the remainder of this chapter that there is no prerotation (i.e. $c_{\theta 1} = 0$).

Conservation of rothalpy

A cornerstone of the analysis of steady, relative flows in rotating systems has, for many years, been the immutable nature of the fluid mechanical property *rothalpy*.

The conditions under which the rothalpy of a fluid is conserved in the flow through impellers and rotors have been closely scrutinised by several researchers. Lyman (1993) reviewed the equations and physics governing the constancy of rothalpy in turbomachine fluid flows and found that *an increase* in rothalpy was possible for steady, viscous flow without heat transfer or body forces. He proved mathematically that the rothalpy increase was generated mainly by the fluid friction acting on the *stationary* shroud of the compressor considered. From his analysis, and put in the simplest terms, he deduced that:

$$h_{02} - h_{01} = (Uc_{\theta})_2 - (Uc_{\theta})_1 + W_f/\dot{m}, \tag{7.4}$$

where $W_f = \dot{m}(I_2 - I_1) = \int \mathbf{n} \cdot \tau \cdot \mathbf{W} \, dA$ is the power loss due to fluid friction on the stationary shroud, $\mathbf{n}$ is a unit normal vector, τ is a viscous stress tensor, $\mathbf{W}$ is the relative velocity vector and dA is an element of the surface area. Lyman did not give any numerical values in support of his analysis.

In the discussion of Lyman's paper, Moore disclosed that earlier viscous flow calculations of the flow in centrifugal flow compressors (see Moore *et al.* 1984) of the power loss in a centrifugal compressor had shown a rothalpy production amounting to 1.2 per cent of the total work input. This was due to the shear work done at the impeller shroud and it was acknowledged to be of the same order of magnitude as the work done overcoming disc friction on the back face of the impeller. Often disc friction is ignored in preliminary design calculations.

A later, careful, order-of-magnitude investigation by Bosman and Jadayel (1996) showed that the change in rothalpy through a centrifugal compressor impeller would be negligible under typical operating conditions. They also believed that it was not possible to *accurately* calculate the change in rothalpy because the effects due to inexact turbulence modelling and truncation error in computation would far exceed those due to non-conservation of rothalpy.

Diffuser

The fluid is decelerated adiabatically from velocity c_2 to a velocity c_3, the static pressure rising from p_2 to p_3 as shown in Figure 7.3. As the volute and outlet diffuser involve some further deceleration it is convenient to group the whole diffusion together as the change of state from point 2 to point 3. As the stagnation enthalpy in steady adiabatic flow without shaft work is constant, $h_{02} = h_{03}$ or $h_2 + \frac{1}{2}c_2^2 = h_3 + \frac{1}{2}c_3^2$. The process 2 to 3 in Figure 7.3 is drawn as irreversible, there being a loss in stagnation pressure $p_{02} - p_{03}$ during the process.

Inlet velocity limitations

The inlet eye is an important critical region in centrifugal pumps and compressors requiring careful consideration at the design stage. If the relative velocity of the inlet flow is too large in pumps, cavitation may result with consequent blade erosion or even reduced performance. In compressors large relative velocities can cause an increase in the impeller total pressure losses. In high-speed centrifugal compressors Mach number effects may become important with high relative velocities in the inlet. By suitable sizing of the eye the maximum relative velocity, or some related parameter, can be minimised to give the optimum inlet flow conditions. As an illustration the following analysis shows a simple optimisation procedure for a low-speed compressor based upon incompressible flow theory. p201

For the inlet geometry shown in Figure 7.1, the absolute eye velocity is assumed to be uniform and axial. The inlet relative velocity is $w_1 = (c_{x1}^2 + U^2)^{1/2}$ which is clearly a maximum at the inducer tip radius r_{s1}. The volume flow rate is

$$Q = c_{x1} A_1 = \pi (r_{s1}^2 - r_{h1}^2)(w_{s1}^2 - \Omega^2 r_{s1}^2)^{1/2}. \tag{7.5}$$

It is worth noticing that with both Q and r_{h1} fixed:

(i) if r_{s1} is made large then, from continuity, the axial velocity is low but the blade speed is high,
(ii) if r_{s1} is made small the blade speed is small but the axial velocity is high.

Both extremes produce large relative velocities and there must exist some optimum radius r_{s1} for which the relative velocity is a minimum.

For maximum volume flow, differentiate eqn. (7.5) with respect to r_{s1} (keeping w_{s1} constant) and equate to zero,

$$\frac{1}{\pi}\frac{\partial Q}{\partial r_{s1}} = 0 = 2r_{s1}(w_{s1}^2 - \Omega^2 r_{s1}^2)^{1/2} - (r_{s1}^2 - r_{h1}^2)\Omega^2 r_{s1}/(w_{s1}^2 - \Omega^2 r_{s1}^2)^{1/2}$$

After simplifying,

$$2(w_{s1}^2 - \Omega^2 r_{s1}^2) = (r_{s1}^2 - r_{h1}^2)\Omega^2,$$

$$\therefore 2c_{x1}^2 = kU_{s1}^2,$$

where $k = 1 - (r_{h1}/r_{s1})^2$ and $U_{s1} = \Omega r_{s1}$. Hence, the optimum inlet velocity coefficient is

$$\phi = c_{x1}/U_{s1} = \cot \beta_{s1} = (k/2)^{1/2}. \tag{7.6}$$

Equation (7.6) specifies the optimum conditions for the inlet velocity triangles in terms of the hub/tip radius ratio. For typical values of this ratio (i.e. $0.3 \leqslant r_{h1}/r_{s1} \leqslant 0.6$) the optimum relative flow angle at the inducer tip β_{s1} lies between 56 deg and 60 deg.

Optimum design of a pump inlet

As discussed in Chapter 1, cavitation commences in a flowing liquid when the decreasing local static pressure becomes approximately equal to the vapour pressure, p_v. To be more precise, it is necessary to assume that gas cavitation is negligible and that sufficient nuclei exist in the liquid to initiate vapour cavitation.

The pump considered in the following analysis is again assumed to have the flow geometry shown in Figure 7.1. Immediately upstream of the impeller blades the static pressure is $p_1 = p_{01} - \frac{1}{2}\rho c_{x1}^2$ where p_{01} is the stagnation pressure and c_{x1} is the axial velocity. In the vicinity of the impeller blades leading edges on the suction surfaces there is normally a rapid velocity increase which produces a further decrease in pressure. At cavitation inception the dynamic action of the blades causes the *local* pressure to reduce such that $p = p_v = p_1 - \sigma_b(1/2\rho w_1^2)$. The parameter σ_b which is the *blade cavitation coefficient* corresponding to the cavitation inception point, depends upon the blade shape and the flow incidence angle. For conventional pumps (see Pearsall 1972) operating normally this coefficient lies in the range $0.2 \leqslant \sigma_b \leqslant 0.4$. Thus, at cavitation inception.

$$p_1 = p_{01} - \tfrac{1}{2}\rho c_{x1}^2 = p_v + \sigma_b(\tfrac{1}{2}\rho w_1^2)$$

$$\therefore gH_s = (p_{01} - p_v)/\rho = \tfrac{1}{2}c_{x1}^2 + \sigma_b(\tfrac{1}{2}w_1^2) = \tfrac{1}{2}c_{x1}^2(1 + \sigma_b) + \tfrac{1}{2}\sigma_b U_{s1}^2$$

where H_s is the net positive suction head introduced earlier and it is implied that this is measured at the shroud radius $r = r_{s1}$.

To obtain the optimum inlet design conditions consider the suction specific speed defined as $\Omega_{ss} = \Omega Q^{1/2}/(gH_s)^{3/4}$, where $\Omega = U_{s1}/r_{s1}$ and $Q = c_{x1}A_1 = \pi k r_{s1}^2 c_{x1}$. Thus,

$$\frac{\Omega_{ss}^2}{\pi k} = \frac{U_{s1}^2 c_{x1}}{\{\tfrac{1}{2}c_{x1}^2(1 + \sigma_b) + \tfrac{1}{2}\sigma_b U_{s1}^2\}^{3/2}} = \frac{\phi}{\{\tfrac{1}{2}(1 + \sigma_b)\phi^2 + \tfrac{1}{2}\sigma_b\}^{3/2}} \tag{7.7}$$

where $\phi = c_{x1}/U_{s1}$. To obtain the condition of maximum Ω_{ss}, eqn. (7.7) is differentiated with respect to ϕ and the result set equal to zero. From this procedure the optimum conditions are found:

$$\phi = \left\{\frac{\sigma_b}{2(1 + \sigma_b)}\right\}^{1/2}, \tag{7.8a}$$

$$gH_s = \tfrac{3}{2}\sigma_b(\tfrac{1}{2}U_{s1}^2), \tag{7.8b}$$

$$\Omega_{ss}^2 = \frac{2\pi k(2/3)^{1.5}}{\sigma_b(1 + \sigma_b)^{0.5}} = \frac{3.420k}{\sigma_b(1 + \sigma_b)^{0.5}}. \tag{7.8c}$$

EXAMPLE 7.1. The inlet of a centrifugal pump of the type shown in Figure 7.1 is to be designed for optimum conditions when the flow rate of water is 25 dm³/s and the impeller rotational speed is 1450 rev/min. The maximum suction specific speed $\Omega_{ss} = 3.0$ (rad) and the inlet eye radius ratio is to be 0.3. Determine

(i) the blade cavitation coefficient,
(ii) the shroud diameter at the eye,
(iii) the eye axial velocity, and
(iv) the NPSH.

Solution. (i) From eqn. (7.8c),

$$\sigma_b^2(1 + \sigma_b) = (3.42\,k)^2/\Omega_{ss}^4 = 0.1196$$

with $k = 1 - (r_{h1}/r_{s1})^2 = 1 - 0.3^2 = 0.91$. Solving iteratively (e.g. using the Newton–Raphson approximation), $\sigma_b = 0.3030$.

(ii) As $Q = \pi k r_{s1}^2 c_{x1}$ and $c_{x1} = \phi\Omega r_{s1}$ then $r_{s1}^3 = Q/(\pi k\Omega\phi)$ and $\Omega = 1450\pi/30 = 151.84$ rad/s.

From eqn. (7.8a), $\phi = \{0.303/(2 \times 1.303)\}^{0.5} = 0.3410$,

$$\therefore r_{s1}^3 = 0.025/(\pi \times 0.91 \times 151.84 \times 0.341) = 1.689 \times 10^{-4},$$

$$\therefore r_{s1} = 0.05528 \text{ m}.$$

The required diameter of the eye is 110.6 mm.

(iii) $c_{x1} = \phi \Omega r_{s1} = 0.341 \times 151.84 \times 0.05528 = 2.862$ m/s.

(iv) From eqn. (7.8b),

$$H_s = \frac{0.75 \sigma_b c_{x1}^2}{g \phi^2} = \frac{0.75 \times 0.303 \times 2.862^2}{9.81 \times 0.341^2} = 1.632 \text{ m}.$$

Optimum design of a centrifugal compressor inlet

To obtain high efficiencies from high pressure ratio compressors it is necessary to limit the relative Mach number at the eye.

The flow area at the eye can be written as

$$A_1 = \pi r_{s1}^2 k, \quad \text{where } k = 1 - (r_{h1}/r_{s1})^2.$$

Hence $A_1 = \pi k U_{s1}^2/\Omega^2$ (7.9)

with $U_{s1} = \Omega r_{s1}$.

With uniform axial velocity the continuity equation is $\dot{m} = \rho_1 A_1 c_{x1}$.

Noting from the inlet velocity diagram (Figure 7.1) that $c_{x1} = w_{s1} \cos \beta_{s1}$ and $U_{s1} = w_{s1} \sin \beta_{s1}$, then, using eqn. (7.9),

$$\frac{\dot{m}\Omega^2}{\rho_1 k \pi} = w_{s1}^3 \sin^2 \beta_{s1} \cos \beta_{s1}. \tag{7.10}$$

For a perfect gas it is most convenient to express the static density ρ_1 in terms of the stagnation temperature T_{01} and stagnation pressure p_{01} because these parameters are usually constant at entry to the compressor. Now,

$$\frac{\rho}{\rho_0} = \frac{p}{p_0}\frac{T_0}{T}.$$

With $C_p T_0 = C_p T + \frac{1}{2}c^2$ and $C_p = \gamma R/(\gamma - 1)$

then $\dfrac{T_0}{T} = 1 + \dfrac{\gamma - 1}{2}M^2 = \dfrac{a_0^2}{a^2}$

where the Mach number, $M = c/(\gamma RT)^{1/2} = c/a$, a_0 and a being the stagnation and local (static) speeds of sound. For isentropic flow,

$$\frac{p}{p_0} = \left(\frac{T}{T_0}\right)^{\gamma/(\gamma-1)}.$$

Thus,

$$\frac{\rho_1}{\rho_0} = \left(\frac{T_1}{T_0}\right)^{1-\gamma/(\gamma-1)} = \left(1 + \frac{\gamma-1}{2}M_1^2\right)^{-1/(\gamma-1)}$$

where

$$\rho_0 = p_0/(RT_0).$$

The absolute Mach number M_1 and the relative Mach number M_{r1} are defined as

$$M_1 = c_{x1}/a_1 = M_{r1} \cos \beta_{s1} \quad \text{and} \quad w_{s1} = M_{r1}a_1.$$

Using these relations together with eqn. (7.10)

$$\frac{\dot{m}\Omega^2 RT_{01}}{k\pi p_{01}} = \frac{M_{r1}^3 a_1^3}{\left[1 + \frac{1}{2}(\gamma - 1)M_1^2\right]^{1/(\gamma-1)}} \sin^2 \beta_{s1} \cos \beta_{s1}$$

Since $a_{01}/a_1 = \left[1 + \frac{1}{2}(\gamma - 1)M_1^2\right]^{1/2}$ and $a_{01} = (\gamma RT_{01})^{1/2}$ the above equation is rearranged to give

$$\frac{\dot{m}\Omega^2}{\pi k \gamma p_{01}(\gamma RT_{01})^{1/2}} = \frac{M_{r1}^3 \sin^2 \beta_{s1} \cos \beta_{s1}}{\left[1 + \frac{1}{2}(\gamma - 1)M_{r1}^2 \cos^2 \beta_{s1}\right]^{1/(\gamma-1)+3/2}} \tag{7.11}$$

This equation is extremely useful and can be used in a number of different ways. For a particular gas and known inlet conditions one can specify values of γ, R, p_{01} and T_{01} and obtain $\dot{m}\Omega^2/k$ as a function of M_{r1} and β_{s1}. By specifying a particular value of M_{r1} as a limit, the optimum value of β_{s1} for maximum mass flow can be found. A graphical procedure is the simplest method of optimising β_{s1} as illustrated below.

Taking as an example air, with $\gamma = 1.4$, eqn. (7.11) becomes

$$f(M_{r1}) = \dot{m}\Omega^2/(\pi k \rho_{01} a_{01}^3) = \frac{M_{r1}^3 \sin^2 \beta_{s1} \cos \beta_{s1}}{\left(1 + \frac{1}{5}M_{r1}^2 \cos^2 \beta_{s1}\right)^4}. \tag{7.11a}$$

The rhs of eqn. (7.11a) is plotted in Figure 7.4 as a function of β_{s1} for $M_{r1} = 0.8$ and 0.9. These curves are a maximum at $\beta_{s1} = 60$ deg (approximately).

Shepherd (1956) considered a more general approach to the design of the compressor inlet which included the effect of a free-vortex prewhirl or prerotation. The effect of prewhirl on the mass flow function is easily determined as follows. From the velocity triangles in Figure 7.5,

$$c_1 = c_x/\cos \alpha_1 = w_1 \cos \beta_1/\cos \alpha_1,$$

$$M_1 = \frac{c_1}{a_1} = \frac{w_1 \cos \beta_1}{a_1 \cos \alpha_1} = M_{r1}\left(\frac{\cos \beta_1}{\cos \alpha_1}\right).$$

Also,

$$U_1 = w_1 \sin \beta_1 + c_1 \sin \alpha_1 = w_1 \cos \beta_1(\tan \beta_1 + \tan \alpha_1),$$

and

$$\dot{m} = \rho_1 A_1 c_{x1},$$

$$\therefore \dot{m} = \frac{\pi k}{\Omega^2} \rho_1 U_1^2 w_1 \cos \beta_1 = \left(\frac{\pi k \rho_1}{\Omega^2}\right) w_1^3 \cos^3 \beta_1(\tan \beta_1 + \tan \alpha_1)^2. \tag{7.11b}$$

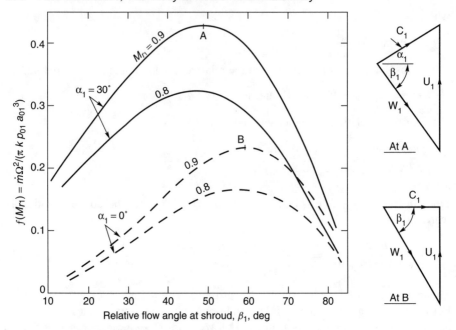

FIG. 7.4. Variation of mass flow function for the inducer of a centrifugal compressor with and without guide vanes ($\gamma = 1.4$). For comparison both velocity triangles are drawn to scale for $M_{r1} = 0.9$ the peak values or curves.

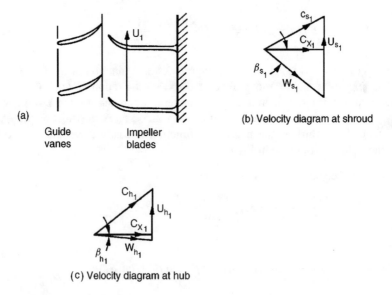

(a)

Guide
vanes

Impeller
blades

(b) Velocity diagram at shroud

(c) Velocity diagram at hub

FIG. 7.5. Effect of free-vortex prewhirl vanes upon relative velocity at impeller inlet.

Thus, using the relations developed earlier for T_{01}/T_1, p_{01}/p_1 and ρ_{01}/ρ_1, we obtain

$$f(M_{r1}) = \frac{\dot{m}\Omega^2}{\pi k \rho_{01} a_{01}^3} = \frac{M_{r1}^3 \cos^3 \beta_1 (\tan \beta_1 + \tan \alpha_1)^2}{\left(1 + \dfrac{\gamma - 1}{2} M_{r1}^2 \cos^2 \beta_1 / \cos^2 \alpha_1\right)^{\frac{1}{\gamma-1}+\frac{3}{2}}}. \tag{7.12}$$

Substituting $\gamma = 1.4$ for air into eqn. (7.12) we get:

$$f(M_{r1}) = \frac{\Omega^2 \dot{m}}{\pi k \rho_{01} a_{01}^3} = \frac{M_{r1}^3 \cos^3 \beta_1 (\tan \beta_1 + \tan \alpha_1)^2}{\left(1 + \frac{1}{5} M_{r1}^2 \cos^2 \beta_1 / \cos^2 \alpha_1\right)^4}. \tag{7.12a}$$

The rhs of eqn. (7.12a) is plotted in Figure 7.4 with $\alpha_{1s} = 30$ deg for $M_{r1} = 0.8$ and 0.9, showing that the peak values of $\dot{m}\Omega^2/k$ are significantly increased and occur at much lower values of β_1.

EXAMPLE 7.2. The inlet of a centrifugal compressor is fitted with free-vortex guide vanes to provide a positive prewhirl of 30 deg at the shroud. The inlet hub/shroud radius ratio is 0.4 and a requirement of the design is that the relative Mach number does not exceed 0.9. The air mass flow is 1 kg/s, the stagnation pressure and temperature are 101.3 kPa and 288 K. For air take $R = 287$ J/(kg K) and $\gamma = 1.4$.

Assuming optimum conditions at the shroud, determine:

(1) the rotational speed of the impeller;
(2) the inlet static density downstream of the guide vanes at the shroud and the axial velocity;
(3) the inducer tip diameter and velocity.

Solution. (1) From Figure 7.4, the peak value of $f(M_{r1}) = 0.4307$ at a relative flow angle $\beta_1 = 49.4$ deg. The constants needed are $a_{01} = \sqrt{(\gamma R T_{01})} = 340.2$ m/s, $\rho_{01} = p_{01}/(R T_{01}) = 1.2255$ kg/m^3 and $k = 1 - 0.4^2 = 0.84$. Thus, from eqn. (7.12a), $\Omega^2 = \pi f k \rho_{01} a_{01}^3 = 5.4843 \times 10^7$. Hence,

$$\Omega = 7405.6 \text{ rad/s} \quad \text{and} \quad N = 70718 \text{ rev/min}.$$

(2) $\rho_1 = \dfrac{\rho_{01}}{\left[1 + \frac{1}{5}(M_{r1} \cos \beta_1)^2\right]^{2.5}} = \dfrac{1.2255}{1.06973^{2.5}} = 0.98464$ kg/m^3.

The axial velocity is determined from eqn. (7.11b):

$$(w_1 \cos \beta_1)^3 = c_x^3 = \frac{\Omega^2 \dot{m}}{\pi k \rho_1 (\tan \beta_1 + \tan \alpha_1)^2} = \frac{5.4843 \times 10^7}{\pi \times 0.84 \times 0.98464 \times 3.0418},$$

$$= 6.9388 \times 10^6,$$

$$\therefore c_x = 190.73 \text{ m/s}.$$

(3) $A_1 = \dfrac{\dot{m}}{\rho_1 c_x} = \pi k r_{s1}^2,$

$$\therefore r_{s1}^2 = \frac{\dot{m}}{\pi \rho_1 c_x k} = \frac{1}{\pi \times 0.98464 \times 190.73 \times 0.84} = 2.0178 \times 10^{-3},$$

$$\therefore r_{s1} = 0.04492 \text{ m} \quad \text{and} \quad d_{s1} = 8.984 \text{ cm},$$

$$U = \Omega r_{s1} = 7405.6 \times 0.04492 = 332.7 \text{ m/s}.$$

Use of prewhirl at entry to impeller

Introducing positive prewhirl (i.e. in the direction of impeller rotation) can give a significant reduction of the inlet Mach number M_{r1} but, as seen from eqn. (2.12c), it reduces the specific work done on the gas. As will be seen later, it is necessary to increase the blade tip speed to maintain the same level of impeller pressure ratio as was obtained without prewhirl.

Prewhirl is obtained by fitting guide vanes upstream of the impeller. One arrangement for doing this is shown in Figure 7.5a. The velocity triangles, Figure 7.5b and c, show how the guide vanes reduce the relative inlet velocity. Guide vanes are designed to produce either a free-vortex or a forced-vortex velocity distribution. In Chapter 6 it was shown that for a free-vortex flow the axial velocity c_x is constant (in the ideal flow) with the tangential velocity c_θ varying inversely with the radius. It was shown by Wallace *et al.* (1975) that the use of free-vortex prewhirl vanes leads to a significant increase in incidence angle at low inducer radius ratios. The use of some forced-vortex velocity distribution does alleviate this problem. Some of the effects resulting from the adoption of various forms of forced-vortex of the type

$$c_\theta = A \left(\frac{r}{r_{s1}} \right)^n$$

have been reviewed by Whitfield and Baines (1990). Figure 7.6a shows, for a particular case in which $\alpha_{1s} = 30$ deg, $\beta_{1s} = 60$ deg and $\beta'_{1s} = 60$ deg, the effect of prewhirl on the variation of the incidence angle, $i = \beta_1 - \beta'_1$ with radius ratio, r/r_{1s}, for various whirl distributions. Figure 7.6b shows the corresponding variations of the absolute flow angle, α_1. It is apparent that a high degree of prewhirl vane twist is required for either a free-vortex design or for the quadratic ($n = 2$) design. The advantage of the quadratic design is the low variation of incidence with radius, whereas it is evident that the free-vortex design produces a wide variation of incidence. Wallace *et al.* (1975) adopted the simple untwisted blade shape ($n = 0$) which proved to be a reasonable compromise.

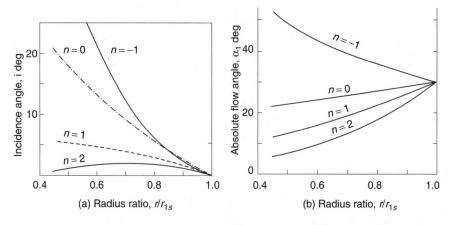

FIG. 7.6. Effect of prewhirl vanes on flow angle and incidence for $\alpha_{1s} = 30$ deg, $\beta_{1s} = 60$ deg and $\beta'_{1s} = 60$ deg. (a) Incidence angle. (b) Inducer flow angle.

Slip factor

Introduction

Even under ideal (frictionless) conditions the relative flow leaving the impeller of a compressor or pump will receive less than perfect guidance from the vanes and the flow is said to *slip*. If the impeller could be imagined as being made with an infinite number of infinitesimally thin vanes, then an ideal flow would be perfectly guided by the vanes and would leave the impeller at the vane angle. Figure 7.7 compares the relative flow angle, β_2, obtained with a finite number of vanes, with the vane angle, β_2'. A *slip factor* may be defined as

$$\sigma = \frac{c_{\theta 2}}{c_{\theta 2}'}, \tag{7.13a}$$

where $c_{\theta 2}$ is the tangential component of the absolute velocity and related to the relative flow angle β_2. The *hypothetical* tangential velocity component $c_{\theta 2}'$ is related to the vane angle β_2'. The *slip velocity* is given by $c_{\theta s} = c_{\theta 2}' - c_{\theta 2}$ so that the slip factor can be written as

$$\sigma = 1 - \frac{c_{\theta s}}{c_{\theta 2}'}. \tag{7.13b}$$

The slip factor is a vital piece of information needed by pump and compressor designers (also by designers of radial turbines) as its accurate estimation enables the correct value of the energy transfer between impeller and fluid to be made. Various attempts to determine values of slip factor have been made and numerous research papers concerned solely with this topic have been published. Wiesner (1967) has given an extensive review of the various expressions used for determining slip factors. Most of the expressions derived relate to radially vaned impellers ($\beta_2' = 0$) or to mixed flow designs, but some are given for backward swept vane (b s v) designs. All of these expressions are derived from inviscid flow theory even though the real flow is far from ideal. However, despite this lack of realism in the flow modelling, the fact remains that good results are still obtained with the various theories.

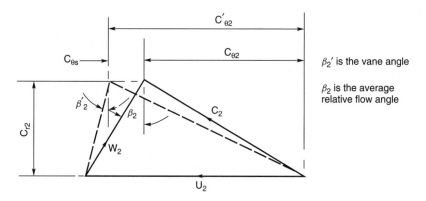

FIG. 7.7. Actual and hypothetical velocity diagrams at exit from an impeller with back swept vanes.

The relative eddy concept

Suppose that an irrotational and frictionless fluid flow is possible which passes through an impeller. If the absolute flow enters the impeller without spin, then at outlet the spin of the absolute flow *must still be zero*. The impeller itself has an angular velocity Ω so that, relative to the impeller, the fluid has an angular velocity of $-\Omega$; this is the termed the *relative eddy*. A simple explanation for the slip effect in an impeller is obtained from the idea of a relative eddy.

At outlet from the impeller the relative flow can be regarded as a through-flow on which is superimposed a relative eddy. The net effect of these two motions is that the average relative flow emerging from the impeller passages is at an angle to the vanes and in a direction opposite to the blade motion, as indicated in Figure 7.8. This is the basis of the various theories of slip.

Slip factor correlations

One of the earliest and simplest expressions for the slip factor was obtained by Stodola (1927). Referring to Figure 7.9 the *slip velocity*, $c_{\theta s} = c'_{\theta 2} - c_{\theta 2}$, is considered to be the product of the relative eddy and the radius $d/2$ of a circle which can be inscribed within the channel. Thus $c_{\theta s} = \Omega d/2$. If the number of vanes is denoted by Z then an approximate expression, $d \simeq (2\pi r_2/Z)\cos\beta'_2$ can be written if Z is not small. Since $\Omega = U_2/r_2$ then

$$c_{\theta s} = \frac{\pi U_2 \cos \beta'_2}{Z}. \tag{7.13c}$$

Now as $c'_{\theta 2} = U_2 - c_{r2}\tan\beta'_2$ the Stodola slip factor becomes

$$\sigma = \frac{c_{\theta 2}}{c'_{\theta 2}} = 1 - \frac{c_{\theta s}}{U_2 - c_{r2}\tan\beta'_2} \tag{7.14}$$

or,

$$\sigma = 1 - \frac{(\pi/Z)\cos\beta'_2}{1 - \phi_2\tan\beta'_2} \tag{7.15}$$

where $\phi_2 = c_{r2}/U_2$.

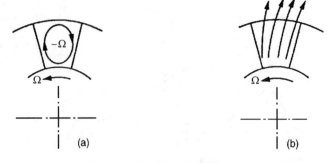

FIG. 7.8. (a) Relative eddy without any throughflow. (b) Relative flow at impeller exit (throughflow added to relative eddy).

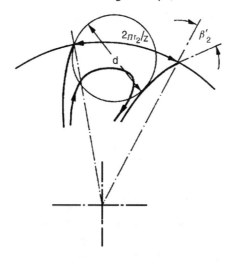

FIG. 7.9. Flow model for Stodola slip factor.

A number of more refined (mathematically exact) solutions have been evolved of which the most well known are those of Busemann, discussed at some length by Wislicenus (1947) and Stanitz (1952) mentioned earlier. The volume of mathematical work required to describe these theories is too extensive to justify inclusion here and only a brief outline of the results is presented.

Busemann's theory applies to the special case of two-dimensional vanes curved as logarithmic spirals as shown in Figure 7.10 Considering the geometry of the vane element shown it should be an easy task for the student to prove that,

$$\gamma = \tan \beta' \ln(r_2/r_1) \tag{7.17a}$$

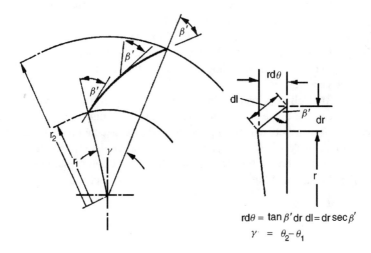

FIG. 7.10. Logarithmic spiral vane. Vane angle β' is constant for all radii.

that the ratio of vane length to equivalent blade pitch is

$$\frac{l}{s} = \frac{Z}{2\pi \cos \beta'} \ln \left(\frac{r_2}{r_1} \right) \tag{7.17b}$$

and that the equivalent pitch is

$$s = \frac{2\pi(r_2 - r_1)}{Z \ln(r_2/r_1)}.$$

The equi-angular or logarithmic spiral is the simplest form of radial vane system and has been frequently used for *pump impellers* in the past. The Busemann slip factor can be written as

$$\sigma_B = (A - B\phi_2 \tan \beta_2')/(1 - \phi_2 \tan \beta_2'), \tag{7.16}$$

where both A and B are functions of r_2/r_1, β_2' and Z. For typical pump and compressor impellers the dependence of A and B on r_2/r_1 is negligible when the equivalent l/s exceeds unity. From eqn. (7.17b) the requirement for $l/s \geq 1$, is that the radius ratio must be sufficiently large, i.e.

$$r_2/r_1 \geqslant \exp(2\pi \cos \beta'/Z). \tag{7.17c}$$

This criterion is often applied to other than logarithmic spiral vanes and then β_2' is used instead of β'. Radius ratios of typical centrifugal pump impeller vanes normally exceed the above limit. For instance, blade outlet angles of impellers are usually in the range $50 \leqslant \beta_2' \leqslant 70$ deg with between 5 and 12 vanes. Taking representative values of $\beta_2' = 60$ deg and $Z = 8$ the rhs of eqn. (7.17c) is equal to 1.48 which is not particularly large for a pump.

So long as these criteria are obeyed the value of B is constant and practically equal to unity for all conditions. Similarly, the value of A is independent of the radius ratio r_2/r_1 and depends on β_2' and Z only. Values of A given by Csanady (1960) are shown in Figure 7.11 and may also be interpreted as the value of σ_B for zero through flow ($\phi_2 = 0$).

The exact solution of Busemann makes it possible to check the validity of approximate methods of calculation such as the Stodola expression. By putting $\phi_2 = 0$ in eqns. (7.15) and (7.16) a comparison of the Stodola and Busemann slip factors at the zero through flow condition can be made. The Stodola value of slip comes close to the exact correction if the vane angle is within the range $50 \leqslant \beta_2' \leqslant 70$ deg and the number of vanes exceeds 6.

Stanitz (1952) applied relaxation methods of calculation to solve the potential flow field between the blades (blade-to-blade solution) of eight impellers with blade tip angles β_2' varying between 0 and 45 deg. His main conclusions were that the computed slip velocity $c_{\theta s}$ was independent of vane angle β_2' and depended only on blade spacing (number of blades). He also found that compressibility effects did not affect the slip factor. Stanitz's expression for slip velocity is,

$$c_{\theta s} = 0.63 U_2 \pi / Z \tag{7.18}$$

and the corresponding slip factor σ_s using eqn. (7.14) is

$$\sigma_s = 1 - \frac{0.63\pi/Z}{1 - \phi_2 \tan \beta_2'}. \tag{7.18a}$$

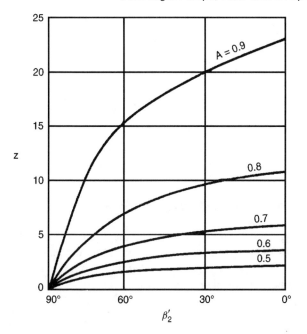

FIG. 7.11. Head correction factors for centrifugal impellers (adapted from Csanady (1960)).

For radial vaned impellers this becomes $\sigma_s = 1 - 0.63\pi/Z$ but is often written for convenience and initial approximate calculations as $\sigma_s = 1 - 2/Z$.

Ferguson (1963) has usefully compiled values of slip factor found from several theories for a number of blade angles and blade numbers and compared them with known experimental values. He found that for pumps, with β_2' between 60 deg and 70 deg, the Busemann or Stodola slip factors gave fairly good agreement with experimental results. For radial vaned impellers on the other hand, the Stanitz expression, eqn. (7.18a) agreed very well with experimental observations. For intermediate values of β_2' the Busemann slip factor gave the most consistent agreement with experiment.

Wiesner (1967) reviewed all the available methods for calculating values of slip factor and compared them with values obtained from tests. He concluded from all the material presented that Busemann's procedure was still the most generally applicable predictor for determining the basic slip factor of centripetal impellers. Wiesner obtained the following simple empirical expression for the slip velocity:

$$c_{\theta s} = \frac{U_2\sqrt{\cos\beta_2'}}{Z^{0.7}}. \tag{7.19a}$$

and the corresponding slip factor

$$\sigma_w = 1 - \frac{\sqrt{\cos\beta_2'}/Z^{0.7}}{(1 - \phi_2 \tan\beta_2')}, \tag{7.19b}$$

which, according to Wiesner, fitted the Busemann results "extremely well over the whole range of practical blade angles and number of blades".

The above equation is applicable to a limiting mean radius ratio for the impeller given by the empirical expression:

$$\varepsilon = \left(\frac{r_1}{r_2}\right)_{lim} = \exp\left(\frac{-8.16\cos\beta_2'}{Z}\right). \tag{7.19c}$$

For values of $r_1/r_2 > \varepsilon$ the slip factor is determined from the empirical expression:

$$\sigma_w' = \sigma_w\left[1 - \left(\frac{r_1/r_2 - \varepsilon}{1 - \varepsilon}\right)^3\right]. \tag{7.19d}$$

Head increase of a centrifugal pump

The actual delivered head H measured as the *head difference* between the inlet and outlet flanges of the pump and sometimes called the *manometric head*, is less than the ideal head H_i defined by eqn. (7.3b) by the amount of the internal losses. The hydraulic efficiency of the pump is defined as

$$\eta_h = \frac{H}{H_i} = \frac{gH}{U_2 c_{\theta 2}}. \tag{7.20}$$

From the velocity triangles of Figure 7.2

$$c_{\theta 2} = U_2 - c_{r2}\tan\beta_2.$$

Therefore $H = \eta_h U_2^2(1 - \phi_2\tan\beta_2)/g$ (7.20a)

where $\phi_2 = c_{r2}/U_2$ and β_2 is the actual averaged relative flow angle at impeller outlet.

With the definition of slip factor, $\sigma = c_{\theta 2}/c_{\theta 2}'$, H can, more usefully, be directly related to the impeller vane outlet angle, as

$$H = \eta_h \sigma U_2^2(1 - \phi_2\tan\beta_2')/g. \tag{7.20b}$$

In general, centrifugal pump impellers have between five and twelve vanes inclined backwards to the direction of rotation, as suggested in Figure 7.2, with a vane tip angle β_2' of between 50 and 70 deg. A knowledge of blade number, β_2' and ϕ_2 (usually small and of the order 0.1) generally enables σ to be found using the Busemann formula. The effect of slip, it should be noted, causes the relative flow angle β_2 to become larger than the vane tip angle β_2'.

EXAMPLE 7.3 A centrifugal pump delivers 0.1 m³/s of water at a rotational speed of 1200 rev/min. The impeller has seven vanes which lean backwards to the direction of rotation such that the vane tip angle β_2' is 50 deg. The impeller has an external diameter of 0.4 m, an internal diameter of 0.2 m and an axial width of 31.7 mm. Assuming that the diffuser efficiency is 51.5%, that the impeller head losses are 10% of the ideal head rise and that the diffuser exit is 0.15 m in diameter, estimate the slip factor, the manometric head and the hydraulic efficiency.

Solution. Equation (7.16) is used for estimating the slip factor. Since $\exp(2\pi \cos \beta_2' /Z) = \exp(2\pi \times 0.643/7) = 1.78$, is less than $r_2/r_1 = 2$, then $B = 1$ and $A \simeq 0.77$, obtained by replotting the values of A given in Figure 7.11 for $\beta_2' = 50$ deg and interpolating.

The vane tip speed, $U_2 = \pi N D_2/60 = \pi \times 1200 \times 0.4/60 = 25.13$ m/s.

The radial velocity, $c_{r2} = Q/(\pi D_2 b_2) = 0.1/(\pi \times 0.4 \times 0.0317)$

$$= 2.51 \text{ m/s}.$$

Hence the Busemann slip factor is

$$\sigma_B = (0.77 - 0.1 \times 1.192)/(1 - 0.1 \times 1.192) = 0.739.$$

Hydraulic losses occur in the impeller and in the diffuser. The kinetic energy leaving the diffuser is not normally recovered and must contribute to the total loss, H_L. From inspection of eqn. (2.45b), the loss in head in the diffuser is $(1 - \eta_D)(c_2^2 - c_3^2)/(2g)$. The head loss in the impeller is $0.1 \times U_2 c_{\theta2}/g$ and the exit head loss is $c_3^2/(2g)$. Summing the losses,

$$H_L = 0.485(c_2^2 - c_3^2)/(2g) + 0.1 \times U_2 c_{\theta2}/g + c_3^2/(2g).$$

Determining the velocities and heads needed,

$$c_{\theta2} = \sigma_B U_2(1 - \phi_2 \tan \beta_2') = 0.739 \times 25.13 \times 0.881 = 16.35 \text{ m/s}.$$

$$H_i = U_2 c_{\theta2}/g = 25.13 \times 16.35/9.81 = 41.8 \text{ m}.$$

$$c_2^2/(2g) = (16.35^2 + 2.51^2)/19.62 = 13.96 \text{ m}.$$

$$c_3 = 4Q/(\pi d^2) = 0.4/(\pi \times 0.15^2) = 5.65 \text{ m/s}.$$

Therefore $c_3^2/(2g) = 1.63 \text{ m}.$

Therefore $H_L = 4.18 + 0.485(13.96 - 1.63) + 1.63 = 11.8 \text{ m}.$

The manometric head is

$$H = H_i - H_L = 41.8 - 11.8 = 30.0 \text{ m}$$

and the hydraulic efficiency

$$\eta_h = H/H_i = 71.7\%.$$

Performance of centrifugal compressors

Determining the pressure ratio

Consider a centrifugal compressor having zero inlet swirl, compressing a perfect gas. With the usual notation the energy transfer is

$$\Delta W = \dot{W}_c/\dot{m} = h_{02} - h_{01} = U_2 c_{\theta2}.$$

The overall or total-to-total efficiency η_c is

$$\eta_c = \frac{h_{03ss} - h_{01}}{h_{03} - h_{01}} = \frac{C_p T_{01}(T_{03ss}/T_{01} - 1)}{h_{02} - h_{01}}$$
$$= C_p T_{01}(T_{03ss}/T_{01} - 1)/(U_2 c_{\theta 2}). \tag{7.21}$$

Now the overall pressure ratio is

$$\frac{p_{03}}{p_{01}} = \left(\frac{T_{03ss}}{T_{01}}\right)^{\gamma/(\gamma - 1)}. \tag{7.22}$$

Substituting eqn. (7.21) into eqn. (7.22) and noting that $C_p T_{01} = \gamma R T_{01}/(\gamma - 1) = a_{01}^2/(\gamma - 1)$, the pressure ratio becomes

$$\frac{p_{03}}{p_{01}} = \left[1 + \frac{(\gamma - 1)\eta_c U_2 c_{r2} \tan \alpha_2}{a_{01}^2}\right]^{\gamma/(\gamma - 1)} \tag{7.23}$$

From the velocity triangle at impeller outlet (Figure 7.1)

$$\phi_2 = c_{r2}/U_2 = (\tan \alpha_2 + \tan \beta_2)^{-1}$$

and, therefore,

$$\frac{p_{03}}{p_{01}} = \left[1 + \frac{(\gamma - 1)\eta_c U_2^2 \tan \alpha_2}{a_{01}^2(\tan \alpha_2 + \tan \beta_2)}\right]^{\gamma/(\gamma - 1)}. \tag{7.24a}$$

This formulation is useful if the flow angles can be specified. Alternatively, and more usefully, as $c_{\theta 2} = \sigma c'_{\theta 2} = \sigma(U_2 - c_{r2} \tan \beta'_2)$, then

$$\frac{p_{03}}{p_{01}} = [1 + (\gamma - 1)\eta_c \sigma(1 - \phi_2 \tan \beta'_2)M_u^2]^{\gamma/(\gamma - 1)} \tag{7.24b}$$

where $M_u = U_2/a_{01}$, is now defined as a blade Mach number.

It is of interest to calculate the variation of the pressure ratio of a radially vaned ($\beta'_2 = 0$) centrifugal air compressor to show the influence of blade speed and efficiency on the performance. With $\gamma = 1.4$ and $\sigma = 0.9$ (i.e. using the Stanitz slip factor, $\sigma = 1 - 1.98/Z$ and assuming $Z = 20$, the results evaluated are shown in Figure 7.12. It is clear that both the efficiency and the blade speed have a strong effect on the pressure ratio. In the 1970s the limit on blade speed due centrifugal stress was about 500 m/s and efficiencies seldom exceeded 80 per cent giving, with a slip factor of 0.9, radial vanes and an inlet temperature of 288 K, a pressure ratio just above 5. In recent years significant improvements in the performance of centrifugal compressors have been obtained, brought about by the development of computer-aided design and analysis techniques. According to Whitfield and Baines (1990) the techniques employed consist of "a judicious mix of empirical correlations and detailed modelling of the flow physics". It is possible to use these computer packages and arrive at a design solution without any real appreciation of the flow phenomena involved. In *all* compressors the basic flow process is one of diffusion; boundary layers are prone to separate and the flow is extremely complex. With separated wakes in the flow, unsteady flow downstream of the impeller can occur. It must

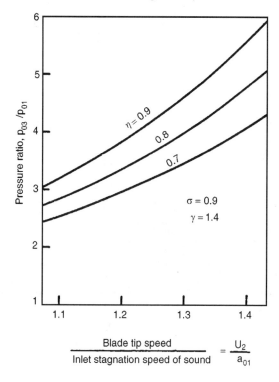

FIG. 7.12. Variation of pressure ratio with blade speed for a radial-bladed compressor ($\beta_2' = 0$) at various values of efficiency.

be stressed that a broad understanding of the flow processes within a centrifugal compressor is still a vital requirement for the more advanced student and for the further progress of new design methods.

A characteristic of all high performance compressors is that as the design pressure ratio has increased, so the range of mass flow between surge and choking has diminished. In the case of the centrifugal compressor, choking can occur when the Mach number entering the diffuser passages is just in excess of unity. This is a severe problem which is aggravated by shock-induced separation of the boundary layers on the vanes which worsens the problem of flow blockage.

Effect of backswept vanes

Came (1978) and Whitfield and Baines (1990) have commented upon the trend towards the use of higher pressure ratios from single-stage compressors leading to more highly stressed impellers. The increasing use of back swept vanes and higher blade tip speeds result in higher direct stress in the impeller and bending stress in the non-radial vanes. However, new methods of computing the stresses in impellers are being implemented (Calvert and Swinhoe 1977), capable of determining both the direct and the bending stresses caused by impeller rotation.

The effect of using back swept impeller vanes on the pressure ratio is shown in Figure 7.13 for a range of blade Mach number. It is evident that the use of back

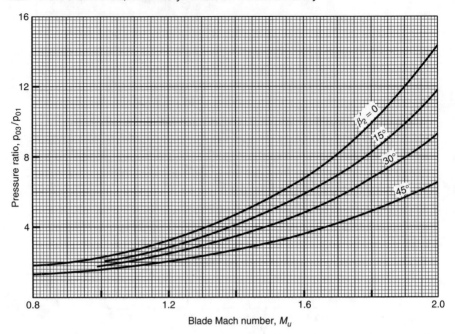

Fɪɢ. 7.13. Variation of pressure ratio vs blade Mach number of a centrifugal compressor for selected back sweep angles ($\gamma = 1.4$, $\eta_c = 0.8$, $\sigma = 0.9$, $\phi_2 = 0.375$).

sweep of the vanes at a given blade speed causes a loss in pressure ratio. In order to maintain a given pressure ratio it would be necessary to increase the design speed which, it has been noted already, increases the blade stresses.

With high blade tip speeds the absolute flow leaving the impeller may have a Mach number well in excess of unity. As this Mach number can be related to the Mach number at entry to the diffuser vanes, it is of some advantage to be able to calculate the former.

Assuming a perfect gas the Mach number at impeller exit M_2 can be written as

$$M_2^2 = \frac{c_2^2}{a_2^2} = \frac{c_2^2}{T_{01}} \cdot \frac{T_{01}}{T_2} \cdot \frac{T_2}{a_2^2} = \frac{c_2^2}{a_{01}^2} \frac{T_{01}}{T_2}, \tag{7.25}$$

since $a_{01}^2 = \gamma R T_{01}$ and $a_2^2 = \gamma R T_2$.

Referring to the outlet velocity triangle, Figure (7.7)

$$c_2^2 = c_{r2}^2 + c_{\theta 2}^2 = c_{r2}^2 + (\sigma c'_{\theta 2})^2,$$

where

$$c'_{\theta 2} = U_2 - c_{r2} \tan \beta'_2,$$

$$\left(\frac{c_2}{U_2} \right)^2 = \phi_2^2 + \sigma^2 (1 - \phi_2 \tan \beta'_2)^2. \tag{7.26}$$

From eqn. (7.2), assuming that rothalpy remains essentially constant,

$$h_2 - h_1 = \frac{1}{2}(U_2^2 - U_1^2) + \frac{1}{2}(w_1^2 - w_2^2),$$

$$\therefore h_2 = (h_1 + \frac{1}{2}w_1^2 - \frac{1}{2}U_1^2) + \frac{1}{2}(U_2^2 - w_2^2) = h_{01} + \frac{1}{2}(U_2^2 - w_2^2)$$

hence,

$$\frac{T_2}{T_{01}} = 1 + \frac{(U_2^2 - w_2^2)}{a_{01}^2/(\gamma - 1)} = 1 + (\gamma - 1)M_u^2\left(1 - \frac{w_2^2}{U_2^2}\right), \quad (7.27)$$

since $h_{01} = C_p T_{01} = a_{01}^2/(\gamma - 1)$.

From the exit velocity triangle, Figure 7.7,

$$w_2^2 = c_{r2}^2 + (U_2 - c_{\theta 2})^2 = c_{r2}^2 + (U_2 - \sigma c_{\theta 2}')^2$$

$$= c_{r2}^2 + [U_2 - \sigma(U_2 - c_{r2}\tan\beta_2')]^2,$$

$$1 - \left(\frac{w_2}{U_2}\right)^2 = 1 - \phi_2^2 - [1 - \sigma(1 - \phi_2\tan\beta_2')]^2. \quad (7.28)$$

Substituting eqns. (7.26), (7.27) and (7.28) into eqn. (7.25), we get:

$$M_2^2 = \frac{M_u^2\left[\sigma^2\left(1 - \phi_2\tan\beta_2'\right)^2 + \phi_2^2\right]}{1 + \frac{1}{2}(\gamma - 1)M_u^2\{1 - \phi_2^2\left[1 - \sigma(1 - \phi_2\tan\beta_2')\right]^2\}}. \quad (7.29)$$

Although eqn. (7.29) at first sight looks complicated it reduces into an easily managed form when constant values are inserted. Assuming the same values used previously, i.e. $\gamma = 1.4$, $\sigma = 0.9$, $\phi_2 = 0.375$ and $\beta_2' = 0, 15, 30$ and 45 deg, the solution for M_2 can be written as

$$M_2 = \frac{AM_u}{\sqrt{(1 + BM_u^2)}}, \quad (7.29a)$$

where the constants A and B are as shown in Table 7.1, and, from which the curves of M_2 against M_u in Figure 7.14 have been calculated.

According to Whitfield and Baines (1990) the two most important aerodynamic parameters at impeller exit are the magnitude and direction of the absolute Mach number M_2. If M_2 has too high a value, the process of efficient flow deceleration within the diffuser itself is made more difficult leading to high friction losses as well as the increased possibility of shock losses. If the flow angle α_2 is large the flow path in the vaneless diffuser will be excessively long resulting in high friction losses

TABLE 7.1. Constants used to evaluate M_2

Constant	β_2' (degrees)			
	0	15	30	45
A	0.975	0.8922	0.7986	0.676
B	0.2	0.199	0.1975	0.1946

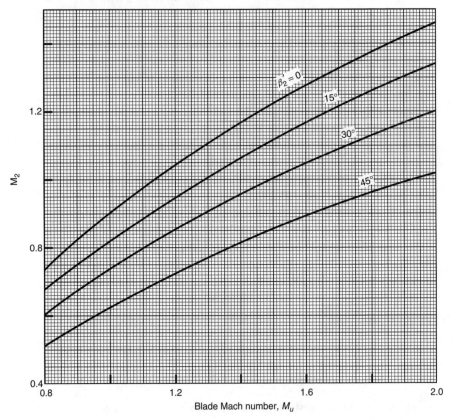

FIG. 7.14. Variation of impeller exit Mach number vs blade Mach number of a centrifugal compressor for selected back sweep angles ($\gamma = 1.4$, $\sigma = 0.9$, $\phi_2 = 0.375$).

and possible stall and flow instability. Several researchers (e.g. Rodgers and Sapiro 1972) have shown that the optimum flow angle is in the range $60° < \alpha_2 < 70°$.

Backswept vanes give a reduction of the impeller discharge Mach number, M_2, at any given tip speed. A designer making the change from radial vanes to back swept vanes will incur a reduction in the design pressure ratio if the vane tip speed remains the same. To recover the original pressure ratio the designer is forced to increase the blade tip speed which increases the discharge Mach number. Fortunately, it turns out that this increase in M_2 is rather less than the reduction obtained by the use of backsweep.

Illustrative Exercise. Consider a centrifugal compressor design which assumes the previous design data (Figures 7.13 and 7.14), together with $\beta_2' = 0°$ and a blade speed such that $M_u = 1.6$. From Figure 7.13 the pressure ratio at this point is 6.9 and, from Figure 7.14, the value of $M_2 = 1.27$. Choosing an impeller with a back sweep angle, $\beta_2' = 30°$, the pressure ratio is 5.0 from Figure 7.13 at the same value of M_u. So, to restore the original pressure ratio of 6.9 the blade Mach number must be increased to $M_u = 1.8$. At this new condition a value of $M_2 = 1.125$ is obtained from Figure 7.14, a significant reduction from the original value.

The absolute flow angle can now be found from the exit velocity triangle, Figure 7.7:

$$\tan\alpha_2 = \frac{c_{\theta 2}}{c_{r2}} = \frac{\sigma(U_2 - c_{r2}\tan\beta_2')}{c_{r2}} = \sigma\left(\frac{1}{\phi_2} - \tan\beta_2'\right).$$

With $\sigma = 0.9$, $\phi_2 = 0.375$ then, for $\beta_2' = 0°$, the value of $\alpha_2 = 67.38°$. Similarly, with $\beta_2' = 30°$, the value of $\alpha_2 = 62°$, i.e. both values of α_2 are within the prescribed range.

Kinetic energy leaving the impeller

According to Van den Braembussche (1985) "the kinetic energy available at the diffuser inlet easily amounts to more than 50 per cent of the total energy added by the impeller". From the foregoing analysis it is not so difficult to determine whether or not this statement is true. If the magnitude of the kinetic energy is so large then the importance of efficiently converting this energy into pressure energy can be appreciated. The conversion of the kinetic energy to pressure energy is considered in the following section on diffusers.

The fraction of the kinetic energy at impeller exit to the specific work input is

$$r_E = \tfrac{1}{2}c_2^2/\Delta W, \tag{7.30}$$

where

$$\Delta W = \sigma U_2^2(1 - \phi_2\tan\beta_2') \text{ and } \left(\frac{c_2}{U_2}\right)^2 = \left(\frac{c_2}{a_2}\cdot\frac{a_2}{a_{01}}\cdot\frac{a_{01}}{U_2}\right)^2$$

$$= \left(\frac{M_2}{M_u}\right)^2\left(\frac{a_2}{a_{02}}\cdot\frac{a_{02}}{a_{01}}\right)^2. \tag{7.31}$$

Define the total-to-total efficiency of the impeller as

$$\eta_1 = \frac{h_{02ss} - h_{01}}{h_{02} - h_{01}} = \frac{h_{01}\left(\dfrac{T_{02ss}}{T_{01}} - 1\right)}{h_{02} - h_{01}} = \frac{h_{01}\left(p_R^{(\gamma-1)/\gamma} - 1\right)}{\Delta W}.$$

where p_r is the total-to-total pressure ratio across the impeller, then

$$\left(\frac{a_{02}}{a_{01}}\right)^2 = \frac{T_{02}}{T_{01}} = 1 + \frac{\Delta T_0}{T_{01}} = 1 + \frac{\Delta W}{C_p T_{01}} = 1 + \frac{1}{\eta_1}(p_r^{(\gamma-1)/\gamma} - 1), \tag{7.32}$$

$$\left(\frac{a_{02}}{a_2}\right)^2 = \frac{T_{02}}{T_2} = 1 + \tfrac{1}{2}(\gamma - 1)M_2^2. \tag{7.33}$$

Substituting eqns. (7.30), (7.31) and (7.32) into eqn. (7.30) we get

$$r_E = \frac{c_2^2/U_2^2}{2\sigma(1 - \phi_2\tan\beta_2')} = \frac{(M_2/M_u)^2\left[1 + \dfrac{1}{\eta_1}(p_r^{(\gamma-1)/\gamma} - 1)\right]}{2\sigma(1 - \phi_2\tan\beta_2')[1 + \tfrac{1}{2}(\gamma - 1)M_2^2]}. \tag{7.34}$$

Exercise. Determine r_E assuming that $\beta'_2 = 0$, $\sigma = 0.9$, $\eta_I = 0.8$, $p_r = 4$ and $\gamma = 1.4$.

N B. It is very convenient to assume that Figures 7.13 and 7.14 can be used to derive the values of the Mach numbers M_u and M_2. From Figure 7.13 we get $M_u = 1.3$ and from Figure 7.14, $M_2 = 1.096$. Substituting values into eqn. (7.34),

$$r_E = \frac{1}{2 \times 0.9} \left(\frac{1.096}{1.3}\right)^2 \frac{\left[1 + \frac{1}{0.8}(4^{1/35} - 1)\right]}{1 + \frac{1}{5} \times 1.096^2} = 0.512.$$

Calculations of r_E at other pressure ratios and sweepback angles show that its value remains about 0.51 provided that σ and η_1 do not change.

EXAMPLE 7.4. Air at a stagnation temperature of 22°C enters the impeller of a centrifugal compressor in the axial direction. The rotor, which has 17 radial vanes, rotates at 15,000 rev/min. The stagnation pressure ratio between diffuser outlet and impeller inlet is 4.2 and the overall efficiency (total-to-total) is 83%. Determine the impeller tip radius and power required to drive the compressor when the mass flow rate is 2 kg/s and the mechanical efficiency is 97%. Given that the air density at impeller outlet is 2 kg/m^3 and the axial width at entrance to the diffuser is 11 mm, determine the absolute Mach number at that point. Assume that the slip factor $\sigma_s = 1 - 2/Z$, where Z is the number of vanes.

(For air take $\gamma = 1.4$ and $R = 0.287$ kJ/(kg K).)

Solution. From eqn. (7.1a) the specific work is

$$\Delta W = h_{02} - h_{01} = U_2 c_{\theta 2} = \sigma_s U_2^2$$

since $c_{\theta 1} = 0$. Combining eqns. (7.20) and (7.21) with the above and rearranging gives

$$U_2^2 = \frac{C_p T_{01}(r^{(\gamma-1)/\gamma} - 1)}{\sigma_s \eta_c}$$

where $r = p_{03}/p_{01} = 4.2$; $C_p = \gamma R/(\gamma - 1) = 1.005$ kJ/kg k; $\sigma_s = 1 - 2/17 = 0.8824$.

Therefore $U_2^2 = \dfrac{1005 \times 295(4.2^{0.286} - 1)}{0.8824 \times 0.83} = 20.5 \times 10^4$.

Therefore $U_2 = 452$ m/s.
The rotational speed is

$$\Omega = 15,000 \times 2\pi/60 = 1570 \text{ rad/s}.$$

Thus, the impeller tip radius is

$$r_t = U_2/\Omega = 452/1570 = 0.288 \text{ m}.$$

The actual shaft power is obtained from

$$\dot{W}_{act} = \dot{W}_c/\eta_m = \dot{m}\Delta W/\eta_m = 2 \times 0.8824 \times 452^2/0.97$$
$$= 373 \text{ kW}.$$

Although the absolute Mach number at the impeller tip can be obtained almost directly from eqn. (7.28) it may be instructive to find it from

$$M_2 = \frac{c_2}{a_2} = \frac{c_2}{(\gamma R T_2)^{1/2}}$$

where $\quad c_2 = (c_{\theta 2}^2 + c_{r2}^2)^{1/2}$

$$c_{r2} = \dot{m}/(\rho_2 2\pi r_t b_2) = 2/(2 \times 2\pi \times 0.288 \times 0.011) = 50.3 \, \text{m/s}$$

$$c_{\theta 2} = \sigma_s U_2 = 400 \, \text{m/s}.$$

Therefore $\quad c_2 = \sqrt{(400^2 + 50.3^2)} = 402.5 \, \text{m/s}.$

Since $\quad h_{02} = h_{01} + \Delta W$

$$h_2 = h_{01} + \Delta W - \tfrac{1}{2}c_2^2.$$

Therefore $\quad T_2 = T_{01} + (\Delta W - \tfrac{1}{2}c_2^2)/C_p = 295 + (18.1 - 8.1)10^4/1005$

$$= 394.5 \, \text{K}.$$

Hence,

$$M_2 = \frac{402.5}{\sqrt{(402 \times 394.5)}} = 1.01.$$

The diffuser system

Centrifugal compressors and pumps are, in general, fitted with either a vaneless or a vaned diffuser to transform the kinetic energy at impeller outlet into static pressure.

Vaneless diffusers (see p 6)

The simplest concept of diffusion in a radial flow machine is one where the swirl velocity is reduced by an increase in radius (conservation of angular momentum) and the radial velocity component is controlled by the radial flow area. From continuity, since $\dot{m} = \rho A c_r = 2\pi r b \rho c_r$, where b is the width of passage, then

$$c_r = \frac{r_2 b_2 \rho_2 c_{r2}}{rb\rho}. \tag{7.30}$$

Assuming the flow is frictionless in the diffuser, the angular momentum is constant and $c_\theta = c_{\theta 2} r_2/r$. Now the tangential velocity component c_θ is usually very much larger than the radial velocity component c_r; therefore, the ratio of inlet to outlet diffuser velocities c_2/c_3 is approximately r_3/r_2. Clearly, to obtain useful reductions in velocity, vaneless diffusers must be large. This may not be a disadvantage in industrial applications where weight and size may be of secondary importance compared with the cost of a vaned diffuser. A factor in favour of vaneless diffusers is the wide operating range obtainable, vaned diffusers being more sensitive to flow variation because of incidence effects.

For a parallel-walled radial diffuser in incompressible flow, the continuity of mass flow equation requires that rc_r is constant. Assuming that rc_θ remains constant, then

(how about Compressible flow ?

the absolute flow angle $\alpha_2 = \tan^{-1}(c_\theta/c_r)$ is also constant as the fluid is diffused outwards. Under these conditions the flow path is a *logarithmic spiral*. The relationship between the change in the circumferential angle $\Delta\theta$ and the radius ratio of the flow in the diffuser can be found from consideration of an element of the flow geometry shown in Figure 7.15. For an increment in radius dr we have, $r\,d\theta = dr\tan\alpha_2$ which, upon integration, gives:

$$\Delta\theta = \theta_3 - \theta_2 = \tan\alpha_2 \ln\left(\frac{r_3}{r_2}\right). \tag{7.31}$$

Values of $\Delta\theta$ are shown in Figure 7.16 plotted against r_3/r_2 for several values of α_2. It can be readily seen that when $\alpha_2 > 70°$, rather long flow paths are implied, friction losses will be significant and the diffuser efficiency will be low.

Vaned diffusers

In the vaned diffuser the vanes are used to remove the swirl of the fluid at a higher rate than is possible by a simple increase in radius, thereby reducing the length of flow path and diameter. The vaned diffuser is advantageous where small size is important.

There is a clearance between the impeller and vane leading edges amounting to about $0.04D_2$ for pumps and between $0.1D_2$ to $0.2D_2$ for compressors. This space

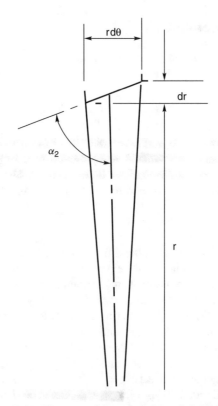

FIG. 7.15. Element of flow path in radial diffuser.

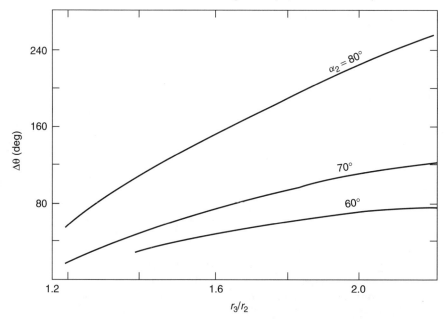

FIG. 7.16. Flow path data for paralled-walled radial diffuser (incompressible flow).

constitutes a vaneless diffuser and its functions are (i) to reduce the circumferential pressure gradient at the impeller tip, (ii) to smooth out velocity variations between the impeller tip and vanes, and (iii) to reduce the Mach number (for compressors) at entry to the vanes.

The flow follows an approximately logarithmic spiral path to the vanes after which it is constrained by the diffuser channels. For rapid diffusion the axis of the channel is straight and tangential to the spiral as shown. The passages are generally designed on the basis of simple channel theory with an equivalent angle of divergence of between 8 deg and 12 deg to control separation. (See remarks in Chapter 2 on straightwalled diffuser efficiency.)

In many applications of the centrifugal compressor, size is important and the outside diameter must be minimised. With a vaned diffuser the channel length can be crucial when considering the final size of the compressor. Clements and Artt (1988) considered this and performed a series of experiments aimed at determining the optimum diffuser channel length to width ratio, L/W. They found that, on the compressor they tested, increasing L/W. beyond 3.7 did not produce any improvement in the performance, the pressure gradient at that point having reached zero. Another significant result found by them was that the pressure gradient in the diffuser channel when $L/W > 2.13$ was no greater than that which could be obtained in a vaneless diffuser. Hence, removing completely that portion of the diffuser after this point would yield the same pressure recovery as with the full diffuser.

The number of diffuser vanes can also have a direct bearing on the efficiency and surge margin of the compressor. It is now widely accepted that surge occurs at higher flow rates when vaned diffusers are used than when a simple vaneless diffuser design is adopted. Came and Herbert (1980) quoted an example where a reduction of the number of diffuser vanes from 29 to 13 caused a significant improvement in the surge margin. Generally, it is accepted that it is better to have fewer diffuser vanes than impeller vanes in order to achieve a wide range of surge-free flow.

With several adjacent diffuser passages sharing the gas from one impeller passage, the uneven velocity distribution from that passage results in alternate diffuser passages being either starved or choked. This is an unstable situation leading to flow reversal in the passages and to surge of the compressor. When the number of diffuser passages is *less* than the number of impeller passages a more uniform total flow results.

Choking in a compressor stage

When the through flow velocity in a passage reaches the speed of sound at some cross-section, the flow *chokes*. For the stationary inlet passage this means that no further increase in mass flow is possible, either by decreasing the back pressure or by increasing the rotational speed. Now the choking behaviour of rotating passages differs from that of stationary passages, making separate analyzes for the inlet, impeller and diffuser a necessity. For each component a simple, one-dimensional approach is used assuming that all flow processes are adiabatic and that the fluid is a perfect gas.

Inlet

Choking takes place when $c^2 = a^2 = \gamma RT$. Since $h_0 = h + \frac{1}{2}c^2$, then $C_p T_0 = C_p T + \frac{1}{2}\gamma RT$ and

$$\frac{T}{T_0} = \left(1 + \frac{\gamma R}{2C_p}\right)^{-1} = \frac{2}{\gamma + 1}.$$
(7.32)

Assuming the flow in the inlet is isentropic,

$$\frac{\rho}{\rho_0} = \frac{p}{p_0}\frac{T_0}{T} = \left[1 + \frac{1}{2}(\gamma - 1)M^2\right]^{1-\gamma/(\gamma-1)}$$

and when $c = a$, $M = 1$, so that

$$\frac{\rho}{\rho_0} = \left(\frac{2}{\gamma + 1}\right)^{1/(\gamma-1)}$$
(7.33)

Substituting eqns. (7.31), (7.32) into the continuity equation, $\dot{m}/A = \rho c = \rho(\gamma RT)^{1/2}$, then

$$\frac{\dot{m}}{A} = \rho_0 a_0 \left(\frac{2}{\gamma + 1}\right)^{(\gamma+1)/2(\gamma-1)}$$
(7.34)

Thus, since ρ_0, a_0 refer to inlet stagnation conditions which remain unchanged, the mass flow rate at choking is constant.

Impeller

In the rotating impeller passages, flow conditions are referred to the factor $I = h + \frac{1}{2}(w^2 - U^2)$, which is constant according to eqn. (7.2). At the impeller inlet and for the special case $c_{\theta 1} = 0$, note that $I_1 = h_1 + \frac{1}{2}c_1^2 = h_{01}$. When choking occurs in the impeller passages it is the *relative velocity* w which equals the speed of sound at some section. Now $w^2 = a^2 = \gamma RT$ and $T_{01} = T + (\gamma RT/2C_p) - (U^2/2C_p)$, therefore

$$\frac{T}{T_{01}} = \left(\frac{2}{\gamma+1}\right)\left(1 + \frac{U^2}{2C_pT_{01}}\right). \tag{7.34}$$

Assuming isentropic flow, $\rho/\rho_{01} = (T/T_{01})^{1/(\gamma-1)}$. Using the continuity equation,

$$\frac{\dot{m}}{A} = \rho_{01}a_{01}\left(\frac{T}{T_{01}}\right)^{(\gamma+1)/2(\gamma-1)}$$

$$= \rho_{01}a_{01}\left[\frac{2}{\gamma+1}\left(1 + \frac{U^2}{2C_pT_{01}}\right)\right]^{(\gamma+1)/2(\gamma-1)}$$

$$= \rho_{01}a_{01}\left[\frac{2 + (\gamma-1)U^2/a_{01}^2}{\gamma+1}\right]^{(\gamma+1)/2(\gamma-1)} \tag{7.36}$$

If chocking occurs in the rotating passages, eqn. (7.36) indicates that the mass flow is dependent on the blade speed. As the speed of rotation is increased the compressor can accept a *greater* mass flow, unless choking occurs in some other component of the compressor. That the choking flow in an impeller can vary, depending on blade speed, may seem at first rather surprising; the above analysis gives the *reason* for the variation of the choking limit of a compressor.

Diffuser

The relation for the choking flow, eqn. (7.34) holds for the diffuser passages, it being noted that stagnation conditions now refer to the diffuser and not the inlet. Thus

$$\frac{\dot{m}}{A_2} = \rho_{02}a_{02}\left(\frac{2}{\gamma+1}\right)^{(\gamma+1)/2(\gamma-1)}. \tag{7.37}$$

Clearly, stagnation conditions at diffuser inlet are dependent on the impeller process. To find how the choking mass flow limit is affected by blade speed it is necessary to refer back to inlet stagnation conditions.

Assuming a radial bladed impeller of efficiency η_i then,

$$T_{02s} - T_{01} = \eta_i(T_{02} - T_{01}) = \eta_i\sigma U_2^2/C_p.$$

Hence

$$p_{02}/p_{01} = (T_{02s}/T_{01})^{\gamma/(\gamma-1)} = [1 + \eta_i\sigma U_2^2/C_pT_{01})]^{\gamma/(\gamma-1)}$$

and

$$T_{02}/T_{01} = [1 + \sigma U_2^2/(C_p T_{01})].$$

Now

$$\rho_{02}a_{02} = \rho_{01}a_{01}(\rho_{02}/\rho_{01})(a_{02}/a_{01})$$
$$= \rho_{01}a_{01}[p_{02}/p_{01}(T_{01}/T_{02})^{1/2}],$$

therefore,

$$\frac{\dot{m}}{A_2} = \rho_{01}a_{01}\frac{[1 + (\gamma - 1)\eta_i\sigma U_2^2/a_{01}^2]^{\gamma/(\gamma-1)}}{[1 + (\gamma - 1)\sigma U_2^2/a_{01}^2]^{1/2}}\left(\frac{2}{\gamma + 1}\right)^{(\gamma+1)/2(\gamma-1)}. \tag{7.38}$$

In this analysis it should be noted that the diffuser process has been assumed to be isentropic but the impeller has not. Eqn. (7.38) indicates that the choking mass flow can be varied by changing the impeller rotational speed.

References

Bosman, C. and Jadayel, O. C. (1996). A quantified study of rothalpy conservation in turbomachines. *Int. J. Heat and Fluid Flow.*, **17**, No. 4, 410–17.

Calvert, W. J. and Swinhoe, P. R. (1977). Impeller computer design package, part VI - the application of finite element methods to the stressing of centrifugal impellers. NGTE Internal Note, National Gas Turbine Establishment, Pyestock, Farnborough.

Came, P. (1978). The development, application and experimental evaluation of a design procedure for centrifugal compressors. *Proc Instn. Mech. Engrs.*, **192**, No. 5, 49–67.

Came, P. M. and Herbert, M. V. (1980). Design and experimental performance of some high pressure ratio centrifugal compressors. *AGARD Conference Proc.* No. 282.

Campbell, K. and Talbert, J. E. (1945). Some advantages and limitations of centrifugal and axial aircraft compressors. *S.A.E. Journal (Transactions)*, **53**, 10.

Cheshire, L. J. (1945). The design and development of centrifugal compressors for aircraft gas turbines. *Proc. Instn. Mech. Engrs. London*, **153**; reprinted by A.S.M.E. (1947), Lectures on the development of the British gas turbine jet.

Clements, W. W. and Artt, D. W. (1988). The influence of diffuser channel length to width ratio on the efficiency of a centrifugal compressor. *Proc. Instn Mech Engrs.*, **202**, No. A3, 163–9.

Csanady, G. T. (1960) Head correction factors for radial impellers. *Engineering*, **190**.

Dean, R. C., Jr. (1973). The centrifugal compressor. Creare Inc. Technical Note TN183.

Ferguson, T. B. (1963). The *Centrifugal Compressor Stage*. Butterworth.

Gas Turbine News (1989). International Gas Turbine Institute. November 1989. ASME.

Hawthorne, Sir William (1978). Aircraft propulsion from the back room. *Aeronautical J.*, March 93–108.

Hess, H. (1985). Centrifugal compressors in heat pumps and refrigerating plants. *Sulzer Tech. Rev.*, 3/1985, 27–30.

Lyman, F. A. (1993). On the conservation of rothalpy in turbomachines. *J. of Turbomachinery Trans. Am. Soc. Mech. Engrs.*, **115**, 520–6.

Moore, J., Moore, J. G. and Timmis, P. H. (1984). Performance evaluation of centrifugal compressor impellers using three-dimensional viscous flow calculations. *J. Eng. Gas Turbines Power. Trans Am. Soc. Mech. Engrs.*, **106**, 475–81.

Moult, E. S. and Pearson, H. (1951). The relative merits of centrifugal and axial compressors for aircraft gas turbines. *J. Roy. Aero. Soc.*, **55**.

Palmer, D. L. and Waterman, W. F. (1995). Design and development of an advanced two-stage centrifugal compressor. *J. Turbomachinery Trans. Am. Soc. Mech. Engrs.*, **117**, 205–12.

Pearsall, I. S. (1972). *Cavitation*. M&B Monograph ME/10. Mills & Boon.

Schorr, P. G., Welliver, A. D. and Winslow, L. J. (1971). Design and development of small, high pressure ratio, single stage centrifugal compressors. *Advanced Centrifugal Compressors*. Am. Soc. Mech. Engrs.

Shepherd, D. G. (1956). *Principles of Turbomachinery*. Macmillan.

Stanitz, J. D. (1952). Some theoretical aerodynamic investigations of impellers in radial and mixed flow centrifugal compressors. *Trans. A.S.M.E.*, **74**, 4.

Stodola, A. (1945). *Steam and Gas Turbines*. Vols. I and II. McGraw-Hill (reprinted, Peter Smith).

Rodgers, C. and Sapiro, L. (1972). Design considerations for high pressure ratio centrifugal compressors. Am. Soc. Mech. Engrs. Paper 72-GT-91.

Van den Braembussche, R. (1985). Design and optimisation of centrifugal compressors. In *Thermodynamics and Fluid Mechanics of Turbomachinery* (A. S. Üçer, P. Stow and Ch. Hirsch, eds) pp. 829–85 Martinus Nijhoff.

Wallace, F. J., Whitfield, A. and Atkey, R. C. (1975). Experimental and theoretical performance of a radial flow turbocharger compressor with inlet prewhirl. *Proc.Instn. Mech Engrs.*, **189**, 177–86.

Whitfield, A. and Baines, N. C. (1990). *Design of Radial Turbomachines*. Longman.

Wiggins, J. O. (1986). The "axi-fuge" – a novel compressor. *J. Turbomachinery Trans. Am. Soc. Mech. Engrs.*, **108**, 240–3.

Wiesner, F. J. (1967). A review of slip factors for centrifugal compressors. *J. Eng. Power. Trans Am. Soc. Mech. Engrs.*, **89**, 558–72.

Wislicenus, G. F. (1947). *Fluid Mechanics of Turbomachinery*. McGraw-Hill.

Problems

NOTE. In problems 2 to 6 assume γ and R are 1.4 and 287 J/(kg°C) respectively. In problems 1 to 4 assume the stagnation pressure and stagnation temperature at compressor entry are 101.3 kPa and 288 K respectively.)

1. A cheap radial-vaned centrifugal fan is required to provide a supply of pressurised air to a furnace. The specification requires that the fan produce a total pressure rise equivalent to 7.5 cm of water at a volume flow rate of 0.2 m³/s. The fan impeller is fabricated from 30 thin sheet metal vanes, the ratio of the passage width to circumferential pitch at impeller exit being specified as 0.5 and the ratio of the radial velocity to blade tip speed as 0.1.

Assuming that the overall isentropic efficiency of the fan is 0.75 and that the slip can be estimated from Stanitz's expression, eqn. (7.18a), determine

(1) the vane tip speed;
(2) the rotational speed and diameter of the impeller;
(3) the power required to drive the fan if the mechanical efficiency is 0.95;
(4) the specific speed.

For air assume the density is 1.2 kg/m³.

2. The air entering the impeller of a centrifugal compressor has an absolute axial velocity of 100 m/s. At rotor exit the relative air angle measured from the radial direction is 26° 36′, the radial component of velocity is 120 m/s and the tip speed of the radial vanes is 500 m/s. Determine the power required to drive the compressor when the air flow rate is 2.5 kg/s and the mechanical efficiency is 95%. If the radius ratio of the impeller eye is 0.3, calculate a suitable inlet diameter assuming the inlet flow is incompressible. Determine the overall total pressure ratio of the compressor when the total-to-total efficiency is 80%, assuming the velocity at exit from the diffuser is negligible.

3. A centrifugal compressor has an impeller tip speed of 366 m/s. Determine the absolute Mach number of the flow leaving the radial vanes of the impeller when the radial component of velocity at impeller exit is 30.5 m/s and the slip factor is 0.90. Given that the flow area at impeller exit is 0.1 m² and the total-to-total efficiency of the impeller is 90%, determine the mass flow rate.

4. The eye of a centrifugal compressor has a hub/tip radius ratio of 0.4, a maximum relative flow Mach number of 0.9 and an absolute flow which is uniform and completely axial. Determine the optimum speed of rotation for the condition of maximum mass flow given that the mass flow rate is 4.536 kg/s. Also, determine the outside diameter of the eye and the ratio of axial velocity/blade speed at the eye tip. Figure 7.4 may be used to assist the calculations.

5. An experimental centrifugal compressor is fitted with free-vortex guide vanes in order to reduce the relative air speed at inlet to the impeller. At the outer radius of the eye, air leaving the guide-vanes has a velocity of 91.5 m/s at 20 deg to the axial direction. Determine the inlet relative Mach number, assuming frictionless flow through the guide vanes, and the impeller total-to-total efficiency.

Other details of the compressor and its operating conditions are:

Impeller entry tip diameter, 0.457 m
Impeller exit tip diameter, 0.762 m
Slip factor 0.9
Radial component of velocity at impeller exit, 53.4 m/s
Rotational speed of impeller, 11 000 rev/min
Static pressure at impeller exit, 223 kPa (abs.)

6. A centrifugal compressor has an impeller with 21 vanes, which are radial at exit, a vaneless diffuser and no inlet guide vanes. At inlet the stagnation pressure is 100 kPa abs. and the stagnation temperature is 300 K.

(i) Given that the mass flow rate is 2.3 kg/s, the impeller tip speed is 500 m/s and the mechanical efficiency is 96%, determine the driving power on the shaft. Use eqn. (7.18a) for the slip factor.

(ii) Determine the total and static pressures at diffuser exit when the velocity at that position is 100 m/s. The total to total efficiency is 82%.

(iii) The reaction, which may be defined as for an axial flow compressor by eqn. (5.10b), is 0.5, the absolute flow speed at impeller entry is 150 m/s and the diffuser efficiency is 84%. Determine the total and static pressures, absolute Mach number and radial component of velocity at the impeller exit.

(iv) Determine the total-to-total efficiency for the impeller.

(v) Estimate the inlet/outlet radius ratio for the diffuser assuming the conservation of angular momentum.

(vi) Find a suitable rotational speed for the impeller given an impeller tip width of 6 mm.

7. A centrifugal pump is used to raise water against a static head of 18.0 m. The suction and delivery pipes, both 0.15 m diameter, have respectively, friction head losses amounting to 2.25 and 7.5 times the dynamic head. The impeller, which rotates at 1450 rev/min, is 0.25 m diameter with 8 vanes, radius ratio 0.45, inclined backwards at $\beta_2' = 60$ deg. The axial width of the impeller is designed so as to give constant radial velocity at all radii and is 20 mm at impeller exit. Assuming an hydraulic efficiency of 0.82 and an overall efficiency of 0.72, determine

(i) the volume flow rate;

(ii) the slip factor using Busemann's method;

(iii) the impeller vane inlet angle required for zero incidence angle;

(iv) the power required to drive the pump.

8. A centrifugal pump delivers $50 \, \mathrm{dm}^3/\mathrm{s}$ of water at an impeller speed of $1450 \, \mathrm{rev/min}$. The impeller has eight vanes inclined backwards to the direction of rotation with an angle at the tip of $\beta_2' = 60°$. The diameter of the impeller is twice the diameter of the shroud at inlet and the magnitude of the radial component of velocity at impeller exit is equal to that of the axial component of velocity at the inlet. The impeller entry is designed for the optimum flow condition to resist cavitation (see eqn. (7.8)), has a radius ratio of 0.35 and the blade shape corresponds to a well tested design giving a cavitation coefficient $\sigma_b = 0.3$.

Assuming that the hydraulic efficiency is 70 per cent and the mechanical efficiency is 90 per cent, determine:

(1) the diameter of the inlet;

(2) the net positive suction head;

(3) the impeller slip factor using Wiesner's formula;

(4) the head developed by the pump;

(5) the power input.

Also calculate values for slip factor using the equations of Stodola and Busemann, comparing the answers obtained with the result found from Wiesner's equation.

9. (a) Write down the advantages and disadvantages of using free-vortex guide vanes upstream of the impeller of a high pressure ratio centrifugal compressor. What other sorts of guide vanes can be used and how do they compare with free-vortex vanes?

(b) The inlet of a centrifugal air compressor has a shroud diameter of 0.2 m and a hub diameter of 0.105 m. Free-vortex guide vanes are fitted in the duct upstream of the impeller so that the flow on the shroud at the impeller inlet has a relative Mach number, $M_{r1} = 1.0$, an absolute flow angle of $\alpha_1 = 20°$ and a relative flow angle $\beta_1 = 55°$. At inlet the stagnation conditions are $288 \, \mathrm{K}$ and $10^5 \, \mathrm{Pa}$.

Assuming frictionless flow into the inlet, determine:

(1) the rotational speed of the impeller;

(2) the air mass flow.

(c) At exit from the radially vaned impeller, the vanes have a radius of 0.16 m and a design point slip factor of 0.9. Assuming an impeller efficiency of 0.9, determine:

(1) the shaft power input;

(2) the impeller pressure ratio.

CHAPTER 8

Radial Flow Gas Turbines

I like work; it fascinates me, I can sit and look at it for hours. (JEROME
K. JEROME, *Three Men in a Boat.*)

Introduction

The radial flow turbine has had a long history of development being first conceived
for the purpose of producing hydraulic power over 170 years ago. A French engineer,
Fourneyron, developed the first commercially successful hydraulic turbine (*c.* 1830)
and this was of the *radial-outflow* type. A *radial-inflow* type of hydraulic turbine
was built by Francis and Boyden in the U.S.A. (*c.* 1847) which gave excellent
results and was highly regarded. This type of machine is now known as the *Francis
turbine*, a simplified arrangement of it being shown in Figure 1.1. It will be observed
that the flow path followed is from the radial direction to what is substantially an
axial direction. A flow path in the reverse direction (radial-outflow), for a single
stage turbine anyway, creates several problems one of which (discussed later) is
low specific work. However, as pointed out by Shepherd (1956) radial-outflow
steam turbines comprising many stages have received considerable acceptance in
Europe. Figure 8.1 from Kearton (1951), shows diagrammatically the *Ljungström
steam turbine* which, because of the tremendous increase in specific volume· of
steam, makes the radial-outflow flow path virtually imperative. A unique feature of
the Ljungstroöm turbine is that it does not have any stationary blade rows. The two
rows of blades comprising each of the stages rotate in opposite directions so that
they can both be regarded as rotors.

The inward-flow radial (IFR) turbine covers tremendous ranges of power, rates of
mass flow and rotational speeds, from very large Francis turbines used in hydroelectric power generation and developing hundreds of megawatts down to tiny closed
cycle gas turbines for space power generation of a few kilowatts.

The IFR turbine has been, and continues to be, used extensively for powering
automotive turbocharges, aircraft auxiliary power units, expansion units in gas lique-
faction and other cryogenic systems and as a component of the small (10 kW) gas
turbines used for space power generation (Anon. 1971). It has been considered
for primary power use in automobiles and in helicopters. According to Huntsman
(1992), studies at Rolls-Royce have shown that a cooled, high efficiency IFR turbine
could offer significant improvement in performance as the gas generator turbine of a
high technology turboshaft engine. What is needed to enable this type of application
are some small improvements in current technology levels! However, designers of
this new required generation of IFR turbines face considerable problems, particularly

236

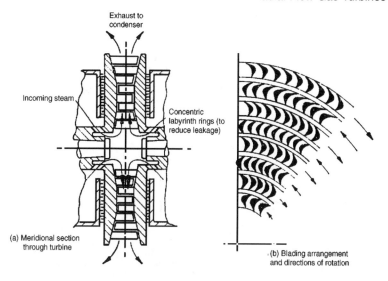

FIG. 8.1. Ljungström type outward flow radial turbine (adapted from Kearton[2]).

in the development of advanced techniques of rotor cooling or of ceramic, shock-resistant rotors.

As indicated later in this chapter, over a limited range of specific speed, IFR turbines provide an efficiency about equal to that of the best axial-flow turbines. The significant advantages offered by the IFR turbine compared with the axial-flow turbine is the greater amount of work that can be obtained per stage, the ease of manufacture and its superior ruggedness.

Types of inward flow radial turbine

In the centripetal turbine energy is transferred from the fluid to the rotor in passing from a large radius to a small radius. For the production of positive work the product of Uc_θ at entry to the rotor must be greater than Uc_θ at rotor exit (eqn. (2.12b)). This is usually arranged by imparting a large component of tangential velocity at rotor entry, using single or multiple nozzles, and allowing little or no swirl in the exit absolute flow.

Cantilever turbine

Figure 8.2a shows a *cantilever* IFR turbine where the blades are limited to the region of the rotor tip, extending from the rotor in the *axial* direction. In practice the cantilever blades are usually of the impulse type (i.e. low reaction), by which it is implied that there is little change in relative velocity at inlet and outlet of the rotor. There is no fundamental reason why the blading should not be of the reaction type. However, the resulting expansion through the rotor would require an increase in flow area. This extra flow area is extremely difficult to accommodate in a small radial distance, especially as the radius decreases through the rotor row.

Aerodynamically, the cantilever turbine is similar to an axial-impulse turbine and can even be designed by similar methods. Figure 8.2b shows the velocity triangles

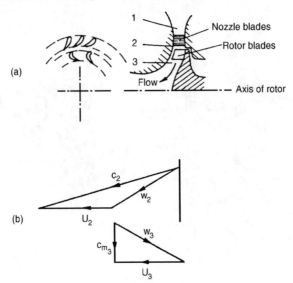

FIG. 8.2. Arrangement of cantilever turbine and velocity triangles at the design point.

at rotor inlet and outlet. The fact that the flow is radially inwards hardly alters the design procedure because the blade radius ratio r_2/r_3 is close to unity anyway.

The 90 degree IFR turbine

Because of its higher structural strength compared with the cantilever turbine, the 90 degree IFR turbine is the preferred type. Figure 8.3 shows a typical layout of a 90 degree IFR turbine; the inlet blade angle is generally made zero, a fact dictated by the material strength and often high gas temperature. The rotor vanes are subject to high stress levels caused by the centrifugal force field, together with a pulsating and often unsteady gas flow at high temperatures. Despite possible performance gains the use of non-radial (or swept) vanes is generally avoided, mainly because of the additional stresses which arise due to bending. Nevertheless, despite this difficulty, Meitner and Glassman (1983) have considered designs using sweptback vanes in assessing ways of increasing the work output of IFR turbines.

From station 2 the rotor vanes extend radially inward and turn the flow into the axial direction. The exit part of the vanes, called the *exducer*, is curved to remove most if not all of the absolute tangential component of velocity. The 90 degree IFR turbine or centripetal turbine is very similar in appearance to the centrifugal compressor of Chapter 7 but with the flow direction and blade motion reversed.

The fluid discharging from the turbine rotor may have a considerable velocity c_3 and an axial diffuser (see Chapter 2) would normally be incorporated to recover most of the kinetic energy, $\frac{1}{2}c_3^2$, which would otherwise be wasted. In hydraulic turbines (discussed in Chapter 9) a diffuser is invariably used and is called the *draught tube*.

In Figure 8.3 the velocity triangles are drawn to suggest that the inlet relative velocity, w_2, is *radially* inward, i.e. zero incidence flow, and the absolute flow

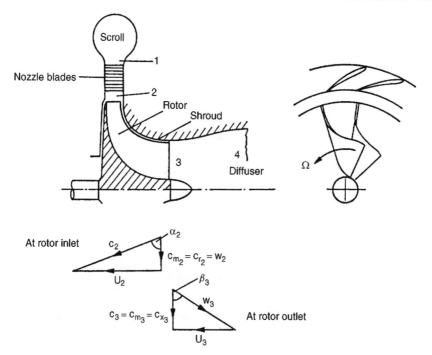

FIG. 8.3. Layout and velocity diagrams for a 90 deg inward flow radial turbine at the nominal design point.

at rotor exit, c_3, is axial. This configuration of the velocity triangles, popular with designers for many years, is called the *nominal design* condition and will be considered in some detail in the following pages. Following this the so-called *optimum efficiency design* will be explained.

Thermodynamics of the 90 deg IFR turbine

The complete adiabatic expansion process for a turbine comprising a nozzle blade row, a radial rotor followed by a diffuser corresponding to the layout of Figure 8.3, is represented by the Mollier diagram shown in Figure 8.4. In the turbine, frictional processes cause the entropy to increase in all components and these irreversibilities are implied in Figure 8.4.

Across the nozzle blades the stagnation enthalpy is assumed constant, $h_{01} = h_{02}$ and, therefore, the static enthalpy drop is,

$$h_1 - h_2 = \tfrac{1}{2}(c_2^2 - c_1^2) \tag{8.1}$$

corresponding to the static pressure change from p_1 to the lower pressure p_2. The *ideal* enthalpy change $(h_1 - h_{2s})$ is between these *same* two pressures but at constant entropy.

In Chapter 7 it was shown that the rothalpy, $I = h_{0rel} - \tfrac{1}{2}U^2$, is constant for an adiabatic irreversible flow process, relative to a rotating component. For the rotor

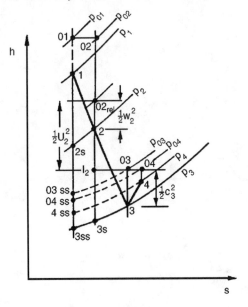

FIG. 8.4. Mollier diagram for a 90 deg inward flow radial turbine and diffuser (at the design point).

of the 90 deg IFR turbine,

$$h_{02\,rel} - \tfrac{1}{2}U_2^2 = h_{03\,rel} - \tfrac{1}{2}U_3^2$$

Thus, as $h_{0\,rel} = h + \tfrac{1}{2}w^2$,

$$h_2 - h_3 = \tfrac{1}{2}[(U_2^2 - U_2^2) - (w_2^2 - w_3^2)] \tag{8.2}$$

In this analysis the reference point 2 (Figure 8.3) is taken to be at the inlet radius r_2 of the rotor (the blade tip speed $U_2 = \Omega r_2$). This implies that the nozzle irreversibilities are lumped together with any friction losses occurring in the annular space between nozzle exit and rotor entry (usually scroll losses are included as well).

Across the diffuser the stagnation enthalpy does not change, $h_{03} = h_{04}$, but the static enthalpy *increases* as a result of the velocity diffusion. Hence,

$$h_4 - h_3 = \tfrac{1}{2}(c_3^2 - c_4^2) \tag{8.3}$$

The specific work done by the fluid on the rotor is

$$\Delta W = h_{01} - h_{03} = U_2 c_{\theta 2} - U_3 c_{\theta 3} \tag{8.4}$$

As $h_{01} = h_{02}$,

$$\Delta W = h_{02} - h_{03} = h_2 - h_3 + \tfrac{1}{2}(c_2^2 - c_3^2)$$

$$= \tfrac{1}{2}[(U_2^2 - U_3^2) - (w_2^2 - w_3^2) + (c_2^2 - c_3^2)] \tag{8.4a}$$

after substituting eqn. (8.2).

Basic design of the rotor

Each term in eqn. (8.4a) makes a contribution to the specific work done on the rotor. A significant contribution comes from the first term, namely $\frac{1}{2}(U_2^2 - U_1^2)$, and is the main reason why the inward flow turbine has such an advantage over the outward flow turbine where the contribution from this term would be negative. For the axial flow turbine, where $U_2 = U_1$, of course no contribution to the specific work is obtained from this term. For the second term in eqn. (8.4a) a positive contribution to the specific work is obtained when $w_3 > w_2$. In fact, accelerating the relative velocity through the rotor is a most useful aim of the designer as this is conducive to achieving a low loss flow. The third term in eqn. (8.4a) indicates that the absolute velocity at rotor inlet should be larger than at rotor outlet so as to increase the work input to the rotor. With these considerations in mind the general shape of the velocity diagram shown in Figure 8.3 results.

Nominal Design

The *nominal design* is defined by a relative flow of zero incidence at rotor inlet (i.e. $w_2 = c_{r2}$) and an absolute flow at rotor exit which is axial (i.e. $c_3 = c_{x3}$). Thus, from eqn. (8.4), with $c_{\theta3} = 0$ and $c_{\theta2} = U_2$, the specific work for the nominal design is simply

$$\Delta W = U_2^2. \tag{8.4b}$$

EXAMPLE 8.1. The rotor of an IFR turbine, which is designed to operate at the nominal condition, is 23.76 cm in diameter and rotates at 38 140 rev/min. At the design point the absolute flow angle at rotor entry is 72 deg. The rotor mean exit diameter is one half of the rotor diameter and the relative velocity at rotor exit is twice the relative velocity at rotor inlet.

Determine the relative contributions to the specific work of each of the three terms in eqn. (8.4a).

Solution. The blade tip speed is $U_2 = \pi N D_2/60 = \pi \times 38\,140 \times 0.2376/60 = 474.5$ m/s.

Referring to Figure 8.3, $w_2 = U_2 \cot \alpha_2 = 154.17$ m/s, and $c_2 = U_2 \sin \alpha_2 = 498.9$ m/s.

$$c_3^2 = w_3^2 - U_3^2 = (2 \times 154.17)^2 - (\tfrac{1}{2} \times 474.5)^2 = 38\,786\,\text{m}^2/\text{s}^2.$$

Hence, $(U_2^2 - U_3^2) = U_2^2(1 - 1/4) = 168\,863\,\text{m}^2/\text{s}^2$, $w_3^2 - w_2^2 = 3 \times w_2^2 = 71\,305\,\text{m}^2/\text{s}^2$ and $c_2^2 - c_3^2 = 210\,115\,\text{m}^2/\text{s}^2$. Thus, summing the values of the three terms and dividing by two, we get $\Delta W = 225\,142\,\text{m}^2/\text{s}^2$.

The fractional inputs from each of the three terms are: for the U^2 terms, 0.375; for the w^2 terms, 0.158; for the c^2 terms, 0.467.

Finally, as a numerical check, the specific work is, $\Delta W = U_2^2 = 474.5^2 = 225\,150$ m²/s² which, apart from some rounding erors, agrees with the above computations.

Spouting velocity

The term *spouting velocity* c_0 (originating from hydraulic turbine practice) is defined as that velocity which has an associated kinetic energy equal to the isentropic enthalpy drop from turbine inlet stagnation pressure p_{01} to the final exhaust pressure. The exhaust pressure here can have several interpretations depending upon whether total or static conditions are used in the related efficiency definition and upon whether or not a diffuser is included with the turbine. Thus, when *no* diffuser is used

$$\tfrac{1}{2}c_0^2 = h_{01} - h_{03ss} \tag{8.5a}$$

or,

$$\tfrac{1}{2}c_0^2 = h_{01} - h_{3ss} \tag{8.5b}$$

for the total and static cases respectively.

In an *ideal* (frictionless) radial turbine with complete recovery of the exhaust kinetic energy, and with $c_{\theta 2} = U_2$,

$$\Delta W = U_2^2 = \tfrac{1}{2}c_0^2$$

$$\therefore \quad \frac{U^2}{c_0} = 0.707$$

At the best efficiency point of actual (frictional) 90 deg IFR turbines it is found that this velocity ratio is, generally, in the range $0.68 < U_2/c_0 < 0.71$.

Nominal design point efficiency

Referring to Figure 8.4, the total-to-static efficiency in the absence of a diffuser, is defined as

$$\eta_{ts} = \frac{h_{01} - h_{03}}{h_{01} - h_{3ss}} = \frac{\Delta W}{\Delta W + \tfrac{1}{2}c_3^2 + (h_3 - h_{3s}) + (h_{3s} - h_{3ss})} \tag{8.6}$$

The passage enthalpy losses can be expressed as a fraction (ζ) of the exit kinetic energy relative to the nozzle row and the rotor, i.e.

$$h_3 - h_{3s} = \tfrac{1}{2}w_3^2 \zeta_R \tag{8.7a}$$

$$h_{3s} - h_{3ss} = \tfrac{1}{2}c_2^2 \zeta_N (T_3/T_2) \tag{8.7b}$$

for the rotor and nozzles respectively. It is noted that for a constant pressure process, $ds = dh/T$, hence the approximation,

$$h_{3s} - h_{3ss} = (h_2 - h_{2s})(T_3/T_2)$$

Substituting for the enthalpy losses in eqn. (8.6),

$$\eta_{ts} = [1 + \tfrac{1}{2}(c_3^2 + w_3^2 \zeta_R + c_2^2 \zeta_N T_3/T_2)/\Delta W]^{-1} \tag{8.8}$$

From the design point velocity triangles, Figure 8.3,

$$c_2 = U_2 \operatorname{cosec} a_2, \quad w_3 = U_3 \operatorname{cosec} \beta_3, \quad c_3 = U_3 \cot \beta_3, \quad \Delta W = U_2^2.$$

Thus substituting all these expressions in eqn. (8.8) and noting that $U_3 = U_2 r_3/r_2$, then

$$\eta_{ts} = \left[1 + \frac{1}{2} \left\{ \zeta_N \frac{T_3}{T_2} \operatorname{cosec}^2 \alpha_2 + \left(\frac{r_3}{r_2} \right)^2 (\zeta_R \operatorname{cosec}^2 \beta_3 + \cot^2 \beta_3) \right\} \right]^{-1} \quad (8.9)$$

Usually r_3 and β_3 are taken to apply at the arithmetic mean radius, i.e. $r_3 = \frac{1}{2}(r_{3t} + r_{3h})$. The temperature ratio (T_3/T_2) in eqn. (8.9) can be obtained as follows.

At the nominal design condition, referring to the velocity triangles of Figure 8.3, $w_3^2 - U_3^2 = c_3^2$, and so eqn. (8.2) can be rewritten as

$$h_2 - h_3 = \frac{1}{2}(U_2^2 - w_2^2 + c_3^2). \quad (8.2a)$$

This particular relationship, in the form $I_2 = h_{02\,\text{rel}} - \frac{1}{2}U_2^2 = h_{03}$ can be easily identified in Figure 8.4.

Again, referring to the velocity triangles, $w_2 = U_2 \cot a_2$ and $c_3 = U_3 \cot \beta_3$, a useful alternative form to eqn. (8.2a) is obtained,

$$h_2 - h_3 = \frac{1}{2}U_2^2[(1 - \cot^2 a_2) + (r_3/r_2)\cot^2 \beta_3], \quad (8.2b)$$

where U_3 is written as $U_2 r_3/r_2$. For a perfect gas the temperature ratio T_3/T_2 can be easily found. Substituting $h = C_p T = \gamma R T / (\gamma - 1)$ in eqn. (8.2b)

$$1 - \frac{T_3}{T_2} = \frac{1}{2}U_2^2 \frac{(\gamma - 1)}{\gamma R T_2} \left[1 - \cot^2 \alpha_2 + \left(\frac{r_3}{r_2} \right)^2 \cot^2 \beta_3 \right]$$

$$\therefore \frac{T_3}{T_2} = 1 - \frac{1}{2}(\gamma - 1) \left(\frac{U_2}{\alpha_2} \right)^2 \left[1 - \cot^2 \alpha_2 + \left(\frac{r_3}{r_2} \right)^2 \cot^2 \beta_3 \right], \quad (8.2c)$$

where $a_2 = (\gamma R T_2)^{1/2}$ is the sonic velocity at temperature T_2.

Generally this temperature ratio will only have a very minor effect upon the numerical value of η_{ts} and so it is often ignored in calculations. Thus,

$$\eta_{ts} \simeq \left[1 + \frac{1}{2} \left\{ \zeta_N \operatorname{cosec}^2 \alpha_2 + \left(\frac{r_{3av}}{r_2} \right)^2 (\zeta_R \operatorname{cosec}^2 \beta_{3av} + \cot^2 \beta_{3av}) \right\} \right]^{-1} \quad (8.9a)$$

is the expression normally used to determine the total-to-static efficiency. An alternative form for η_{ts} can be obtained by rewriting eqn. (8.6) as

$$\eta_{ts} = \frac{h_{01} - h_{03}}{h_{01} - h_{3ss}} = \frac{(h_{01} - h_{3ss}) - (h_{03} - h_3) - (h_3 - h_{3s}) - (h_{3s} - h_{3ss})}{(h_{01} - h_{3ss})}$$

$$= 1 - (c_3^2 + \zeta_N c_2^2 + \zeta_R w_3^2)/c_0^2 \quad (8.10)$$

where the spouting velocity c_0 is defined by,

$$h_{01} - h_{3ss} = \frac{1}{2}c_0^2 = C_p T_{01}[1 - (p_3/p_{01})^{(\gamma-1)/\gamma}] \quad (8.11)$$

A simple connection exists between total-to-total and total-to-static efficiency which can be obtained as follows. Writing

$$\Delta W = U_2^2 = \eta_{ts} \Delta W_{is} = \eta_{ts}(h_{01} - h_{3ss})$$

then,

$$\eta_{tt} = \frac{\Delta W}{\Delta W_{is} - \frac{1}{2}c_3^2} = \frac{1}{\dfrac{1}{\eta_{ts}} - \dfrac{c_3^2}{2\Delta W}}$$

$$\therefore \ \frac{1}{\eta_{ts}} = \frac{1}{\eta_{ts}} - \frac{c_3^2}{2\Delta W}$$

$$= \frac{1}{\eta_{ts}} - \frac{1}{2}\left(\frac{r_{3av}}{r_2} \cot \beta_{3av}\right)^2 \tag{8.12}$$

EXAMPLE 8.2. Performance data from the CAV type 01 radial turbine (Benson *et al.* 1968) operating at a pressure ratio p_{01}/p_3 of 1.5 with zero incidence relative flow onto the rotor, is presented in the following form:

$$\dot{m}\sqrt{T_{01}}/p_{01} = 1.44 \times 10^{-5}, \ \text{ms(deg. K)}^{1/2}$$

$$N/\sqrt{T_{01}} = 2410, \ \text{(rev/min)/(deg. K)}^{1/2}$$

$$\tau/p_{01} = 4.59 \times 10^{-6}, \ \text{m}^3$$

where τ is the torque, corrected for bearing friction loss. The principal dimensions and angles, etc. are given as follows:

Rotor inlet diameter,	72.5 mm
Rotor inlet width,	7.14 mm
Rotor mean outlet diameter,	34.4 mm
Rotor outlet annulus width,	20.1 mm
Rotor inlet angle,	0 deg
Rotor outlet angle,	53 deg
Number of rotor blades,	10
Nozzle outlet diameter,	74.1 mm
Nozzle outlet angle,	80 deg
Nozzle blade number,	15

The turbine is "cold tested" with air heated to 400 K (to prevent condensation erosion of the blades). At nozzle outlet an estimate of the flow angle is given as 71 deg and the corresponding enthalpy loss coefficient is stated to be 0.065. Assuming that the absolute flow at rotor exit is without swirl and uniform, and the relative flow leaves the rotor without any deviation, determine the total-to-static and overall efficiencies of the turbine, the rotor enthalpy loss coefficient and the rotor relative velocity ratio.

Solution. The data given is obtained from an actual turbine test and, even though the bearing friction loss has been corrected, there is an additional reduction in the specific work delivered due to disk friction and tip leakage losses, etc. The rotor speed $N = 2410\sqrt{400} = 48\,200\,\text{rev/min}$, the rotor tip speed $U_2 = \pi N D_2/60 = 183\,\text{m/s}$ and hence the specific work done by the rotor $\Delta W = U_2^2 = 33.48\,\text{kJ/kg}$. The corresponding isentropic total-to-static enthalpy drop is

$$h_{01} - h_{3ss} = C_p T_{01}[1 - (p_3/p_{01})^{(\gamma-1)/\gamma}]$$
$$= 1.005 \times 400[1 - (1/1.5)^{1/3.5}] = 43.97\,\text{kJ/kg}$$

Thus, the total-to-static efficiency is

$$\eta_{ts} = \Delta W/(h_{01} - h_{3ss}) = 76.14\%$$

The actual specific work output to the shaft, after allowing for the bearing friction loss, is

$$\Delta W_{act} = \tau\Omega/\dot{m} = \left(\frac{\tau}{p_{01}}\right)\frac{N}{\sqrt{T_{01}}}\left(\frac{p_{01}}{\dot{m}\sqrt{T_{01}}}\right)\frac{\pi}{30}T_{01}$$
$$= 4.59 \times 10^{-6} \times 2410 \times \pi \times 400/(30 \times 1.44 \times 10^{-5})$$
$$= 32.18\,\text{kJ/kg}$$

Thus, the turbine overall total-to-static efficiency is

$$\eta_0 = \Delta W_{act}/(h_{01} - h_{3ss}) = 73.18\%$$

By rearranging eqn. (8.9a) the rotor enthalpy loss coefficient can be obtained:

$$\zeta_R = \{2(1/\eta_{ts} - 1) - \zeta_N\,\text{cosec}^2\,\alpha_2\}(r_2/r_{3av})^2\sin^2\beta_{3av} - \cos^2\beta_{3av}$$
$$= \{2(1/0.7613 - 1) - 0.065 \times 1.1186\} \times 4.442 \times 0.6378$$
$$- 0.3622$$
$$= 1.208$$

At rotor exit c_3 is assumed to be uniform and axial. From the velocity triangles, Figure 8.3,

$$c_3 = U_3 \cot\beta_3 = U_{3av}\cot\beta_{3av} = \text{constant}$$
$$w_3^2 = U_3^2 + c_3^2$$
$$= U_{3av}^2\left[\left(\frac{r_3}{r_{3av}}\right)^2 + \cot^2\beta_{3av}\right]$$
$$w_{2av} = U_2 \cot\alpha_2$$

ignoring blade to blade velocity variations. Hence,

$$\frac{w_3}{w_{2av}} = \frac{r_{3av}}{r_2}\tan\alpha_2\left[\left(\frac{r_3}{r_{3av}}\right)^2 + \cot^2\beta_{3av}\right]^{1/2}. \tag{8.13}$$

The lowest value of this relative velocity ratio occurs when r_3 is least, i.e. $r_3 = r_{3h} = (34.4 - 20.1)/2 = 7.15 \, \text{mm}$, so that

$$\left(\frac{w_3}{w_{2av}}\right)_{min} = 0.475 \times 2.904[0.415^2 + 0.7536^2]^{1/2} = 1.19.$$

The relative velocity ratio corresponding to the mean exit radius is,

$$\frac{w_{3av}}{w_{2av}} = 0.475 \times 2.904[1 + 0.7536^2]^{1/2} = 1.73.$$

It is worth commenting that higher total-to-static efficiencies have been obtained in other small radial turbines operating at higher pressure ratios. Rodgers (1969) has suggested that total-to-static efficiencies in excess of 90% for pressure ratios up to five to one can be attained. Nusbaum and Kofskey (1969) reported an experimental value of 88.8% for a small radial turbine (fitted with an outlet diffuser, admittedly!) at a pressure ratio p_{01}/p_4 of 1.763. In the design point exercise given above the high rotor enthalpy loss coefficient and the corresponding relatively low total-to-static efficiency may well be related to the low relative velocity ratio determined on the hub. Matters are probably worse than this as the calculation is based only on a simple one-dimensional treatment. In determining velocity ratios across the rotor, account should also be taken of the effect of blade to blade velocity variation (outlined in this chapter) as well as viscous effects. The number of vanes in the rotor (ten) may be insufficient on the basis of Jamieson's theory* (1955) which suggests 18 vanes (i.e. $Z_{min} = 2\pi \tan \alpha_2$). For this turbine, at lower nozzle exit angles, eqn. (8.13) suggests that the relative velocity ratio becomes even less favourable despite the fact that the Jamieson blade spacing criterion is being approached. (For $Z = 10$, the optimum value of α_2 is about 58 deg.)

Mach number relations

Assuming the fluid is a perfect gas, expressions can be deduced for the important Mach numbers in the turbine. At nozzle outlet the absolute Mach number at the nominal design point is,

$$M_2 = \frac{c_2}{a_2} = \frac{U_2}{a_2} \operatorname{cosec} \alpha_2.$$

Now, $T_2 = T_{01} - c_2^2/(2C_p) = T_{01} - \frac{1}{2}U_2^2 \operatorname{cosec}^2 \alpha_2/C_p.$

$$\therefore \frac{T_2}{T_{01}} = 1 - \frac{1}{2}(\gamma - 1)(U_2/a_{01})^2 \operatorname{cosec}^2 \alpha_2$$

where $a_2 = a_{01}(T_2/T_{01})^{1/2}$. Hence,

$$M_2 = \frac{U_2/a_{01}}{\sin \alpha_2[1 - \frac{1}{2}(\gamma - 1)(U_2/a_{01})^2 \operatorname{cosec}^2 \alpha_2]^{1/2}} \tag{8.14}$$

---

* Included in a later part of this Chapter.

At rotor outlet the relative Mach number at the design point is defined by,

$$M_{r3} = \frac{w_3}{a_3} = \frac{r_3 U_2}{r_2 a_3} \operatorname{cosec} \beta_3.$$

Now,

$$h_3 = h_{01} - (U_2^2 + \tfrac{1}{2}c_3^2) = h_{01} - (U_2^2 + \tfrac{1}{2}U_3^2 \cot^2 \beta_3)$$

$$= h_{01} - U_2^2 \left[1 + \tfrac{1}{2} \left(\frac{r_3}{r_2} \cot \beta_3 \right)^2 \right]$$

$$a_3^2 = a_{01}^2 - (\gamma - 1)U_2^2 \left[1 + \tfrac{1}{2} \left(\frac{r_3}{r_2} \cot \beta_3 \right)^2 \right]$$

$$\therefore M_{r3} = \frac{(U_2/a_{01})(r_3/r_2)}{\sin \beta_3 \left[1 - (\gamma - 1)(U_2/a_{01})^2 \left\{ 1 + \tfrac{1}{2} \left(\frac{r_3}{r_2} \cot \beta_3 \right)^2 \right\} \right]^{1/2}} \qquad (8.15)$$

Loss coefficients in 90 deg IFR turbines

There are a number of ways of representing the losses in the passages of 90 deg IFR turbines and these have been listed and inter-related by Benson (1970). As well as the nozzle and rotor passage losses there is, in addition, a loss at rotor entry at off-design conditions. This occurs when the relative flow entering the rotor is at some angle of incidence to the radial vanes so that it can be called an *incidence loss*. It is often referred to as a "shock loss" but this can be a rather misleading term because, usually, there is no shock wave.

(i) Nozzle loss coefficients

The enthalpy loss coefficient, which normally includes the inlet scroll losses, has already been defined and is,

$$\zeta_N = (h_2 - h_{2s})/(\tfrac{1}{2}c_2^2). \qquad (8.16)$$

Also in use is the *velocity coefficient*,

$$\phi_N = c_2/c_{2s} \qquad (8.17)$$

and the *stagnation pressure loss coefficient*,

$$Y_N = (p_{01} - p_{02})/(p_{02} - p_2) \qquad (8.18a)$$

which can be related, approximately, to ζ_N by

$$Y_N \simeq \zeta_N(1 + \tfrac{1}{2}\gamma M_2^2) \qquad (8.18b)$$

Since, $h_{01} = h_2 + \tfrac{1}{2}c_2^2 = h_{2s} + \tfrac{1}{2}c_{2s}^2$, then $h_2 - h_{2s} = \tfrac{1}{2}(c_{2s}^2 - c_2^2)$ and

$$\zeta_N = \frac{1}{\phi_N^2} - 1. \qquad (8.19)$$

Practical values of ϕ_N for well-designed nozzle rows in normal operation are usually in the range $0.90 \leqslant \phi_N \leqslant 0.97$.

(ii) Rotor loss coefficients

At either the design condition (Figure 8.4), or at the off-design condition dealt with later (Figure 8.5), the rotor passage friction losses can be expressed in terms of the following coefficients.

The enthalpy loss coefficient is,

$$\zeta_R = (h_3 - h_{3s})/(\tfrac{1}{2}w_3^2). \tag{8.20}$$

The velocity coefficient is,

$$\phi_R = w_3/w_{3s} \tag{8.21}$$

which is related to ζ_R by

$$\zeta_R = \frac{1}{\phi_R^2} - 1 \tag{8.22}$$

The normal range of ϕ for well-designed rotors is approximately, $0.70 \leqslant \phi_R \leqslant 0.85$.

Optimum efficiency considerations

According to Abidat *et al.* (1992) the understanding of incidence effects on the rotors of radial and mixed flow turbines is very limited. Normally, IFR turbines are made with radial vanes in order to reduce bending stresses. In most flow analyses that have been published of the IFR turbine, including all earlier editions of this text, it was assumed that the *average* relative flow at entry to the rotor was radial, i.e. the incidence of the relative flow approaching the radial vanes was zero. The following discussion of the flow model will show that this is an over-simplification and the flow angle for optimum efficiency is significantly different from zero incidence. Rohlik (1975) had asserted that "there is some incidence angle that provides *optimum flow conditions* at the rotor-blade leading edge. This angle has a value sometimes as high as 40° with a radial blade."

The flow approaching the rotor is assumed to be in the radial plane with a velocity c_2 and flow angle α_2 determined by the geometry of the nozzles or volute. Once the fluid enters the rotor the process of work extraction proceeds rapidly with reduction in the magnitude of the tangential velocity component and blade speed as the flow radius decreases. Corresponding to these velocity changes there is a high blade loading and an accompanying large pressure gradient across the passage from the pressure side to the suction side (Figure 8.5a).

With the rotor rotating at angular velocity Ω and the entering flow assumed to be irrotational, a counter-rotating vortex (or relative eddy) is created in the relative flow, whose magnitude is $-\Omega$, which conserves the irrotational state. The effect is virtually the same as that described earlier for the flow leaving the impeller of a centrifugal compressor, but in reverse (see Chapter 7 under the heading "Slip factor"). As a result of combining the incoming irrotational flow with the relative

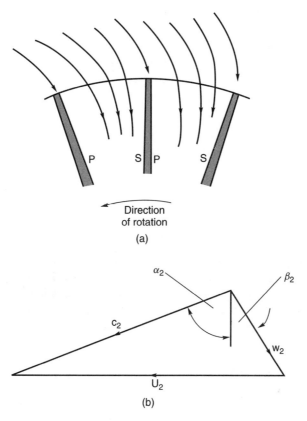

FIG. 8.5. Optimum flow condition at inlet to the rotor. (a) Streamline flow at rotor inlet; p is for pressure surface, s is for suction surface. (b) Velocity diagram for the pitchwise averaged flow.

eddy, the relative velocity on the pressure (or trailing) surface of the vane is reduced. Similarly, on the suction (or leading) surface of the vane it is seem that the relative velocity is increased. Thus, a static pressure gradient exists *across* the vane passage in agreement with the reasoning of the preceding paragraph.

Figure 8.5b indicates the *average* relative velocity w_2, entering the rotor at angle β_2 and giving optimum flow conditions at the vane leading edge. As the rotor vanes in IFR turbines are assumed to be radial, the angle β_2 is an angle of incidence, and as drawn it is numerically positive. Depending upon the number of rotor vanes this angle may be between 20 and 40 degrees. The static pressure gradient across the passage causes a streamline shift of the flow towards the suction surface. Stream-function analyzes of this flow condition show that the streamline pattern properly locates the inlet stagnation point on the vane leading edge so that this streamline is approximately radial (see Figure 8.5a). It is reasoned that only at this flow condition will the fluid move smoothly into the rotor passage. Thus, it is the *averaged* relative flow that is at an angle of incidence β_2 to the vane. Whitfield and Baines (1990) have comprehensively reviewed computational methods used in determining turbomachinery flows, including streamfunction methods.

Wilson and Jansen (1965) appear to have been the first to note that the optimum angle of incidence was virtually identical to the angle of "slip" of the flow leaving the impeller of a radially bladed centrifugal compressor with the same number of vanes as the turbine rotor. Following Whitfield and Baines (1990), an *incidence factor*, λ, is defined, analogous to the slip factor used in centrifugal compressors:

$$\lambda = c_{\theta 2}/U_2.$$

The slip factor most often used in determining the flow angle at rotor inlet is that devised by Stanitz (1952) for radial vaned impellers, so for the incidence factor

$$\lambda = 1 - 0.63\pi/Z \approx 1 - 2/Z. \tag{7.18a}$$

Thus, from the geometry of Figure 8.5b, we obtain

$$\tan \beta_2 = (2/Z)U_2/c_{m2}. \tag{8.23}$$

In order to determine the relative flow angle, β_2, we need to know, at least, the values of the flow coefficient, $\phi_2 = c_{m2}/U_2$ and the vane number Z. A simple method of determining the minimum number of vanes needed in the rotor, due to Jamieson (1955), is given later in this chapter. However, in the next section an optimum efficiency design method devised by Whitfield (1990) provides an alternative way for deriving β_2.

Design for optimum efficiency

Whitfield (1990) presented a general one-dimensional design procedure for the IFR turbine in which, initially, only the required power output is specified. The specific power output is given:

$$\Delta W = \frac{\dot{W}}{\dot{m}} = h_{01} - h_{03} = \frac{\gamma R}{\gamma - 1}(T_{01} - T_{03}) \tag{8.24}$$

and, from this a non-dimensional *power ratio*, S, is defined:

$$S = \Delta W/h_{01} = 1 - T_{03}/T_{01}. \tag{8.25}$$

The power ratio is related to the overall pressure ratio through the total-to-static efficiency:

$$\eta_{ts} = \frac{S}{[1 - (p_3/p_{01})^{(\gamma-1)/\gamma}]}. \tag{8.26}$$

If the power output, mass flow rate and inlet stagnation temperature are specified, then S can be directly calculated but, if only the output power is known, then an iterative procedure must be followed.

Whitfield (1990) chose to develop his procedure in terms of the power ratio S and evolved a new non-dimensional design method. At a later stage of the design when the rate of mass flow and inlet stagnation temperature can be quantified, then the actual gas velocities and turbine size can be determined. Only the first part of Whitfield's method dealing with the rotor design is considered in this chapter.

Solution of Whitfield's design problem

At the design point it is usually assumed that the fluid discharges from the rotor in the axial direction so that with $c_{\theta 3} = 0$, the specific work is

$$\Delta W = U_2 c_{\theta 2}$$

and, combining this with eqns. (8.24) and (8.25), we obtain,

$$U_2 c_{\theta 2}/a_{01}^2 = S/(\gamma - 1), \tag{8.27}$$

where $a_{01} = (\gamma R T_{01})^{1/2}$ is the speed of sound corresponding to the temperature T_{01}. Now, from the velocity triangle at rotor inlet, Figure 8.5b,

$$U_2 - c_{\theta 2} = c_{m2} \tan \beta_2 = c_{\theta 2} \tan \beta_2 / \tan \alpha_2. \tag{8.28}$$

Multiplying both sides of eqn. (8.28) by $c_{\theta 2}/c_{m2}^2$, we get

$$U_2 c_{\theta 2}/c_{m2}^2 - c_{\theta 2}^2/c_{m2}^2 - \tan \alpha_2 \tan \beta_2 = 0.$$

But,

$$U_2 c_{\theta 2}/c_{m2}^2 = (U_2 c_{\theta 2}/c_2^2) \sec^2 \alpha_2^2 = c(1 + \tan^2 \alpha_2^2),$$

which can be written as a quadratic equation for $\tan \alpha_2$:

$$\tan^2 \alpha_2 (c - 1) - b \tan \alpha_2 + c = 0,$$

where, for economy of writing, $c = U_2 c_{\theta 2}/c_2^2$ and $b = \tan \beta_2$. Solving for $\tan \alpha_2$,

$$\tan \alpha_2 = \{b \pm \sqrt{[b^2 + 4c(1 - c)]}\}/[2(c - 1)]. \tag{8.29}$$

For a real solution to exist the radical must be greater than, or equal to, zero; i.e. $b^2 + 4c(1 - c) \geqslant 0$. Taking the zero case and rearranging the terms, another quadratic equation is found, namely

$$c^2 - c - b^2/4 = 0.$$

Hence, solving for c,

$$c = \left(1 \pm \sqrt{1 + b^2}\right)/2 = \tfrac{1}{2}(1 \pm \sec \beta_2) = U_2 c_{\theta 2}/c_2^2. \tag{8.30}$$

From eqn. (8.29) and then eqn. (8.30), the corresponding solution for $\tan \alpha_2$ is

$$\tan \alpha_2 = b/[2(c - 1)] = \tan \beta_2/(-1 \pm \sec \beta_2).$$

The correct choice between these two solutions will give a value for $\alpha_2 > 0$, thus:

$$\tan \alpha_2 = \frac{\sin \beta_2}{1 - \cos \beta_2} \tag{8.31}$$

It is easy to see from Table 8.1 that a simple numerical relation exists between these two parameters, namely

$$\alpha_2 = 90 - \beta_2/2. \tag{8.31a}$$

TABLE 8.1. Variation of α_2 for several values of β_2.

β_2 (deg)	10	20	30	40
α_2 (deg)	85	80	75	70

From eqns. (8.27) and (8.30), after some rearranging, a minimum stagnation Mach number at rotor inlet can be found:

$$M_{02}^2 = c_2^2/a_{01}^2 = \left(\frac{S}{\gamma - 1}\right) \frac{2\cos\beta_2}{1 + \cos\beta_2} \tag{8.32}$$

and the inlet Mach number can be determined using the equation

$$M_2^2 = \left(\frac{c_2}{a_2}\right)^2 = \frac{M_{02}^2}{1 - \frac{1}{2}(\gamma - 1)M_{02}^2} \tag{8.33}$$

assuming that $T_{02} = T_{01}$, the flow through the stator is adiabatic.
   Now, from eqn. (8.28)

$$\frac{c_{\theta 2}}{U_2} = \frac{1}{1 + \tan\beta_2/\tan\alpha_2}.$$

After rearranging eqn. (8.31) to give

$$\tan\beta_2/\tan\alpha_2 = \sec\beta_2 - 1 \tag{8.34}$$

and combining these equations,

$$c_{\theta 2}/U_2 = \cos\beta_2 = 1 - 2/Z. \tag{8.35}$$

Equation (8.35) is a direct relationship between the number of rotor blades and the relative flow angle at inlet to the rotor. Also, from eqn. (8.31a),

$$\cos 2\alpha_2 = \cos(180 - \beta_2) = -\cos\beta_2$$

so that, from the identity $\cos 2\alpha_2 = 2\cos^2\alpha_2 - 1$, we get the result:

$$\cos^2\alpha_2 = (1 - \cos\beta_2)/2 = 1/Z, \tag{8.31b}$$

using also eqn. (8.35).

   EXAMPLE 8.3. An IFR turbine with 12 vanes is required to develop 230 kW from a supply of dry air available at a stagnation temperature of 1050 K and a flow rate of 1 kg/s. Using the optimum efficiency design method and assuming a total-to-static efficiency of 0.81, determine:

(1) the absolute and relative flow angles at rotor inlet;
(2) the overall pressure ratio, p_{01}/p_3;
(3) the rotor tip speed and the inlet absolute Mach number.

Solution. (1) From the gas tables, e.g. Rogers and Mayhew (1995), at $T_{01} = 1050$ K, we can find values for $C_p = 1.1502$ kJ/kg K and $\gamma = 1.333$. Using eqn. (8.25),

$$S = \Delta W / (C_p T_{01}) = 230/(1.15 \times 1050) = 0.2.$$

From Whitfield's eqn. (8.31b),

$$\cos^2 \alpha_2 = 1/Z = 0.083333, \quad \therefore \alpha_2 = 73.22 \text{ deg}$$

and, from eqn. (8.31a), $\beta_2 = 2(90 - \alpha_2) = 33.56$ deg.

(2) Rewriting eqn. (8.26),

$$\frac{p_3}{p_{01}} = \left(1 - \frac{S}{\eta_{ts}}\right)^{\gamma/(\gamma-1)} = \left(1 - \frac{0.2}{0.81}\right)^4 = 0.32165, \quad \therefore \frac{p_{01}}{p_3} = 3.109.$$

(3) Using eqn. (8.32),

$$M_{02}^2 = \left(\frac{S}{\gamma - 1}\right) \frac{2\cos\beta_2}{1 + \cos\beta_2} = \frac{0.2}{0.333} \times \frac{2 \times 0.8333}{1 + 0.8333} = 0.5460$$

$$\therefore M_{02} = 0.7389.$$

Using eqn. (8.33),

$$M_2^2 = \frac{M_{02}^2}{1 - \frac{1}{2}(\gamma - 1)M_{02}^2} = \frac{0.546}{1 - (0.333/2) \times 0.546} = 0.6006 \quad \therefore M_2 = 0.775.$$

To find the rotor tip speed, substitute eqn. (8.35) into eqn. (8.27) to obtain:

$$\left(\frac{U_2^2}{a_{01}^2}\right) \cos\beta_2 = \frac{S}{\gamma - 1}$$

$$\therefore U_2 = a_{01} \sqrt{\frac{S}{(\gamma - 1)\cos\beta_2}} = 633.8\sqrt{\frac{0.2}{0.333 \times 0.8333}} = 538.1 \text{ m/s},$$

where $a_{01} = \sqrt{\gamma R T_{01}} = \sqrt{1.333 \times 287 \, 1050} = 633.8$ m/s, and $T_{02} = T_{01}$ is assumed.

Criterion for minimum number of blades

The following simple analysis of the relative flow in a radially bladed rotor is of considerable interest as it illustrates an important fundamental point concerning blade spacing. From elementary mechanics, the radial and transverse components of acceleration, f_r and f_t respectively, of a particle moving in a radial plane (Figure 8.6a) are:

$$f_r = \dot{w} - \Omega^2 r \tag{8.36a}$$

$$f_t = r\dot{\Omega} + 2\Omega w \tag{8.36b}$$

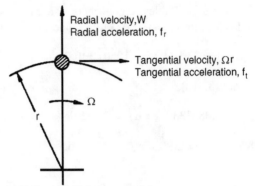

(a) Motion of particle in a radial plane

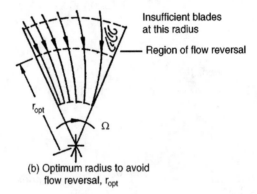

(b) Optimum radius to avoid
flow reversal, r_{opt}

FIG. 8.6. Flow models used in analysis of minimum number of blades.

where w is the radial velocity, $\dot{w} = (dw)/(dt) = w(\partial w)/(\partial r)$ (for steady flow), Ω is the angular velocity and $\dot{\Omega} = d\Omega/dt$ is set equal to zero.

Applying *Newton's second law of motion* to a fluid element (as shown in Figure 6.2) of unit depth, ignoring viscous forces, but putting $c_r = w$, the radial equation of motion is,

$$(p + dp)(r + dr)d\theta - prd\theta - pdrd\theta = -f_r dm$$

where the elementary mass $dm = \rho r d\theta dr$. After simplifying and substituting for f_r from eqn. (8.25a), the following result is obtained,

$$\frac{1}{\rho}\frac{\partial p}{\partial r} + w\frac{\partial w}{\partial r} = \Omega^2 r. \tag{8.37}$$

Integrating eqn. (8.37) with respect to r obtains

$$p/\rho + \tfrac{1}{2}w^2 - \tfrac{1}{2}U^2 = \text{constant} \tag{8.38}$$

which is merely the *inviscid form* of eqn. (8.2).

The torque transmitted to the rotor by the fluid manifests itself as a pressure difference across each radial vane. Consequently, there must be a pressure gradient in the *tangential direction* in the space between the vanes. Again, consider the element of fluid and apply Newton's second law of motion in the tangential direction

$$\mathrm{d}p.\mathrm{d}r = f_t \mathrm{d}m = 2\Omega w(\rho r \mathrm{d}\theta \mathrm{d}r).$$

Hence,

$$\frac{1}{\rho}\frac{\partial p}{\partial \theta} = 2\Omega r w \tag{8.39}$$

which establishes the magnitude of the tangential pressure gradient. Differentiating eqn. (8.38) with respect to θ,

$$\frac{1}{\rho}\frac{\partial p}{\partial \theta} = -w\frac{\partial w}{\partial \theta}. \tag{8.40}$$

Thus, combining eqns. (8.39) and (8.40) gives,

$$\frac{\partial w}{\partial \theta} = -2\Omega r \tag{8.41}$$

This result establishes the important fact that *the radial velocity is not uniform across the passage* as is frequently assumed. As a consequence of this fact the radial velocity on one side of a passage is lower than on the other side. Jamieson (1955), who originated this method, conceived the idea of determining the *minimum* number of blades based upon these velocity considerations.

Let the mean radial velocity be $\overline{w}$ and the angular space between two adjacent blades be $\Delta\theta = 2\pi/Z$ where Z is the number of blades. The maximum and minimum radial velocities are, therefore,

$$w_{\max} = \overline{w} + \tfrac{1}{2}\Delta w = \overline{w} + \Omega r \Delta\theta \tag{8.42a}$$

$$w_{\min} = \overline{w} - \tfrac{1}{2}\Delta w = \overline{w} - \Omega r \Delta\theta \tag{8.42b}$$

using eqn. (8.41).

Making the reasonable assumption that the radial velocity should not drop below zero, (see Figure 8.6b), then the limiting case occurs at the rotor tip, $r = r_2$ with $w_{\min} = 0$. From eqn. (8.42b) with $U_2 = \Omega r_2$, the minimum number of rotor blades is

$$Z_{\min} = 2\pi U_2/\overline{w}_2 \tag{8.43a}$$

At the design condition, $U_2 = \overline{w}_2 \tan\alpha_2$, hence

$$Z_{\min} = 2\pi \tan\alpha_2 \tag{8.43b}$$

Jamieson's result, eqn. (8.43b), is plotted in Figure 8.7 and shows that a large number of rotor vanes are required, especially for high absolute flow angles at rotor inlet. In practice a large number of vanes are not used for several reasons, e.g. excessive flow blockage at rotor exit, a disproportionally large "wetted" surface

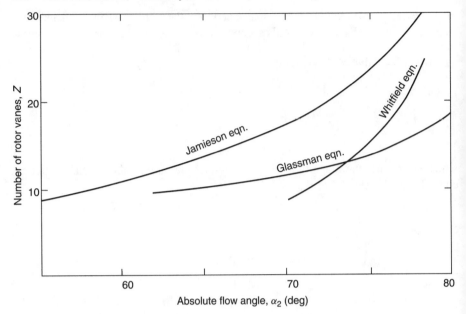

FIG. 8.7. Flow angle at rotor inlet as a function of the number of rotor vanes.

area causing high friction losses, and the weight and inertia of the rotor become relatively high.

Some experimental tests reported by Hiett and Johnston (1964) are of interest in connection with the analysis presented above. With a nozzle outlet angle $\alpha_2 = 77$ deg and a 12 vane rotor, a total-to-static efficiency $\eta_{ts} = 0.84$ was measured at the optimum velocity ratio U_2/c_0. For that magnitude of flow angle, eqn. (8.43b) suggests 27 vanes would be required in order to avoid reverse flow at the rotor tip. However, a second test with the number of vanes increased to 24 produced a gain in efficiency of only 1%. Hiett and Johnston suggested that the criterion for the optimum number of vanes might not simply be the avoidance of local flow reversal but might require a compromise between total pressure losses from this cause and friction losses based upon rotor and blade surface areas.

Glassman (1976) preferred to use an empirical relationship between Z and α_2, namely

$$Z = \frac{\pi}{30}(110 - \alpha_2)\tan\alpha_2, \tag{8.44}$$

as he also considered Jamieson's result, eqn. (8.43b), gave too many vanes in the rotor. Glassman's result, which gives far fewer vanes than Jamieson's is plotted in Figure 8.7. Whitfield's result given in eqn. (8.31b), is not too dissimilar from the result given by Glassman's equation, at least for low vane numbers.

Design considerations for rotor exit

Several decisions need to be made regarding the design of the rotor exit. The flow angle β_3, the meridional velocity to blade tip speed ratio, c_{m3}/U_2, the shroud tip to rotor tip radius ratio, r_{3s}/r_2, and the exit hub to shroud radius ratio, $\nu = r_{3h}/r_{3s}$, all have to be considered. It is assumed that the absolute flow at rotor exit is entirely axial so that the relative velocity can be written:

$$w_3^2 = c_{m3}^2 + U_3^2.$$

If values of c_{m3}/U_2 and r_{3av}/r_2 can be chosen, then the exit flow angle variation can be found for all radii. From the rotor exit velocity diagram in Figure 8.3,

$$\frac{c_{m3}}{U_2} = \frac{r_{3av}}{r_2} \cot \beta_{3av} = \frac{r_3}{r_2} \cot \beta_3. \qquad (8.45)$$

The meridional velocity c_{m3} should be kept small in order to minimise the exhaust energy loss, unless an exhaust diffuser is fitted to the turbine.

Rodgers and Geiser (1987) correlated attainable efficiency levels of IFR turbines against the blade tip speed/spouting velocity ratio, U_2/c_0, and the axial exit flow coefficient, c_{m3}/U_2, and their result is shown in Figure 8.8. From this figure it can be seen that peak efficiency values are obtained with velocity ratios close to 0.7 and with values of exit flow coefficient between 0.2 and 0.3.

Rohlik (1968) suggested that the ratio of mean rotor exit radius to rotor inlet radius, r_{3av}/r_2, should not exceed 0.7 to avoid excessive curvature of the shroud. Also, the exit hub to shroud radius ratio, r_{3h}/r_{3s}, should not be less than 0.4 because of the likelihood of flow blockage caused by closely spaced vanes. Based upon the metal thickness alone it is easily shown that,

$$(2\pi r_{3h}/Z) \cos \beta_{3h} > t_{3h},$$

where t_{3h} is the vane thickness at the hub. It is also necessary to allow more than this thickness because of the boundary layers on each vane. Some of the rather limited test data available on the design of the rotor exit, comes from Rodgers and Geiser (1987), and concerns the effect of rotor radius ratio and blade solidity on turbine efficiency (see Figure 8.9). It is the *relative* efficiency variation, η/η_{opt}, that is depicted as a function of the rotor inlet radius/exit *root mean square* radius ratio,

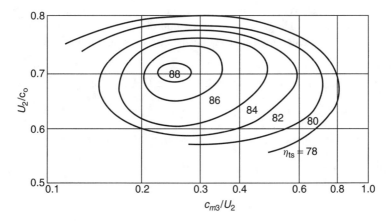

FIG. 8.8. Correlation of attainable efficiency levels of IFR turbines against velocity ratios (adapted from Rodgers and Geiser 1987).

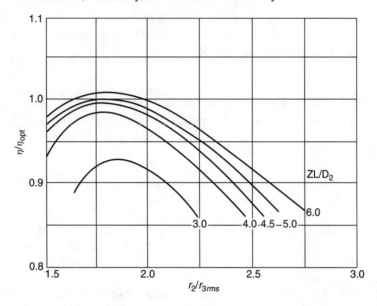

FIG. 8.9. Effects of vane solidity and rotor radius ratio on the efficiency ratio of the IFR turbine (adapted from Rodgers and Geiser 1987).

r_2/r_{3rms}, for various values of a blade solidity parameter, ZL/D_2 (where L is the length of the blade along the mean meridion). This radius ratio is related to the rotor exit hub to shroud ratio, ν, by

$$\frac{r_{3rms}}{r_2} = \frac{r_{3s}}{r_2}\left(\frac{1+\nu^2}{2}\right)^{1/2}$$

From Figure 8.9, for r_2/r_{3rms}, a value between 1.6 and 1.8 appears to be the optimum.

Rohlik (1968) suggested that the ratio of the relative velocity at the mean exit radius to inlet relative velocity, w_{3av}/w_2, should be sufficiently high to assure a low total pressure loss. He gave w_{3av}/w_2 a value of 2.0. The relative velocity at the shroud tip will be greater than that at the mean radius depending upon the radius ratio at rotor exit.

EXAMPLE 8.4. Given the following data for an IFR turbine:

$$c_{m3}/U_2 = 0.25, \nu = 0.4, r_{3s}/r_2 = 0.7 \text{ and } w_{3av}/w_2 = 2.0,$$

determine the ratio of the relative velocity ratio, w_{3s}/w_2 at the shroud.

Solution. As $w_{3s}/c_{m3} = \sec\beta_{3s}$ and $w_{3av}/c_{m3} = \sec\beta_{3av}$, then

$$\frac{w_{3s}}{w_{3av}} = \frac{\sec\beta_{3s}}{\sec\beta_{3av}}$$

$$\frac{r_{3av}}{r_{3s}} = \tfrac{1}{2}(1+\nu) = 0.7 \text{ and } \frac{r_{3av}}{r_2} = \frac{r_{3av}}{r_{3s}}\frac{r_{3s}}{r_2} = 0.7 \times 0.7 = 0.49.$$

From eqn. (8.45):

$$\cot \beta_{3av} = \frac{c_{m3}}{U_2} \frac{r_2}{r_{3av}} = \frac{0.25}{0.49} = 0.5102 \quad \therefore \beta_{3av} = 62.97 \text{ deg}$$

$$\cot \beta_{3s} = \frac{c_{m3}}{U_2} \frac{r_2}{r_{3s}} = \frac{0.25}{0.7} = 0.3571 \quad \therefore \beta_{3s} = 70.35 \text{ deg}$$

$$\therefore \frac{w_{3s}}{w_2} = \frac{w_{3s}}{w_{3av}} \frac{w_{3av}}{w_2} = \frac{\sec \beta_{3s}}{\sec \beta_{3av}} \times 2 = \frac{0.4544}{0.3363} \times 2 = 2.702.$$

The relative velocity ratio will increase progressively from the hub to the shroud.

EXAMPLE 8.5. Using the data and results given in the examples 8.3 and 8.4 together with the additional information that

(a) the static pressure at rotor exit is 100 kPa, and
(b) the nozzle enthalpy loss coefficient, $\zeta_N = 0.06$, determine:
   (1) the diameter of the rotor and its speed of rotation;
   (2) the vane width to diameter ratio, b_2/D_2 at rotor inlet.

Solution. (1) The rate of mass flow is given by

$$\dot{m} = \rho_3 c_{m3} A_3 = \left(\frac{p_3}{RT_3}\right)\left(\frac{c_{m3}}{U_2}\right) U_2 2\pi \left(\frac{r_{3s}}{r_2}\right)^2 (1 - v^2) r_2^2.$$

From eqn. (8.25), $T_{03} = T_{01}(1 - S) = 1050 \times 0.8 = 840 \text{ K}$.

$$T_3 = T_{03} - c_{m3}^2/(2C_p) = T_{03} - \left(\frac{c_{m3}}{U_2}\right)^2 \frac{U_2^2}{2C_p}$$

$$= 840 - 0.25^2 \times 538.1^2/(2 \times 1150.2).$$

Hence, $T_3 = 832.1 \text{ K}$.
   Substituting values into the mass flow equation above,

$$1 = [10^5/(287 \times 832.1)] \times 0.25 \times 538.1 \times 0.7^2 \times \pi \times (1 - 0.4^2) r_2^2$$

$$\therefore r_2^2 = 0.01373 \text{ and } r_2 = 0.1172 \text{ m}, \quad \therefore D_2 = \underline{0.2343 \text{ m}}$$

$$\therefore \Omega = U_2/r_2 = 4591.3 \text{ rad/s } (N = \underline{43\,843 \text{ rev/min}}).$$

(2) The rate of mass flow equation is now written as

$$\dot{m} = \rho_2 c_{m2} A_2, \text{ where } A_2 = 2\pi r_2 b_2 = 4\pi r_2^2 (b_2/D_2)$$

$$\therefore \frac{b_2}{D_2} = \frac{\dot{m}}{4\pi \rho_2 c_{m2} r_2^2}.$$

Solving for the absolute velocity at rotor inlet and its components,

$$c_{\theta 2} = S C_p T_{01}/U_2 = 0.2 \times 1150.2 \times 1050/538.1 = 448.9 \text{ m/s},$$

$$c_{m2} = c_{\theta2}/\tan\alpha_2 = 448.9/3.3163 = 135.4\,\text{m/s},$$

$$c_2 = c_{\theta2}/\sin\alpha_2 = 448.9/0.9574 = 468.8\,\text{m/s}.$$

To obtain a value for the static density, ρ_2, we need to determine T_2 and p_2:

$$T_2 = T_{02} - c_2^2/(2C_p) = 1050 - 468.8^2/(2 \times 1150.2) = 954.5\,\text{K},$$

$$h_{02} - h_2 = \tfrac{1}{2}c_2^2 \text{ and as } \zeta_N = (h_2 - h_{2s})/(\tfrac{1}{2}c_2^2), \text{ then } h_{01} - h_{2s} = \tfrac{1}{2}c_2^2(1 + \zeta_N),$$

$$\therefore \frac{T_{02} - T_{2s}}{T_{02}} = \frac{c_2^2(1 + \zeta_N)}{2C_pT_{02}} = \frac{468.8^2 \times 1.06}{2 \times 1150.2 \times 1050} = 0.096447$$

$$\frac{T_{2s}}{T_{01}} = \left(\frac{p_2}{p_{01}}\right)^{(\gamma-1)/\gamma} = 1 - 0.09645 = 0.90355$$

$$\therefore \frac{p_2}{p_{01}} = \left(\frac{T_{2s}}{T_{01}}\right)^{\gamma/(\gamma-1)} = 0.90355^4 = 0.66652$$

$$\therefore p_2 = 3.109 \times 10^5 \times 0.66652 = 2.0722 \times 10^5\,\text{Pa}$$

$$\frac{b_2}{D_2} = \frac{1}{4\pi}\left(\frac{RT_2}{p_2}\right)\left(\frac{\dot{m}}{c_{m2}r_2^2}\right)$$

$$= \frac{1}{4 \times \pi}\left(\frac{287 \times 954.5}{2.0722 \times 10^5}\right)\frac{1}{135.4 \times 0.01373} = \underline{0.0566}.$$

Incidence losses

At off-design conditions of operation with the fluid entering the rotor at a relative flow angle, β_2, different from the optimum relative flow angle, $\beta_{2,\text{opt}}$, an additional loss due to an effective angle of incidence, $i_2 = \beta_2 - \beta_{2,\text{opt}}$, will be incurred. Operationally, off-design conditions can arise from changes in

(a) the rotational speed of the rotor,
(b) the rate of mass flow,
(c) the setting angle of the stator vanes.

Because of its inertia the speed of the rotor can change only relatively slowly, whereas the flow rate can change very rapidly, as it does in the pulsating flow of turbomachine turbines. The time required to alter the stator vane setting angle will also be relatively long.

Futral and Wasserbauer (1965) defined the incidence loss as equal to the kinetic energy corresponding to the component of velocity normal to the rotor vane at inlet. This may be made clearer by referring to the Mollier diagram and velocity diagrams of Figure 8.10. Immediately *before* entering the rotor the relative velocity is w_2'. Immediately *after* entering the rotor the relative velocity is changed, *hypothetically*, to w_2. Clearly, in reality this change cannot take place so abruptly and will require some finite distance for it to occur. Nevertheless, it is convenient to consider that the

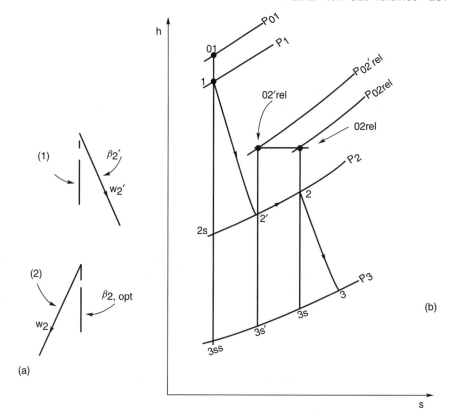

FIG. 8.10. (a) Simple flow model of the relative velocity vector (1) immediately before entry to the rotor, (2) immediately after entry to the rotor. (b) Mollier diagram indicating the corresponding entropy increase, $(s_{3s} - s_{3s'})$, and enthalpy "loss", $(h_2 - h_{2'})$ as a constant pressure process resulting from non-optimum flow incidence.

change in velocity occurs suddenly, at one radius and is the basis of the so-called "shock loss model" used at one time to estimate the incidence loss.

The method used by NASA to evaluate the incidence loss was described by Meitner and Glassman (1980) and (1983) and was based upon a re-evaluation of the experimental data of Kofskey and Nusbaum (1972). They adopted the following equation devised originally by Roelke (1973) to evaluate the incidence losses in axial flow turbines:

$$\Delta h_i = h_2 - h_{2'} = \tfrac{1}{2}w_2^2(1 - \cos^n i_2). \tag{8.46}$$

Based upon data relating to six stators and one rotor, they found values for the exponent n which depended upon whether the incidence was positive or negative. With the present angle convention,

$$n = 2.5 \text{ for } i > 0 \text{ and } n = 1.75 \text{ for } i < 0.$$

Figure 8.11 shows the variation of the incidence loss function $(1 - \cos^n i)$ for a range of the incidence angle i using the appropriate values of n.

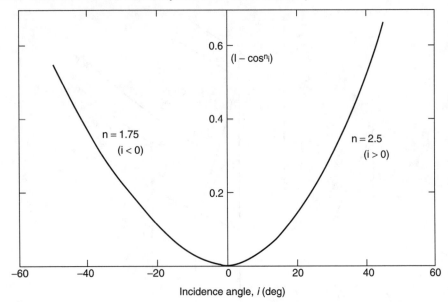

FIG. 8.11. Variation of incidence loss function at rotor inlet as a function of the incidence angle.

EXAMPLE 8.6(a): For the IFR turbine described in Example 8.3, and using the data and results in Example 8.4 and 8.5, deduce a value for the rotor enthalpy loss coefficient, ζ_R, at the optimum efficiency flow condition.

(b) The rotor speed of rotation is now reduced so that the relative flow enters the rotor radially (i.e. at the nominal flow condition). Assuming that the enthalpy loss coefficients, ζ_N and ζ_R remain the same, determine the total-to-static efficiency of the turbine for this off-design condition.

Solution. (a) From eqn. (8.10), solving for ζ_R,

$$\zeta_R = [(1 - \eta_{ts})c_0^2 - c_3^2 - \zeta_N c_2^2]/w_3^2.$$

We need to find values for c_0, c_3, w_3 and c_2.
  From the data,

$$c_3 = c_{m3} = 0.25 \times 538.1 = 134.5 \, \text{m/s}.$$

$$w_{3av} = 2w_2 = 2c_{m2}/\cos \beta_2 = 2 \times 135.4/\cos 33.560 = 324.97 \, \text{m/s}.$$

$$\tfrac{1}{2}c_0^2 = \Delta W/\eta_{ts} = 230 \times 10^3/0.81 = 283.95 \times 10^3$$

$$c_2 = 468.8 \, \text{m/s}.$$

$$\therefore \zeta_R = (2 \times 283.95 \times 10^3 \times 0.19 - 134.5^2 - 0.06 \times 468.8^2)/324.97^2$$

$$= 76,624/105,605$$

$$= 0.7256.$$

(b) Modifying the simplified expression for η_{ts}, eqn. (8.10), to include the incidence loss term given above,

$$\eta_{ts} = 1 - [c_3^2 + \zeta_N c_2^2 + \zeta_R w_3^2 + (1 - \cos^n i_2)w_2^2]/c_0^2.$$

As noted earlier, eqn. (8.10) is an approximation which ignores the weak effect of the temperature ratio T_3/T_2 upon the value of η_{ts}. In this expression $w_2 = c_{m2}$, the relative velocity at rotor entry, $i = -\beta_{2,\text{opt}} = -33.56$ deg. and $n = 1.75$. Hence, $(1 - \cos^{1.75} 33.56) = 0.2732$.

$$\therefore \eta_{ts} = 1 - [134.5^2 + 0.06 \times 468.8^2$$

$$+ 0.7256 \times 324.97^2 + 0.2732 \times 135.4^2]/567\,900$$

$$= 1 - [18\,090 + 13\,186 + 76\,627 + 5\,008]/567\,900$$

$$\therefore \eta_{ts} = 0.801.$$

This example demonstrates that the efficiency reduction when operating at the nominal design state is only one per cent and shows the relative insensitivity of the IFR turbine to operating at this off-design condition. At other off-design conditions the inlet relative velocity w_2 could be much bigger and the incidence loss correspondingly larger.

Significance and application of specific speed

The concept of specific speed N_s has already been discussed in Chapter 1 and some applications of it have been made already. Specific speed is extensively used to describe turbomachinery operating requirements in terms of shaft speed, volume flow rate and ideal specific work (alternatively, power developed is used instead of specific work). Originally, specific speed was applied almost exclusively to *incompressible* flow machines as a tool in the selection of the optimum type and size of unit. Its application to units handling *compressible* fluids was somewhat inhibited, due, it would appear, to the fact that volume flow rate changes through the machine, which raised the awkward question of which flow rate should be used in the specific speed definition. According to Balje (1981), the significant volume flow rate which should be used for turbines is that in the rotor exit, Q_3. This has now been widely adopted by many authorities.

Wood (1963) found it useful to factorise the basic definition of the specific speed equation, eqn. (1.8), in terms of the geometry and flow conditions within the radial-inflow turbine. Adopting the non-dimensional form of specific speed, in order to avoid ambiguities,

$$N_s = \frac{NQ_3^{1/2}}{\Delta h_{0s}^{3/4}} \tag{8.47}$$

where N is in rev/s, Q_3 is in m^3/s and the isentropic total-to-total enthalpy drop Δh_{0s} (from turbine inlet to exhaust) is in J/kg (i.e. m^2/s^2).

For the 90 deg IFR turbine, writing $U_2 = \pi N D_2$ and $\Delta h_{0s} = \frac{1}{2}c_0^2$, eqn. (8.47) can be factorised as follows:

$$N_s = \frac{Q_3^{1/2}}{(\frac{1}{2}c_0^2)^{3/4}}\left(\frac{U_2}{\pi D_2}\right)\left(\frac{U_2}{\pi N D_2}\right)^{1/2}$$

$$= \left(\frac{\sqrt{2}}{\pi}\right)^{3/2}\left(\frac{U_2}{c_0}\right)^{3/2}\left(\frac{Q_3}{N D_2^3}\right)^{1/2} \tag{8.48}$$

For the *ideal* 90 deg. IFR turbine and with $c_{02} = U_2$, it was shown earlier that the blade speed to spouting velocity ratio, $U_2/c_0 = 1/\sqrt{2} = 0.707$. Substituting this value into eqn. (8.34),

$$N_s = 0.18\left(\frac{Q_3}{N D_2^3}\right)^{1/2}, \quad \text{(rev)} \tag{8.48a}$$

i.e. specific speed is directly proportional to the square root of the volumetric flow coefficient.

To obtain some physical significance from eqns. (8.47) and (8.48a), define a *rotor disc area* $A_d = \pi D_2^2/4$ and assume a uniform axial rotor exit velocity c_3 so that $Q_3 = A_3 c_3$, then as

$$N = U_2/(\pi D_2) = \frac{c_0\sqrt{2}}{2\pi D_2}$$

$$\frac{Q_3}{N D_2^3} = \frac{A_3 c_3 2\pi D_2}{\sqrt{2 c_0} D_2^2} = \frac{A_3}{A_d}\frac{c_3}{c_0}\frac{\pi^2}{2\sqrt{2}}$$

Hence,

$$N_s = 0.336\left(\frac{c_3}{c_0}\right)^{1/2}\left(\frac{A_3}{A_d}\right)^{1/2}, \quad \text{(rev)} \tag{8.48b}$$

or,

$$\Omega_s = 2.11\left(\frac{c_3}{c_0}\right)^{1/2}\left(\frac{A_3}{A_d}\right)^{1/2}, \quad \text{(rad)} \tag{8.48c}$$

In an early study of IFR turbine design for maximum efficiency, Rohlik (1968) specified that the ratio of the rotor shroud diameter to rotor inlet diameter should be limited to a maximum value of 0.7 to avoid excessive shroud curvature and that the exit hub to shroud tip ratio was limited to a minimum of 0.4 to avoid excess hub blade blockage and loss. Using this as data, an upper limit for A_3/A_d can be found,

$$\frac{A_3}{A_d} = \left(\frac{D_{3s}}{D_2}\right)^2\left[1 - \left(\frac{D_{3h}}{D_{3s}}\right)^2\right] = 0.7^2 \times (1 - 0.16) = 0.41.$$

Figure 8.12 shows the relationship between Ω_s, the *exhaust energy factor* $(c_3/c_0)^2$ and the area ratio A_3/A_d based upon eqn. (8.48c). According to Wood (1963), the

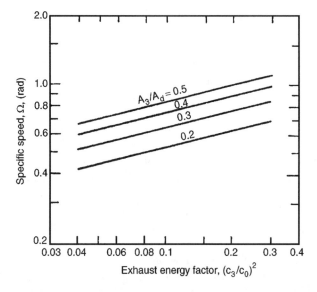

FIG. 8.12. Specific speed function for a 90 deg inward flow radial turbine (adapted from Wood 1963).

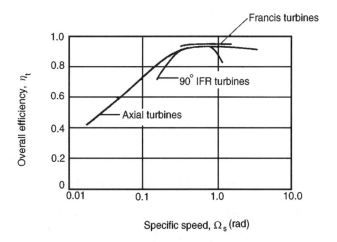

FIG. 8.13. Specific speed-efficiency characteristics for various turbines (adapted from Wood 1963).

limits for the exhaust energy factor in gas turbine practice are $0.04 < (c_3/c_0)^2 < 0.30$, the lower value being apparently a flow stability limit.

The numerical value of specific speed provides a general index of flow capacity relative to work output. Low values of Ω_s are associated with relatively small flow passage area and high values with relatively large flow passage areas. Specific speed has also been widely used as a general indication of achievable efficiency. Figure 8.13 presents a broad correlation of maximum efficiencies for hydraulic and compressible fluid turbines as functions of specific speed. These efficiencies apply to favourable design conditions with high values of flow Reynolds number, efficient diffusers and low leakage losses at the blade tips. It is seen that over a limited range of specific speed the best radial-flow turbines match the best axial-flow turbine efficiency, but from $\Omega_s = 0.03$ to 10, no other form of turbine handling compressible fluids can exceed the peak performance capability of the axial turbine.

Over the fairly limited range of specific speed $(0.3 \leq \Omega_s < 1.0)$ that the IFR turbine can produce a high efficiency, but it is difficult to find a decisive performance advantage in favour of either the axial flow turbine or the radial-flow turbine. New methods of fabrication enable the blades of small axial-flow turbines to be cast integrally with the rotor so that both types of turbine can operate at about the same blade tip speed. Wood (1963) compared the relative merits of axial and radial gas turbines at some length. In general, although weight, bulk and diameter are greater for radial than axial turbines, the differences are not so large and mechanical design compatibility can reverse the difference in a complete gas turbine power plant. The NASA nuclear Brayton cycle space power studies were all been made with 90 deg IFR turbines rather than with axial flow turbines.

The design problems of a small axial-flow turbine were discussed by Dunham and Panton (1973) who studied the cold performance measurements made on a single-shaft turbine of 13 cm diameter, about the same size as the IFR turbines tested by NASA. Tests had been performed with four different rotors to try and determine the effects of aspect ratio, trailing edge thickness, Reynolds number and tip-clearance. One turbine build achieved a total-to-total efficiency of 90 per cent, about equal to that of the best IFR turbine. However, because of the much higher outlet velocity, the total-to-static efficiency of the axial turbine gave a less satisfactory value (84 per cent) than the IFR type which could be decisive in some applications. They also confirmed that the axial turbine tip-clearance were comparatively large, losing two per cent efficiency for every one per cent increase in clearance. The tests illustrated one major design problem of a small axial turbine which was the extreme thinness of the blade trailing edges needed to achieve the efficiencies stated.

Optimum design selection of 90 deg IFR turbines

Rohlik (1968) has examined analytically the performance of 90 deg inward flow radial turbines in order to determine *optimum* design geometry for various applications as characterised by specific speed. His procedure, which extends an earlier treatment of Balje (1981) and Wood (1963) was used to determine the design point losses and corresponding efficiencies for various combinations of nozzle exit flow angle α_2, rotor diameter ratio D_2/D_{3av} and rotor blade entry height to exit diameter ratio, b_2/D_{3av}. The losses taken into account in the calculations are those associated with,

 (i) nozzle blade row boundary layers,
 (ii) rotor passage boundary layers,
 (iii) rotor blade tip clearance,
 (iv) disc windage (on the back surface of the rotor),
 (v) kinetic energy loss at exit.

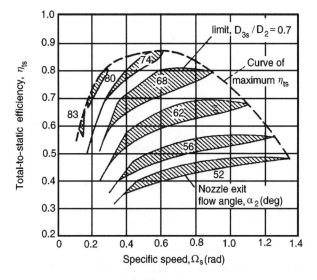

FIG. 8.14. Calculated performance of 90 deg IFR turbine (adapted from Rohlick 1968).

A mean-flowpath analysis was used and the passage losses were based upon the data of Stewart *et al.* (1960). The main constraints in the analysis were:

(a) $w_{3av}/w_2 = 2.0$
(b) $c_{\theta 3} = 0$
(c) $\beta_2 = \beta_{2,opt}$, i.e. zero incidence
(d) $r_{3s}/r_2 = 0.7$
(e) $r_{3h}/r_{3s} = 0.4$.

Figure 8.14 shows the variation in total-to-static efficiency with specific speed (Ω_s) for a selection of nozzle exit flow angles, α_2. For each value of α_2 a hatched area is drawn, inside of which the various diameter ratios are varied. The envelope of maximum η_{ts} is bounded by the constraints $D_{3h}/D_{3s} = 0.4$ in all cases and $D_{3s}/D_2 = 0.7$ for $\Omega_s \geqslant 0.58$ in these hatched regions. This envelope is the *optimum geometry curve* and has a peak η_{ts} of 0.87 at $\Omega_s = 0.58$ rad. An interesting comparison is made by Rohlik with the experimental results obtained by Kofskey and Wasserbauer (1966) on a single 90 deg IFR turbine rotor operated with several nozzle blade row configurations. The peak value of η_{ts} from this experimental investigation also turned out to be 0.87 at a slightly higher specific speed, $\Omega_s = 0.64$ rad.

The distribution of losses for optimum geometry over the specific speed range is shown in Figure 8.15. The way the loss distributions change is a result of the changing ratio of flow to specific work. At low Ω_s all friction losses are relatively large because of the high ratios of surface area to flow area. At high Ω_s the high velocities at turbine exit cause the kinetic energy leaving loss to predominate. Figure 8.16 shows several meridional plane sections at three values of specific speed corresponding to the curve of maximum total-to-static efficiency. The ratio of nozzle exit height to rotor diameter, b_2/D_2, is shown in Figure 8.17, the general rise of this

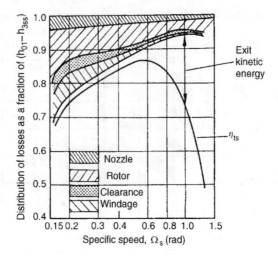

FIG. 8.15. Distribution of losses along envelope of maximum total-to-static efficiency (adapted from Rohlik 1968).

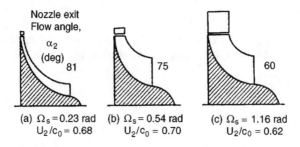

FIG. 8.16. Sections of radial turbines of maximum static efficiency (adapted from Rohlik 1968).

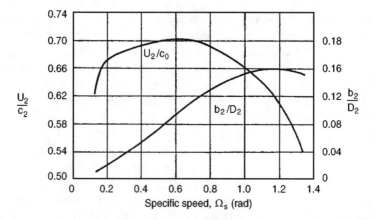

FIG. 8.17. Variation in blade speed/spouting velocity ratio (U_2/c_0) and nozzle blade height/rotor inlet diameter (b_2/D_2) corresponding to maximum total-to-static efficiency with specific speed (adapted from Rohlik 1968).

ratio with increasing Ω_s reflecting the increase in nozzle flow area* accompanying the larger flow rates of higher specific speed. Figure 8.17 also shows the variation of U_2/c_0 with Ω_s along the curve of maximum total-to-static efficiency.

Clearance and windage losses

A clearance gap must exist between the rotor vanes and the shroud. Because of the pressure difference between the pressure and suction surfaces of a vane, a leakage flow occurs through the gap introducing a loss in efficiency of the turbine. The minimum clearance is usually a compromise between manufacturing difficulty and aerodynamic requirements. Often, the minimum clearance is determined by the differential expansion and cooling of components under *transient* operating conditions which can compromise the steady state operating condition. According to Rohlik (1968) the loss in specific work as a result of gap leakage can be determined with the simple proportionality:

$$\Delta h_c = \Delta h_0(c/b_{av}) \tag{8.49}$$

where Δh_0 is the turbine specific work uncorrected for clearance or windage losses and c/b_{av} is the ratio of the gap to average vane height (i.e. $b_{av} = \frac{1}{2}(b_2 + b_3)$). A constant axial and radial gap, $c = 0.25$ mm, was used in the analytical study of Rohlik quoted earlier. According to Rodgers (1969) extensive development on small gas turbines has shown that it is difficult to maintain clearances less than about 0.4 mm. One consequence of this is that as small gas turbines are made progressively smaller the *relative* magnitude of the clearance loss must increase.

The non-dimensional power loss due to windage on the back of the rotor has been given by Shepherd (1956) in the form:

$$\Delta P_w/(\rho_2\Omega^3 D_2^5) = \text{ constant } \times Re^{-1/5}$$

where Ω is the rotational speed of the rotor and Re is a Reynolds number. Rohlik (1968) used this expression to calculate the loss in specific work due to windage,

$$\Delta h_w = 0.56\rho_2 D_2^2(U_2/100)^3/(\dot{m}\ Re) \tag{8.50}$$

where $\dot{m}$ is the total rate of mass flow entering the turbine and the Reynolds number is defined by $Re = U_2 D_2/v_2$, v_2 being the kinematic viscosity of the gas corresponding to the static temperature T_2 at nozzle exit.

Pressure ratio limits of the 90 deg IFR turbine

Every turbine type has pressure ratio limits, which are reached when the flow chokes. Choking usually occurs when the absolute flow at rotor exit reaches sonic velocity. (It can also occur when the *relative* velocity within the rotor reaches sonic conditions.) In the following analysis it is assumed that the turbine first chokes when

---

*The ratio b_2/D_2 is also affected by the pressure ratio and this has not been shown.

the absolute exit velocity c_3 reaches the speed of sound. It is also assumed that c_3 is without swirl and that the fluid is a perfect gas.

For simplicity it is also assumed that the diffuser efficiency is 100% so that, referring to Figure 8.4, $T_{04ss} = T_{03ss}(p_{03} = p_{04})$. Thus, the turbine total-to-total efficiency is,

$$\eta_t = \frac{T_{01} - T_{03}}{T_{01} - T_{03ss}}.$$ (8.51)

The expression for the spouting velocity, now becomes

$$c_0^2 = 2C_p(T_{01} - T_{03ss}),$$

is substituted into eqn. (8.51) to give,

$$\eta_t = \frac{1}{1 - (T_{03ss}/T_{01})} - \frac{2C_p T_{03}}{c_0^2}.$$ (8.52)

The stagnation pressure ratio across the turbine stage is given by $p_{03}/p_{01} = (T_{03ss}/T_{01})^{\gamma/(\gamma-1)}$; substituting this into eqn. (8.52) and rearranging, the exhaust energy factor is,

$$\left(\frac{c_3}{c_0}\right)^2 = \left[\frac{1}{1 - (p_{03}/p_{01})^{(\gamma-1)/\gamma}} - \eta_t\right]\frac{c_3^2}{2C_p T_{03}}.$$ (8.53)

Now $T_{03} = T_3[1 + \frac{1}{2}(\gamma - 1)M_3^2]$ and

$$\frac{c_3^2}{2C_p} = T_{03} - T_{01} = T_3\left(\frac{\gamma - 1}{2}\right)M_3^2,$$

therefore,

$$\frac{c_3^2}{2C_p T_{03}} = \frac{\frac{1}{2}(\gamma - 1)M_3^2}{1 + \frac{1}{2}(\gamma - 1)M_3^2}.$$ (8.54)

With further manipulation of eqn. (8.53) and using eqn. (8.54) the stagnation pressure ratio is expressed explicitly as

$$\left(\frac{p_{01}}{p_{03}}\right)^{(\gamma-1)/\gamma} = \frac{(c_3/c_0)^2 + [\frac{1}{2}(\gamma - 1)M_3^2\eta_t]/[1 + \frac{1}{2}(\gamma - 1)M_3^2]}{(c_3/c_0)^2 - [\frac{1}{2}(\gamma - 1)M_3^2(1 - \eta_t)]/[1 + \frac{1}{2}(\gamma - 1)M_3^2]}.$$ (8.55)

Wood (1963) has calculated the pressure ratio (p_{01}/p_{03}) using this expression, with $\eta_t = 0.9$, $\gamma = 1.4$ and for $M_3 = 0.7$ and 1.0. The result is shown in Figure 8.14. In practice, exhaust choking effectively occurs at nominal values of $M_3 \doteq 0.7$ (instead of at the ideal value of $M_3 = 1.0$) due to non-uniform exit flow.

The kinetic energy ratio $(c_3/c_0)^2$ has a first order effect on the pressure ratio limits of single stage turbines. The effect of any exhaust swirl present would be to lower the limits of choking pressure ratio.

It has been observed by Wood that high pressure ratios tend to compel the use of lower specific speeds. This assertion can be demonstrated by means of Figure 8.12 taken together with Figure 8.18. In Figure 8.12, for a given value of A_3/A_d, Ω_s

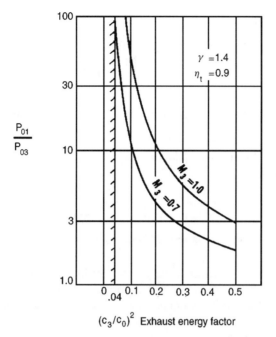

FIG. 8.18. Pressure ratio limit function for a turbine (Wood 1963) (By courtesy of the American Society of Mechanical Engineers).

increases with $(c_3/c_0)^2$ increasing. From Figure 8.18, (p_{01}/p_{03}) decreases with increasing values of $(c_3/c_0)^2$. Thus, for a given value of $(c_3/c_0)^2$, the specific speed must *decrease* as the design pressure ratio is increased.

Cooled 90 deg IFR turbines

The incentive to use higher temperatures in the basic Brayton gas turbine cycle is well known and arises from a desire to increase cycle efficiency and specific work output. In all gas turbines designed for high efficiency a compromise is necessary between the turbine inlet temperature desired and the temperature which can be tolerated by the turbine materials used. This problem can be minimised by using an auxiliary supply of cooling air to lower the temperature of the highly stressed parts of the turbine exposed to the high temperature gas. Following the successful application of blade cooling techniques to axial flow turbines (see, for example, Horlock 1966 or Fullagar 1973), methods of cooling small radial gas turbines have been developed.

According to Rodgers (1969) the most practical method of cooling small radial turbines is by film (or veil) cooling, Figure 8.19, where cooling air is impinged on the rotor and vane tips. The main problem with this method of cooling being its relatively low *cooling effectiveness*, defined by

$$\varepsilon = \frac{T_{01} - (T_m + \Delta T_0)}{T_{01} - (T_{0c} + \Delta T_0)} \tag{8.56}$$

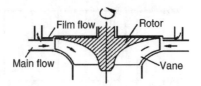

FIG. 8.19. Cross section of film-cooled radial turbine.

where T_m is the rotor metal temperature,

$\Delta T_0 = \frac{1}{2} U_2^2 / C_p$, is half the drop in stagnation temperature of the

gas as a result of doing work on the rotor,

T_{0c} is the stagnation temperature of the cooling air.

Rodgers refers to tests which indicate the possibility of obtaining $\varepsilon = 0.30$ at the rotor tip section with a cooling flow of approximately 10% of the main gas flow. Since the cool and hot streams rapidly mix, effectiveness decreases with distance from the point of impingement. A model study of the heat transfer aspects of film-cooled radial flow gas turbines is given by Metzger and Mitchell (1966).

References

Abidat, M., Chen, H., Baines, N. C. and Firth, M. R. (1992). Design of a highly loaded mixed flow turbine. *J. Power and Energy, Proc. Instn. Mech. Engrs.*, **206**, 95–107.

Anon. (1971). Conceptual design study of a nuclear Brayton turboalternator-compressor. *Contractor Report, General Electric Company. NASA CR-113925.*

Balje, O. E. (1981). *Turbomachines – A guide to Design, Selection and Theory.* Wiley.

Benson, R. S. (1970). A review of methods for assessing loss coefficients in radial gas turbines. *Int. J. Mech. Sci.*, **12**.

Benson, R. S., Cartwright, W. G. and Das, S. K. (1968). An investigation of the losses in the rotor of a radial flow gas turbine at zero incidence under conditions of steady flow. *Proc. Instn. Mech. Engrs. London*, **182**, Pt 3H.

Bridle, E. A. and Boulter, R. A. (1968). A simple theory for the prediction of losses in rotors of inward radial flow turbines. *Proc. Instn. Mech. Engrs. London*, **182**, Pt 3H.

Fullagar, K. P. L. (1973). The design of air cooled turbine rotor blades. *Symposium on Design and Calculation of Constructions Subject to High Temperature*, University of Delft.

Futral, M. J. and Wasserbauer, C. A. (1965). Off-design performance prediction with experimental verification for a radial-inflow turbine. *NASA TN D-2621.*

Glassman, A. J. (1976). Computer program for design and analysis of radial inflow turbines. *NASA TN 8164.*

Hiett, G. F. and Johnson, I. H. (1964). Experiments concerning the aerodynamic performance of inward radial flow turbines. *Proc. Instn. Mech. Engrs.* **178**, Pt 3I.

Horlock, J. H. (1966). *Axial Flow Turbines.* Butterworths. (1973 reprint with corrections, Huntington, New York: Krieger.)

Huntsman, I., Hodson, H. P. and Hill, S. H. (1992). The design and testing of a radial flow turbine for aerodynamic research. *J. Turbomachinery, Trans. Am. Soc. Mech. Engrs.*, **114**, 4.

Jamieson, A. W. H. (1955). The radial turbine. Chapter 9 in *Gas Turbine Principles and Practice* (Sir H. Roxbee-Cox, ed.). Newnes.

Kearton, W. J. (1951). *Steam turbine theory and practice*. (6th edn). Pitman.

Kofskey, M. G. and Nusbaum, W. J. (1972). Effects of specific speed on experimental performance of a radial-inflow turbine. *NASA TN D-6605*.

Kofskey, M. G. and Wasserbauer, C. A. (1966). Experimental performance evaluation of a radial inflow turbine over a range of specific speeds. *NASA TN D-3742*.

Meitner, P. L. and Glassman, A. J. (1983). Computer code for off-design performance analysis of radial-inflow turbines with rotor blade sweep. *NASA TP 2199, AVRADCOM Tech. Report 83-C-4*.

Metzger, D. E. and Mitchell, J. W. (1966). Heat transfer from a shrouded rotating disc with film cooling. *J. Heat Transfer, Trans. Am. Soc. Mech Engrs*, **88**.

Nusbaum, W. J. and Kofskey, M. G. (1969). Cold performance evaluation of 4.97 inch radial-inflow turbine designed for single-shaft Brayton cycle space-power system. *NASA TN D-5090*.

Rodgers, C. (1969). A cycle analysis technique for small gas turbines. Technical Advances in Gas Turbine Design. *Proc. Instn. Mech. Engrs. London*, **183**, Pt 3N.

Rodgers, C. and Geiser, R. (1987). Performance of a high-efficiency radial/axial turbine. *J. of Turbomachinery, Trans. Am. Soc. Mech. Engrs.*, **109**.

Roelke, R. J. (1973). Miscellaneous losses. Chapter 8 in *Turbine Design and Applications* (A.J. Glassman, ed.) NASA SP 290, Vol. 2.

Rogers, G. F. C. and Mayhew, Y. R. (1995). *Thermodynamic and Transport Properties of Fluids* (5th edn). Blackwell.

Rohlik, H. E. (1968). Analytical determination of radial-inflow turbine design geometry for maximum efficiency. *NASA TN D-4384*.

Rohlik, H. E. (1975). Radial-inflow turbines. In *Turbine Design and Applications*. (A. J. Glassman, ed.). NASA SP 290, vol. 3.

Shepherd, D. G. (1956). *Principles of Turbomachinery*. Macmillan.

Stanitz, J. D. (1952). Some theoretical aerodynamic investigations of impellers in radial and mixed flow centrifugal compressors. *Trans. Am. Soc. Mech. Engrs.*, **74**, 4.

Stewart, W. L., Witney, W. J. and Wong, R. Y. (1960). A study of boundary layer characteristics of turbomachine blade rows and their relation to overall blade loss. *J. Basic Eng., Trans. Am. Soc. Mech. Engrs.*, **82**.

Whitfield, A. (1990). The preliminary design of radial inflow turbines. *J. Turbomachinery, Trans. Am. Soc. Mech. Engrs.*, **112**, 50–57.

Whitfield, A. & Baines, N. C. (1990). Computation of internal flows. Chapter 8 in *Design of Radial Turbomachines*. Longman.

Wilson, D. G. and Jansen, W. (1965). The aerodynamic and thermodynamic design of cryogenic radial-inflow expanders. ASME Paper 65 – WA/PID-6, 1–13.

Wood, H. J. (1963). Current technology of radial-inflow turbines for compressible fluids. *J. Eng. Power., Trans. Am. Soc. Mech. Engrs.*, **85**.

Problems

1. A small inward radial flow gas turbine, comprising a ring of nozzle blades, a radial-vaned rotor and an axial diffuser, operates at the nominal design point with a total-to-total efficiency of 0.90. At turbine entry the stagnation pressure and temperature of the gas is 400 kPa and 1,140 K. The flow leaving the turbine is diffused to a pressure of 100 kPa and has negligible final velocity. Given that the flow is just choked at nozzle exit, determine the impeller peripheral speed and the flow outlet angle from the nozzles.

For the gas assume $\gamma = 1.333$ and $R = 287\,\text{J/(kg °C)}$.

2. The mass flow rate of gas through the turbine given in Problem No. 1 is 3.1 kg/s, the ratio of the rotor axial width/rotor tip radius (b_2/r_2) is 0.1 and the nozzle isentropic velocity ratio (ϕ_2) is 0.96. Assuming that the space between nozzle exit and rotor entry is negligible and ignoring the effects of blade blockage, determine:

(i) the static pressure and static temperature at nozzle exit;
(ii) the rotor tip diameter and rotational speed;
(iii) the power transmitted assuming a mechanical efficiency of 93.5%.

3. A radial turbine is proposed as the gas expansion element of a nuclear powered Brayton cycle space power system. The pressure and temperature conditions through the stage at the design point are to be as follows:

Upstream of nozzles, $p_{01} = 699\,\text{kPa}, T_{01} = 1,145\,\text{K};$

Nozzle exit, $p_2 = 527.2\,\text{kPa}, T_2 = 1,029\,\text{K};$

Rotor exit, $p_3 = 384.7\,\text{kPa}, T_3 = 914.5\,\text{K}, T_{03} = 924.7\,\text{K}.$

The ratio of rotor exit mean diameter to rotor inlet tip diameter is chosen as 0.49 and the required rotational speed as 24,000 rev/min. Assuming the relative flow at rotor inlet is radial and the absolute flow at rotor exit is axial, determine:

(i) the total-to-static efficiency of the turbine;
(ii) the rotor diameter;
(iii) the implied enthalpy loss coefficients for the nozzles and rotor row.

The gas employed in this cycle is a mixture of helium and xenon with a molecular weight of 39.94 and a ratio of specific heats of 5/3. The universal gas constant is, $R_0 = 8.314\,\text{kJ/(kg-mol K)}$.

4. A film-cooled radial inflow turbine is to be used in a high performance open Brayton cycle gas turbine. The rotor is made of a material able to withstand a temperature of 1145 K at a tip speed of 600 m/s for short periods of operation. Cooling air is supplied by the compressor which operates at a stagnation pressure ratio of 4 to 1, with an isentropic efficiency of 80%, when air is admitted to the compressor at a stagnation temperature of 288 K. Assuming that the effectiveness of the film cooling is 0.30 and the cooling air temperature at turbine entry is the same as that at compressor exit, determine the maximum permissible gas temperature at entry to the turbine.

Take $\gamma = 1.4$ for the air. Take $\gamma = 1.333$ for the gas entering the turbine. Assume $R = 287\,\text{J/(kg K)}$ in both cases.

5. The radial inflow turbine in Problem 8.3 is designed for a specific speed Ω_s of 0.55 (rad). Determine:

(1) the volume flow rate and the turbine power output;
(2) the rotor exit hub and tip diameters;
(3) the nozzle exit flow angle and the rotor inlet passage width/diameter ratio, b_2/D_2.

6. An inward flow radial gas turbine with a rotor diameter of 23.76 cm is designed to operate with a gas mass flow of 1.0 kg/s at a rotational speed of 38 140 rev/min. At the design condition the inlet stagnation pressure and temperature are to be 300 kPa and 727°C. The turbine is to be "cold" tested in a laboratory where an air supply is available only at the stagnation conditions of 200 kPa and 102°C.

(a) Assuming dynamically similar conditions between those of the laboratory and the projected design determine, for the "cold" test, the equivalent mass flow rate and the speed of rotation. Assume the gas properties are the same as for air.

(b) Using property tables for air, determine the Reynolds numbers for both the hot and cold running conditions. The Reynolds number is defined in this context as:

$$Re = \rho_{01} N D^2 / \mu_{01}$$

where ρ_{01} and μ_{01} are the stagnation density and stagnation viscosity of the air, N is the rotational speed (rev/s) and D is the rotor diameter.

7. For the radial flow turbine described in the previous question and operating at the prescribed "hot" design point condition, the gas leaves the exducer directly to the atmosphere at a pressure of 100 kPa and without swirl. The absolute flow angle at rotor inlet is 72° to the radial direction. The relative velocity w_3 at the the mean radius of the exducer (which is one half of the rotor inlet radius r_2) is twice the rotor inlet relative velocity w_2. The nozzle enthalpy loss coefficient, $\zeta_N = 0.06$.

Assuming the gas has the properties of air with an average value of $\gamma = 1.34$ (this temperature range) and $R = 287$ J/kg K, determine:

(1) the total-to-static efficiency of the turbine;
(2) the static temperature and pressure at the rotor inlet;
(3) the axial width of the passage at inlet to the rotor;
(4) the absolute velocity of the flow at exit from the exducer;
(5) the rotor enthalpy loss coefficient;
(6) the radii of the exducer exit given that the radius ratio at that location is 0.4.

8. One of the early space power systems built and tested for NASA was based on the Brayton cycle and incorporated an IFR turbine as the gas expander. Some of the data available concerning the turbine are as follows:

Total-to total pressure ratio (turbine inlet to turbine exit),	$p_{01}/p_{03} = 1.560$
Total-to-static pressure ratio,	$p_{01}/p_3 = 1.613$
Total temperature at turbine entry,	$T_{01} = 1083$ K
Total pressure at inlet to turbine,	$T_{01} = 91$ kPa
Shaft power output (measured on a dynamometer)	$P_{net} = 22.03$ kW
Bearing and seal friction torque (a separate test),	$\tau_f = 0.0794$ Nm
Rotor diameter,	$D_2 = 15.29$ cm
Absolute flow angle at rotor inlet,	$\alpha_2 = 72°$
Absolute flow angle at rotor exit,	$\alpha_3 = 0°$
The hub to shroud radius ratio at rotor exit,	$r_h/r_t = 0.35$
Ratio of blade speed to jet speed,	$\nu = U_2/c_0 = 0.6958$

(c_0 based on total-to-static pressure ratio)

For reasons of crew safety, an inert gas argon ($R = 208.2$ J/(kg K), ratio of specific heats, $\gamma = 1.667$) was used in the cycle. The turbine design scheme was based on the concept of optimum efficiency.

Determine, for the design point:

(1) the rotor vane tip speed;
(2) the static pressure and temperature at rotor exit;

(3) the gas exit velocity and mass flow rate;
(4) the shroud radius at rotor exit;
(5) the relative flow angle at rotor inlet;
(6) the specific speed.

NB. The volume flow rate to be used in the definition of the specific speed is based on the rotor exit conditions.

CHAPTER 9

Hydraulic Turbines

Hear ye not the hum of mighty workings? (KEATS, *Sonnet No. 14*).
The power of water has changed more in this world than emperors or kings.
(Leonardo da Vinci).

Introduction

To put this chapter into perspective some idea of the scale of hydropower development in the world might be useful before delving into the intricacies of hydraulic turbines. A very detailed and authoritative account of virtually every aspect of hydropower is given by Raabe (1985) and this brief introduction serves merely to illustrate a few aspects of a very extensive subject.

Hydropower is the longest established source for the generation of electric power which, starting in 1880 as a small dc generating plant in Wisconsin, USA, developed into an industrial size plant following the demonstration of the economic transmission of high voltage ac at the Frankfurt Exhibition in 1891. Hydropower now has a worldwide yearly growth rate of about five per cent (i.e. doubling in size every 15 years). In 1980 the worldwide installed generating capacity was 460 GW according to the United Nations (1981) so, by the year 2000, at the above growth rate this should have risen to a figure of about 1220 GW. The main areas with potential for growth are China, Latin America and Africa.

Table 9.1 is an extract of data quoted by Raabe (1985) of the distribution of harnessed and harnessable potential of some of the countries with the biggest usable potential of hydro power. From this list it is seen that the People's Republic of China, the country with the largest harnessable potential in the world had, in 1974, harnessed only 4.22 per cent of this. According to Cotillon (1978), with growth rates of 14.2 per cent up to 1985 and then with a growth rate of eight per cent, the PRC should have harnessed about 26 per cent of its harvestable potential by the year 2000. This would need the installation of nearly 4600 MW per annum of new hydropower plant, and a challenge to the makers of turbines around the world! One scheme in the PRC, under construction since 1992 and scheduled for completion in 2009, is the Xanxia (Three Gorges) project on the Yangtse which has a planned installed capacity of 25 000 MW, and which would make it the biggest hydropower plant in the world.

Features of hydropower plants

The initial cost of hydropower plants may be much higher than those of thermal power plants. However, the present value of total costs (which includes those of

277

TABLE 9.1. Distribution of harnessed and harnessable potential of hydroelectric power.

Country	Usable potential, TWh	Amount of potential used, TWh	Percentage of usable potential
1 China (PRC)	1320	55.6	4.22
2 Former USSR	1095	180	16.45
3 USA	701.5	277.7	39.6
4 Zaire	660	4.3	0.65
5 Canada	535.2	251	46.9
6 Brazil	519.3	126.9	24.45
7 Malaysia	320	1.25	0.39
8 Columbia	300	13.8	4.6
9 India	280	46.87	16.7
Sum 1–9	5731	907.4	15.83
Other countries	4071	843	20.7
Total	9802.4	1750.5	17.8

TABLE 9.2. Features of hydroelectric powerplants.

Advantages	Disadvantages
Technology is relatively simple and proven. High efficiency. Long useful life. No thermal phenomena apart from those in bearings and generator.	Number of favourable sites limited and only available in some countries. Problems with cavitation and water hammer.
Small operating, maintenance and replacement costs.	High initial cost especially for low head plants compared with thermal power plants.
No air pollution. No thermal pollution of water.	Inundation of the reservoirs and displacement of the population. Loss of arable land. Facilitates sedimentation upstream and erosion downstream of a barrage.

fuel) is, in general, lower in hydropower plants. Raabe (1985) listed the various advantages and disadvantages of hydropower plants and a brief summary of these is given in Table 9.2.

Hydraulic turbines

Early history of hydraulic turbines

The hydraulic turbine has a long period of development, its oldest and simplest form being the waterwheel, first used in ancient Greece and subsequently adopted throughout medieval Europe for the grinding of grain, etc. It was a French engineer, Benoit Fourneyron, who developed the first commercially successful hydraulic turbine (circa 1830). Later Fourneyron built turbines for industrial purposes that achieved a speed of 2300 rev/min, developing about 50 kW at an efficiency of over 80 per cent.

The American engineer James B. Francis designed the first *radial-inflow* hydraulic turbine which became widely used, gave excellent results and was highly regarded. In its original form it was used for heads of between 10 and 100 m. A simplified form of this turbine is shown in Figure 1.1d. It will be observed that the flow path followed is essentially from a radial direction to an axial direction.

The Pelton wheel turbine, named after its American inventor Lester A. Pelton, was brought into use in the second half of the nineteenth century. This is an impulse turbine in which water is piped at high pressure to a nozzle where it expands completely to atmospheric pressure. The emerging jet impacts onto the blades (or buckets) of the turbine producing the required torque and power output. A simplified diagram of a Pelton wheel turbine is shown in Figure 1.1f. The head of water used originally was between about 90 m and 900 m (modern versions operate up to heads of 2000 m).

The increasing need for more power during the early years of the twentieth century also led to the invention of a turbine suitable for small heads of water, i.e. 3 m to 9 m, in river locations where a dam could be built. It was in 1913 that Viktor Kaplan revealed his idea of the propeller (or Kaplan) turbine, see Figure 1.1e, which acts like a ship's propeller but in reverse At a later date Kaplan improved his turbine by means of swivelable blades which improved the efficiency of the turbine in accordance with the prevailing conditions (i.e. the available flow rate and head).

Flow regimes for maximum efficiency

Although there are a large number of turbine types in use, only the three mentioned above and variants of them are considered in this book. The efficiencies of the three types are shown in Figure 9.1 as functions of the power specific speed, Ω_{sp} which from eqn. (1.9), is

$$\Omega_{sp} = \frac{\Omega\sqrt{P/\rho}}{(gH_E)^{5/4}} \tag{9.1}$$

where P is the power delivered by the shaft, H_E is the effective head at turbine entry and Ω is the rotational speed in rad/s.

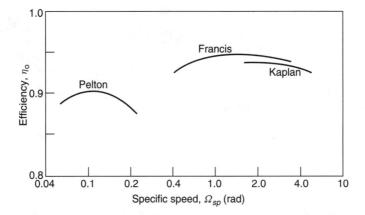

FIG. 9.1. Typical design point efficiencies of Pelton, Francis and Kaplan turbines.

TABLE 9.3. Operating ranges of hydraulic turbines.

	Pelton turbine	Francis turbine	Kaplan turbine
Specific speed (rad)	$0.05 - 0.4$	$0.4 - 2.2$	$1.8 - 5.0$
Head (m)	100–1770	20–900	6–70
Maximum power (MW)	500	800	300
Optimum efficiency, per cent	90	95	94
Regulation method	Needle valve and deflector plate	Stagger angle of guide vanes	Stagger angle of rotor bades

NB. Values shown in the table are only a rough guide and are subject to change.

   The regimes of these turbine types are of some significance to the designer as they indicate the most suitable choice of machine for an application once the specific speed has been determined. In general low specific speed machines correspond to low volume flow rates and high heads, whereas high specific speed machines correspond to high volume flow rates and low heads. Table 9.3 summarises the normal operating ranges for the specific speed, the effective head, the maximum power and best efficiency for each type of turbine.
   According to the experience of Sulzer Hydro Ltd., of Zurich, the application ranges of the various types of turbines and turbine pumps (including some not mentioned

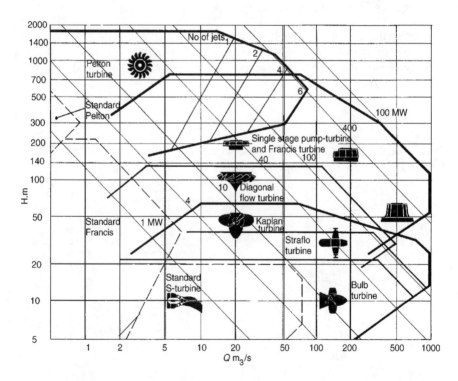

FIG. 9.2. Application ranges for various types of hydraulic turbomachinery, as a plot of Q vs H with lines of constant power determined assuming $\eta_0 = 0.8$. (Courtesy Sulzer Hydro Ltd., Zurich).

here) are plotted in Figure 9.2 on a ln Q vs ln H diagram, and reflect the present state of the art of hydraulic turbomachinery design. Also in Figure 9.2 lines of constant power output are conveniently shown and have been calculated as the product $\eta \rho g Q H$, where the efficiency η is accorded the value of 0.8 throughout the chart.

Capacity of large Francis turbines

The size and capacity of some of the recently built Francis turbines is a source of wonder, they seem so enormous! The size and weight of the runners cause special problems getting them to the site, especially when rivers have to be crossed and the bridges are inadequate.

The largest installation now in North America is at La Grande on James Bay in eastern Canada where 22 units each rated at 333 MW have a total capacity of 7326 MW. For the record, the Itaipu hydroelectric plant on the Paraná river (between Brazil and Paraguay), dedicated in 1982, has the greatest capacity of 12 870 MW in full operation (with a planned value of 21 500 MW) using 18 Francis turbines each sized at over 700 MW.

The efficiency of large Francis turbines has gradually risen over the years and now is about 95 per cent. An historical review of this progress has been given by Danel (1959). There seems to be little prospect of much further improvement in efficiency as skin friction, tip leakage and exit kinetic energy from the diffuser now apparently account for the remaining losses. Raabe (1985) has given much attention to the statistics of the world's biggest turbines. It would appear at the present time that the largest hydroturbines in the world are the three vertical shaft Francis turbines installed at Grand Coulee III on the Columbia River, Washington, USA. Each of these leviathans has been uprated to 800 MW, with the delivery (or effective) head, $H = 87$ m, $N = 85.7$ rev/min, the runner having a diameter of $D = 9.26$ m and weighing 450 ton. Using this data in eqn. (9.1) it is easy to calculate that the power specific speed is 1.74 rad.

The Pelton turbine

This is the only hydraulic turbine of the impulse type now in common use. It is an efficient machine and it is particularly suited to high head applications. The rotor consists of a circular disc with a number of blades (usually called "buckets") spaced around the periphery. One or more nozzles are mounted in such a way that each nozzle directs its jet along a tangent to the circle through the centres of the buckets. There is a "splitter" or ridge which splits the oncoming jet into two equal streams so that, after flowing round the inner surface of the bucket, the two streams depart from the bucket in a direction nearly opposite to that of the incoming jet.

Figure. 9.3 shows the runner of a Pelton turbine and Figure 9.4 shows a six-jet vertical axis Pelton turbine. Considering one jet impinging on a bucket, the appropriate velocity diagram is shown in Figure 9.5. The jet velocity at entry is c_1 and the blade speed is U so that the relative velocity at entry is $w_1 = c_1 - U$. At exit from the bucket one half of the jet stream flows as shown in the velocity diagram, leaving with a relative velocity w_2 and at an angle β_2 to the original direction of flow. From the velocity diagram the much smaller absolute exit velocity c_2 can be determined.

FIG. 9.3. Pelton turbine runner (Courtesy Sulzer Hydro Ltd, Zurich).

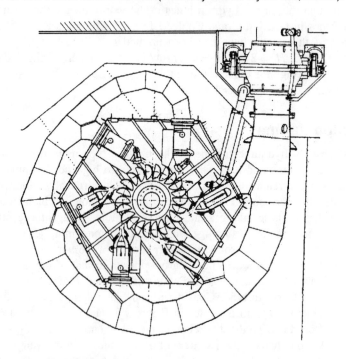

FIG. 9.4. Six-jet vertical shaft Pelton turbine, horizontal section. Power rating 174.4 MW, runner diameter 4.1 m, speed 300 rev/min, head 587 m. (Courtesy Sulzer Hydro Ltd., Zurich).

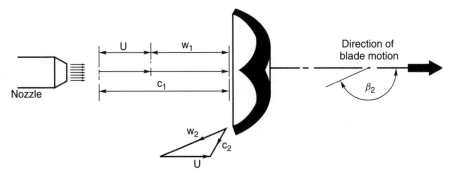

FIG. 9.5. The Pelton wheel showing the jet impinging onto a bucket and the relative and absolute velocities of the flow (only one-half of the emergent velocity diagram is shown).

From Euler's turbine equation, eqn. (2.12b), the specific work done by the water is

$$\Delta W = U_1 c_{\theta 1} - U_2 c_{\theta 2}.$$

For the Pelton turbine, $U_1 = U_2 = U$, $c_{\theta 1} = c_1$ so we get

$$\Delta W = U[U + w_1 - (U + w_2 \cos \beta_2)] = U(w_1 - w_2 \cos \beta_2)$$

in which the value of $c_{\theta 2} < 0$, as defined in Figure 9.5, i.e. $c_{\theta 2} = U + w_2 \cos \beta_2$.

The effect of friction on the fluid flowing inside the bucket will cause the relative velocity at outlet to be less than the value at inlet. Writing $w_2 = kw_1$, where $k < 1$, then,

$$\Delta W = Uw_1(1 - k \cos \beta_2) = U(c_1 - U)(1 - k \cos \beta_2). \tag{9.2}$$

An efficiency η_R for the runner can be defined as the specific work done ΔW divided by the incoming kinetic energy, i.e.

$$\eta_R = \Delta W / (\tfrac{1}{2} c_1^2) \tag{9.3}$$

$$= 2U(c_1 - U)(1 - k \cos \beta_2)/c_1^2$$

$$\therefore \eta_R = 2\nu(1 - \nu)(1 - k \cos \beta_2) \tag{9.4}$$

where the blade speed to jet speed ratio, $\nu = U/c_1$.

In order to find the optimum efficiency, differentiate eqn. (9.4) with respect to the blade speed ratio, i.e.

$$\frac{d\eta_R}{d\nu} = 2 \frac{d}{d\nu}(\nu - \nu^2)(1 - k \cos \beta_2)$$

$$= 2(1 - 2\nu)(1 - k \cos \beta_2) = 0.$$

Therefore, the maximum efficiency of the runner occurs when $\nu = 0.5$, i.e. $U = c_1/2$. Hence,

$$\eta_{R\,max} = (1 - k \cos \beta_2)/2. \tag{9.5}$$

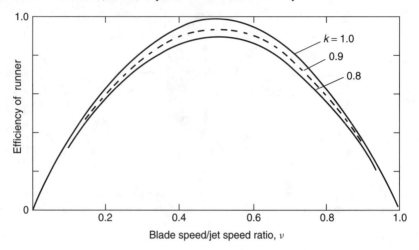

FIG. 9.6. Theoretical variation of runner efficiency for a Pelton wheel with blade speed to jet speed ratio for several values of friction factor k.

Figure. 9.6 shows the variation of the runner efficiency with blade speed ratio for assumed values of $k = 0.8, 0.9$ and 1.0 with $\beta_2 = 165$ deg. In practice the value of k is usually found to be between 0.8 and 0.9.

A simple hydroelectric scheme

The layout of a Pelton turbine hydroelectric scheme is shown in Figure 9.7. The water is delivered from a constant level reservoir at an elevation z_R (above sea level) and flows via a pressure tunnel to the penstock head, down the penstock to the turbine nozzles emerging onto the buckets as a high speed jet. In order to reduce the deleterious effects of large pressure surges, a *surge tank* is connected to the flow close to the penstock head which acts so as to damp out transients. The elevation of the nozzles is z_N and the gross head, $H_G = z_R - z_N$.

Controlling the speed of the Pelton turbine

The Pelton turbine is usually directly coupled to an electrical generator which must run at synchronous speed. With large size hydroelectric schemes supplying electricity to a national grid it is essential for both the voltage and the frequency to closely match the grid values. To ensure that the turbine runs at constant speed despite any load changes which may occur, the rate of flow Q is changed. A spear (or needle) valve, Figure 9.8a, whose position is controlled by means of a servomechanism, is moved axially within the nozzle to alter the diameter of the jet. This works well for very gradual changes in load. However, when a sudden loss in load occurs a more rapid response is needed. This is accomplished by temporarily deflecting the jet with a deflector plate so that some of the water does not reach the buckets, Figure 9.8b. This acts to prevent overspeeding and allows time for the slower acting spear valve to move to a new position.

It is vital to ensure that the spear valve *does move slowly* as a sudden reduction in the rate of flow could result in serious damage to the system from pressure surges

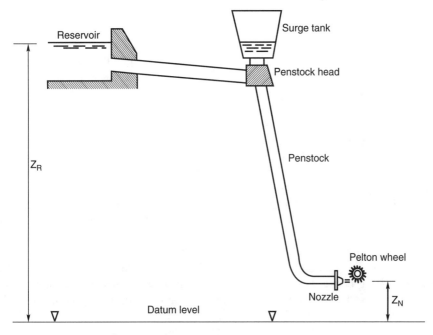

FIG. 9.7. Pelton turbine hydroelectric scheme.

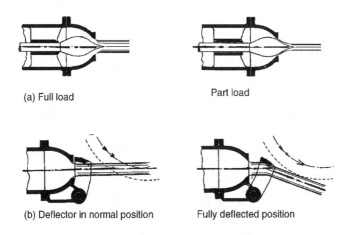

(a) Full load   Part load

(b) Deflector in normal position   Fully deflected position

FIG. 9.8. Methods of regulating the speed of a Pelton turbine: (a) with a spear (or needle) valve; (b) with a deflector plate.

(called "water hammer"). If the spear valve did close quickly, all the kinetic energy of the water in the penstock would be absorbed by the elasticity of the supply pipeline (penstock) and the water, creating very large stresses which would reach their greatest intensity at the turbine inlet where the pipeline is already heavily stressed. The surge chamber, shown in Figure 9.7, has the function of absorbing and dissipating some of the pressure and energy fluctuations created by too rapid a closure of the needle valve.

Sizing the penstock

It is shown in elementary textbooks on fluid mechanics, e.g. Shames (1992), Douglas *et al.* (1995), that the loss in head with incompressible, steady, turbulent flow in pipes of circular cross-section is given by Darcy's equation:

$$H_f = \frac{2flV^2}{gd} \tag{9.6}$$

where f is the friction factor, l is the length of the pipe, d is the pipe diameter and V is the mass average velocity of the flow in the pipe. It is assumed, of course, that the pipe is running full. The value of the friction factor has been determined for various conditions of flow and pipe surface roughness and the results are usually presented in what is called a "Moody diagram". This diagram gives values of f as a function of pipe Reynolds number for varying levels of relative roughness of the pipe wall.

The penstock (the pipeline bringing the water to the turbine) is long and of large diameter and this can add significantly to the total cost of a hydroelectric power scheme. Using Darcy's equation it is easy to calculate a suitable pipe diameter for such a scheme if the friction factor is known and an estimate can be made of the allowable head loss. Logically, this head loss would be determined on the basis of the cost of materials, etc. needed for a large diameter pipe and compared with the value of the useful energy lost from having too small a pipe. A commonly used compromise for the loss in head in the supply pipes is to allow $H_f \leqslant 0.1 H_G$.

A summary of various factors on which the "economic diameter" of a pipe can be determined is given by Raabe (1985).

From eqn. (9.6), substituting for the velocity, $V = 4Q/(\pi d^2)$, we get

$$H_f = \left(\frac{32fl}{\pi^2 g} \right) \frac{Q^2}{d^5}. \tag{9.7}$$

EXAMPLE 9.1. Water is supplied to a turbine at the rate $Q = 2.272 \, \text{m}^3/\text{s}$ by a single penstock 300 m long. The allowable head loss due to friction in the pipe amounts to 20 m. Determine the diameter of the pipe if the friction factor $f = 0.1$.

Solution. Rearranging eqn. (9.7):

$$d^5 = \frac{32fl}{gH_f} \left(\frac{Q}{\pi} \right)^2 = \frac{32 \times 0.01 \times 300}{9.81 \times 20} \left(\frac{2.272}{\pi} \right)^2$$

$$= 0.2559$$

$$\therefore d = 0.7614 \, \text{m}.$$

Energy losses in the Pelton turbine

Having accounted for the energy loss due to friction in the penstock, the energy losses in the rest of the hydroelectric scheme must now be considered. The effective head, H_E, (or delivered head) at entry to the turbine is the gross head minus the friction head loss, H_f, i.e.

$$H_E = H_G - H_f = z_R - z_N - H_f$$

and the spouting (or ideal) velocity, c_0, is

$$c_0 = \sqrt{2gH_E}.$$

The pipeline friction loss H_f is regarded as an external loss and is not included in the losses attributed to the turbine system. The efficiency of the turbine is measured against the ideal total head H_E.

The nozzle velocity coefficient, K_N, is

$$K_N = \frac{\text{actual velocity at nozzle exit}}{\text{spouting velocity at nozzle exit}} = \frac{c_1}{c_0}.$$

Values of K_N are normally around 0.98 to 0.99.

Other energy losses occur in the nozzles and also because of windage and friction of the turbine wheel. Let the loss in head in the nozzle be ΔH_N then the head available for conversion into power is

$$H_E - \Delta H_N = c_1^2/(2g). \tag{9.8}$$

$$\text{nozzle efficiency, } \eta_N = \frac{\text{energy at nozzle exit}}{\text{energy at nozzle inlet}} = \frac{c_1^2}{2gH_E} \tag{9.9}$$

Equation (2.23) is an expression for the hydraulic efficiency of a turbine which, in the present notation and using eqns. (9.3) and (9.9), becomes

$$\eta_h = \frac{\Delta W}{gH_E} = \left(\frac{\Delta W}{\frac{1}{2}c_1^2}\right)\left(\frac{\frac{1}{2}c_1^2}{gH_E}\right) = \eta_R\eta_N. \tag{9.10}$$

The efficiency η_R only represents the effectiveness of converting the kinetic energy of the jet into the mechanical energy of the runner. Further losses occur as a result of bearing friction and "windage" losses inside the casing of the runner. In large Pelton turbines efficiencies of around 90 per cent may be achieved but, in smaller units, a much lower efficiency is usually obtained.

The overall efficiency

In Chapter 2 the overall efficiency was defined as

$$\eta_0 = \frac{\text{mechanical energy available at output shaft in unit time}}{\text{maximum energy difference possible for the fluid in unit time}}$$

$$\eta_0 = \eta_m\eta_h = \eta_m\eta_R\eta_N$$

where η_m is the mechanical efficiency.

The external losses, bearing friction and windage, are chiefly responsible for the energy deficit between the runner and the shaft. An estimate of the effect of the windage loss can be made using the following simple flow model in which the specific energy loss is assumed to be proportional to the square of the blade speed, i.e.

$$\text{loss/unit mass flow} = KU^2$$

where K is a dimensionless constant of proportionality.

The overall efficiency can now be written as

$$\eta_0 = \frac{\Delta W - KU^2}{gH_E} = \eta_h - \frac{KU^2}{gH_g} = \eta_h - 2K\left(\frac{U}{c_1}\right)^2\left(\frac{c_1^2}{2gH_E}\right)$$

$$\therefore \eta_0 = \eta_R\eta_N - 2K\eta_N v^2 = \eta_N(\eta_R - 2Kv^2). \tag{9.11}$$

Hence, the mechanical efficiency is,

$$\eta_m = 1 - 2Kv^2/\eta_R. \tag{9.12}$$

It can be seen that according to eqn. (9.12), as the speed ratio is reduced towards zero, the mechanical efficiency increases and approaches unity. As there must be *some* bearing friction at all speeds, however small, an additional term is needed in the loss equation of the form $Ac_0^2 + kU^2$, where A is another dimensionless constant. The solution of this is left for the student to solve.

The variation of the overall efficiency based upon eqn. (9.11) is shown in Figure 9.9 for several values of K. It is seen that the peak efficiency:

(1) is progressively reduced as the value of K is increased;
(2) occurs at lower values of v than the optimum determined for the runner.

Thus, this evaluation of overall efficiency demonstrates the reason why experimental results obtained of the performance of Pelton turbines always yields a peak efficiency at a value of $v < 0.5$.

Typical performance of a Pelton turbine *under conditions of constant head and speed* is shown in Figure 9.10 in the form of the variation of overall efficiency against load ratio. As result of a change in the load the output of the turbine must then be regulated by a change in the setting of the needle valve in order to keep the turbine speed constant. The observed almost constant value of the efficiency over most of the load range is the result of the *hydraulic losses* reducing in proportion to the power output. However, as the load ratio is reduced to even lower values, the windage and bearing friction losses, which have not diminished, assume a relatively greater importance and the overall efficiency rapidly diminishes towards zero.

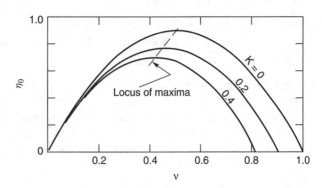

FIG. 9.9. Variation of overall efficiency of a Pelton turbine with speed ratio for several values of windage coefficient, K.

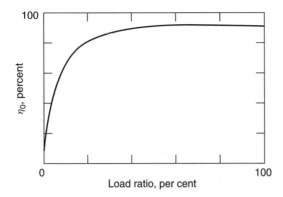

Fig. 9.10. Pelton turbine overall efficiency variation with load under constant head and constant speed conditions.

EXAMPLE 9.2. A Pelton turbine is driven by two jets, generating 4.0 MW at 375 rev/min. The effective head at the nozzles is 200 m of water and the nozzle velocity coefficient, $K_N = 0.98$. The axes of the jets are tangent to a circle 1.5 m in diameter. The relative velocity of the flow across the buckets is decreased by 15 per cent and the water is deflected through an angle of 165 deg.

Neglecting bearing and windage losses, determine:

(1) the runner efficiency;
(2) the diameter of each jet;
(3) the power specific speed.

Solution. (1) The blade speed is:

$$U = \Omega r = (375 \times \pi/30) \times 1.5/2$$

$$= 39.27 \times 1.5/2 = 29.45 \text{ m/s}.$$

The jet speed is:

$$c_1 = K_N \sqrt{2gHe} = 0.98 \times \sqrt{(2 \times 9.81 \times 200)} = 61.39 \text{ m/s}$$

$$\therefore v = U/c_1 = 0.4798.$$

The efficiency of the runner is obtained from eqn. (9.4):

$$\eta_R = 2 \times 0.4798 \times (1 - 0.4798)(1 - 0.85 \times \cos 165°)$$

$$= 0.9090.$$

(2) The "theoretical" power is $P_{th} = P/\eta_R = 4.0/0.909 = 4.40 \text{ MW}$ where $P_{th} = \rho g Q H_e$

$$\therefore Q = P_{th}/(\rho g H_e) = 4.4 \times 10^6/(9810 \times 200) = 2.243 \text{ m}^3/\text{s}.$$

Each jet must have a flow area of,

$$A_j = \frac{Q}{2c_1} = 2.243/(2 \times 61.39) = 0.01827 \text{ m}^2.$$

$$\therefore d_j = 0.1525 \text{ m}.$$

(3) Substituting into eqn. (9.1), the power specific speed is,

$$\Omega_{sp} = 39.27 \times \left(\frac{4.0 \times 10^6}{10^3} \right)^{1/2} /(9.81 \times 200)^{5/4}$$

$$= 0.190 \, \text{rad}.$$

Reaction turbines

The primary features of the reaction turbine are:

(1) only part of the overall pressure drop has occurred up to turbine entry, the remaining pressure drop takes place in the turbine itself;
(2) the flow completely fills all of the passages in the runner, unlike the Pelton turbine where, for each jet, only one or two of the buckets at a time are in contact with the water;
(3) pivotable guide vanes are used to control and direct the flow;
(4) a draft tube is normally added on to the turbine exit; it is considered as an integral part of the turbine.

The pressure of the water gradually decreases as it flows through the runner and it is the reaction from this pressure change which earns this type of turbine its appellation.

The Francis turbine

The majority of Francis turbines are arranged so that the axis is vertical (some smaller machines can have horizontal axes). Figure 9.11 illustrates a section through a vertical shaft Francis turbine with a runner diameter of 5 m, a head of 110 m and a power rating of nearly 200 MW. Water enters via a spiral casing called a *volute* or *scroll* which surrounds the runner. The area of cross-section of the volute decreases along the flow path in such a way that the flow velocity remains constant. From the volute the flow enters a ring of stationary guide vanes which direct it onto the runner at the most appropriate angle.

In flowing through the runner the angular momentum of the water is reduced and work is supplied to the turbine shaft. At the design condition the absolute flow leaves the runner axially (although a small amount of swirl may be countenanced) into the *draft tube* and, finally, the flow enters the *tail race*. It is essential that the exit of the draft tube is submerged below the level of the water in the tail race in order that the turbine remains full of water. The draft tube also acts as a diffuser; by careful design it can ensure maximum recovery of energy through the turbine by significantly reducing the exit kinetic energy.

Figure. 9.12 shows a section through part of a Francis turbine together with the velocity triangles at inlet to and exit from the runner at mid-blade height. At inlet to the guide vanes the flow is in the radial/tangential plane, the absolute velocity is c_1 and the absolute flow angle is α_1. Thus,

$$\alpha_1 = \tan^{-1}(c_{\theta 1}/c_{r 1}). \tag{9.13}$$

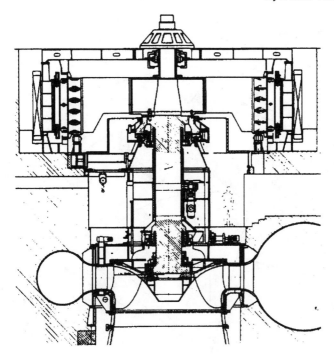

FIG. 9.11. Vertical shaft Francis turbine: runner diameter 5 m, head 110 m, power 200 MW (courtesy Sulzer Hydro Ltd, Zurich).

The flow is turned to angle α_2 and velocity c_2, the absolute condition of the flow at entry to the runner. By vector subtraction the relative velocity at entry to the runner is found, i.e. $\mathbf{w_2} = \mathbf{c_2} - \mathbf{U_2}$. The relative flow angle β_2 at inlet to the runner is defined as

$$\beta_2 = \tan^{-1}\left[(c_{\theta 2} - U_2)/c_{r2}\right]. \tag{9.14}$$

Further inspection of the velocity diagrams in Figure 9.12 reveals that the direction of the velocity vectors approaching both guide vanes and runner blades are tangential to the camber lines at the leading edge of each row. This is the ideal flow condition for "shockless" low loss entry, although an incidence of a few degrees may be beneficial to output without a significant extra loss penalty. At vane outlet some deviation from the blade outlet angle is to be expected (see Chapter 3). For these reasons, in all problems concerning the direction of flow, it is clear that it is the angle of the fluid flow which is important and not the vane angle as is often quoted in other texts.

At outlet from the runner the flow plane is simplified as though it was actually in the radial/tangential plane. This simplification will not affect the subsequent analysis of the flow but it must be conceded that some component of velocity in the axial direction does exist at runner outlet.

The water leaves the runner with a relative flow angle β_3 and a relative flow velocity w_3. The absolute velocity at runner exit is found by vector addition, i.e.

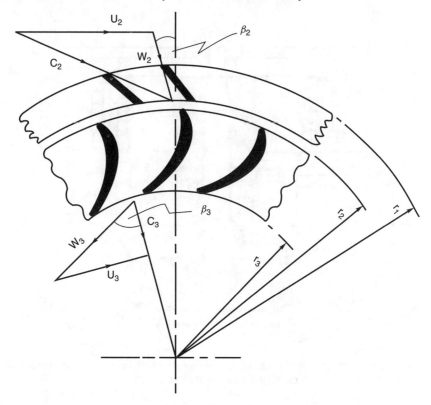

FIG. 9.12. Sectional sketch of blading for a Francis turbine showing velocity diagrams at runner inlet and exit.

$c_3 = w_3 + U_3$. The relative flow angle, β_3, at runner exit is given by

$$\beta_3 = \tan^{-1}\left[(c_{\theta 3} + U_3)/c_{r3}\right]. \tag{9.15}$$

In this equation it is assumed that some residual swirl velocity $c_{\theta 3}$ is present (c_{r3} is the radial velocity at exit from the runner). In most simple analyses of the Francis turbine it is assumed that there is no exit swirl. Detailed investigations have shown that some extra *counter-swirl* (i.e. acting so as to increase Δc_θ) at the runner exit does increase the amount of work done by the fluid without a significant reduction in turbine efficiency.

When a Francis turbine is required to operate at part load, the power output is reduced by swivelling the guide vanes to restrict the flow, i.e. Q is reduced, while the blade speed is maintained constant. Figure. 9.13 compares the velocity triangles at full load and at part load from which it will be seen that the relative flow at runner entry is at a high incidence and at runner exit the absolute flow has a large component of swirl. Both of these flow conditions give rise to high head losses. Figure. 9.14 shows the variation of hydraulic efficiency for several types of turbine, including the Francis turbine, over the full load range at constant speed and constant head.

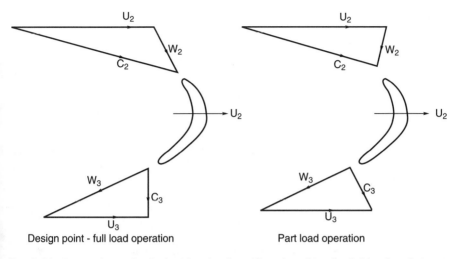

FIG. 9.13. Comparison of velocity triangles for a Francis turbine for full load and at part load operation.

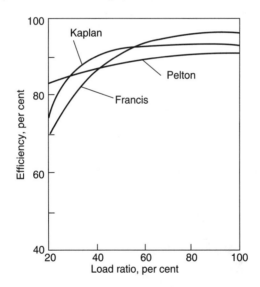

FIG. 9.14. Variation of hydraulic efficiency for various types of turbine over a range of loading, at constant speed and constant head.

It is of interest to note the effect that swirling flow has on the performance of the following diffuser. The results of an extensive experimental investigation made by McDonald *et al.* (1971), showed that swirling inlet flow *does not* affect the performance of conical diffusers which are well designed and give unseparated or only slightly separated flow when the flow through them is entirely axial. Accordingly, part load operation of the turbine is unlikely to give adverse diffuser performance.

Basic equations

Euler's turbine equation, eqn. (2.12b), in the present notation, is written as

$$\Delta W = U_2 c_{\theta 2} - U_3 c_{\theta 3}. \tag{9.16}$$

If the flow at runner exit is without swirl then the equation reduces to

$$\Delta W = U_2 c_{\theta 2}. \tag{9.16a}$$

The effective head for all reaction turbines, H_E, is the total head available at the turbine inlet *relative to the surface of the tailrace*. At entry to the runner the energy available is equal to the sum of the kinetic, potential and pressure energies, i.e.

$$g(H_E - \Delta H_N) = \frac{p_2 - p_a}{\rho} + \tfrac{1}{2} c_2^2 + g z_2, \tag{9.17}$$

where ΔH_N is the loss of head due to friction in the volute and guide vanes and p_2 is the *absolute* static pressure at inlet to the runner.

At runner outlet the energy in the water is further reduced by the amount of specific work ΔW and by friction work in the runner, $g\Delta H_R$ and this remaining energy equals the sum of the pressure potential and kinetic energies, i.e.

$$g(H_E - \Delta H_N - \Delta H_R) - \Delta W = \tfrac{1}{2} c_3^2 + p_3/\rho - p_a/\rho + g z_3 \tag{9.18}$$

where p_3 is the *absolute* static pressure at runner exit.

By differencing eqns. (9.17) and (9.18), the specific work is obtained

$$\Delta W = (p_{02} - p_{03})/\rho - g\Delta H_R + g(z_2 - z_3) \tag{9.19}$$

where p_{02} and p_{03} are the absolute total pressures at runner inlet and exit.

Figure 9.15 shows the draft tube in relation to a vertical-shaft Francis turbine. The most important dimension in this diagram is the vertical distance ($z = z_3$) between the

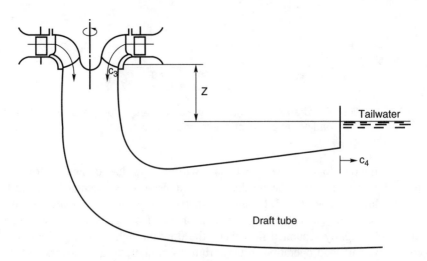

FIG. 9.15. Location of draft tube in relation to vertical shaft Francis turbine.

exit plane of the runner and the free surface of the tailrace. The energy equation between the exit of the runner and the tailrace can now be written as

$$p_3/\rho + \tfrac{1}{2}c_3^2 + gz - g\Delta H_{DT} = \tfrac{1}{2}c_4^2 + p_a/\rho, \qquad (9.20)$$

where ΔH_{DT} is the loss in head in the draft tube and c_4 is the exit velocity.

The hydraulic efficiency is given by

$$\eta_H = \frac{\Delta W}{gH_E} = \frac{U_2 c_{\theta 2} - U_3 c_{\theta 3}}{gH_E} \qquad (9.21)$$

and, if $c_{\theta 3} = 0$, then

$$\eta_H = \frac{U_2 c_{\theta 2}}{gH_E}. \qquad (9.21a)$$

The overall efficiency is given by $\eta_0 = \eta_m \eta_H$. For large machines the mechanical losses are relatively small and $\eta_m \approx 100$ per cent and so $\eta_0 \approx \eta_H$.

For the Francis turbine the ratio of the runner speed to the spouting velocity, $v = U/c_0$, is not as critical for high efficiency operation as it is for the Pelton turbine and, in practice, it lies within a fairly wide range, i.e. $0.6 \leqslant v \leqslant 0.9$. In most applications of Francis turbines the turbine drives an alternator and its speed must be maintained constant. The regulation at part load operation is achieved by varying the angle of the guide vanes. The guide vanes are pivoted and, by means of a gearing mechanism, the setting can be adjusted to the optimum angle. However, operation at part load causes a whirl velocity component to be set up downstream of the runner causing a reduction in efficiency. The strength of the vortex can be such that cavitation can occur along the axis of the draft tube (see remarks on cavitation later in this chapter).

EXAMPLE 9.3. In a vertical-shaft Francis turbine the available head at the inlet flange of the turbine is 150 m and the vertical distance between the runner and the tailrace is 2.0 m. The runner tip speed is 35 m/s, the meridional velocity of the water through the runner is constant and equal to 10.5 m/s, the flow leaves the runner without whirl and the velocity at exit from the draft tube is 3.5 m/s. The hydraulic energy losses estimated for the turbine are as follows:

$$\Delta H_N = 6.0\,\text{m}, \ \Delta H_R = 10.0\,\text{m}, \ \Delta H_{DT} = 1.0\,\text{m}.$$

Determine:

(1) the pressure head (relative to the tailrace) at inlet to and at exit from the runner;
(2) the flow angles at runner inlet and at guide vane exit;
(3) the hydraulic efficiency of the turbine.

If the flow discharged by the turbine is 20 m³/s and the power specific speed of the turbine is 0.8 (rad), determine the speed of rotation and the diameter of the runner.

Solution. From eqn. (9.20)

$$\frac{p_3 - p_a}{\rho g} = H_3 = \frac{1}{2g}(c_4^2 - c_3^2) + \Delta H_{DT} - z.$$

NB. The head H_3 is relative to the tailrace.

$$\therefore H_3 = \frac{(3.5^2 - 10.5^2)}{2 \times 9.81} + 1 - 2 = -6.0\,\text{m},$$

i.e. the pressure at runner outlet is *below* atmospheric pressure, a matter of some importance when we come to consider the subject of cavitation later in this chapter. From eqn. (9.18),

$$H_2 = H_E - \Delta H_N - c_2^2/(2g) = 150 - 6 - 38.73^2/(2 \times 9.81) = 67.22\,\text{m}.$$

From eqn. (9.18),

$$\Delta W = g(H_E - \Delta H_N - \Delta H_{R-z}) - \tfrac{1}{2}c_3^2 - gH_3$$

$$= 9.81 \times (150 - 6 - 10 - 2) - 10.5^2/2 + 9.81 \times 6 = 1298.7\,\text{m}^2/\text{s}^2$$

$$\therefore c_{\theta 2} = \Delta W/U_2 = 1298.7/35 = 37.1\,\text{m/s}$$

$$\alpha_2 = \tan^{-1}\left(\frac{c_{\theta 2}}{c_{r2}}\right) = \tan^{-1}\left(\frac{37.1}{10.5}\right) = 74.2\,\text{deg}$$

$$\beta_2 = \tan^{-1}\left(\frac{c_{\theta 2}}{c_{r2}}\right) = \tan^{-1}\left(\frac{37.1 - 35}{10.5}\right) = 11.31\,\text{deg}.$$

The hydraulic efficiency is

$$\eta_H = \frac{\Delta W}{gH_E} = 1298.7/(9.81 \times 150) = 0.8826.$$

From the definition of the power specific speed, eqn. (9.1),

$$\Omega = \frac{\Omega_{SP}(gH_E)^{5/4}}{\sqrt{Q\Delta W}} = \frac{0.8 \times 9114}{\sqrt{20 \times 1298.7}} = 45.24\,\text{rad/s}.$$

Thus, the rotational speed is $N = 432$ rev/min and the runner diameter is

$$D_2 = 2U_2/\Omega = 70/45.24 = 1.547\,\text{m}.$$

The Kaplan turbine

This type of turbine evolved from the need to generate power from much lower pressure heads than are normally employed with the Francis turbine. To satisfy large power demands very large volume flow rates need to be accommodated in the Kaplan turbine, i.e. the product QH_E is large. The overall flow configuration is from radial to axial. Figure 9.16 is a part sectional view of a Kaplan turbine in which the flow enters from a volute into the inlet guide vanes which impart a degree of swirl to the flow determined by the needs of the runner. The flow leaving the guide vanes is forced by the shape of the passage into an axial direction and the swirl becomes essentially a free vortex, i.e.

$rc_\theta = $ a constant.

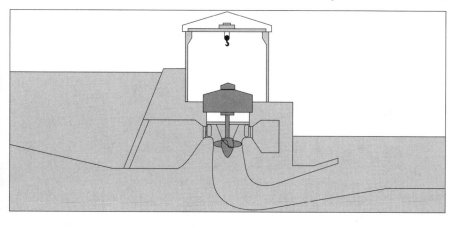

Fɪɢ. 9.16. Part section of a Kaplan turbine *in situ* (courtesy Sulzer Hydro Ltd, Zurich).

The vanes of the runner are similar to those of an axial-flow turbine rotor but designed with a twist suitable for the free-vortex flow at entry and an axial flow at outlet. Because of the very high torque that must be transmitted and the large length of the blades, strength considerations impose the need for large blade chords. As a result, pitch/chord ratios of 1.0 to 1.5 are commonly used by manufacturers and, consequently, the number of blades is small, usually 4, 5 or 6. The Kaplan turbine incorporates one essential feature not found in other turbine rotors and that is the setting of the stagger angle can be controlled. At part load operation the setting angle of the runner vanes is adjusted automatically by a servo mechanism to maintain optimum efficiency conditions. This adjustment requires a complementary adjustment of the inlet guide vane stagger angle in order to maintain an absolute axial flow at exit from the runner.

Basic equations

Most of the equations presented for the Francis turbine also apply to the Kaplan (or propeller) turbine, apart from the treatment of the runner. Figure 9.17 shows the velocity triangles and part section of a Kaplan turbine drawn for the mid-blade height. At exit from the runner the flow is shown leaving the runner without a whirl velocity, i.e. $c_{\theta 3} = 0$ and constant axial velocity. The theory of free-vortex flows was expounded in Chapter 6 and the main results as they apply to an incompressible fluid are given here. The runner blades will have a fairly high degree of twist, the amount depending upon the strength of the circulation function K and the magnitude of the axial velocity. Just upstream of the runner the flow is assumed to be a free-vortex and the velocity components are accordingly:

$$c_{\theta 2} = K/r \quad c_x = \text{a constant.}$$

The relations for the flow angles are

$$\tan \beta_2 = U/c_x - \tan \alpha_2 = \Omega r/c_x - K/(rc_x) \tag{9.22a}$$

$$\tan \beta_3 = U/c_x = \Omega r/c_x. \tag{9.22b}$$

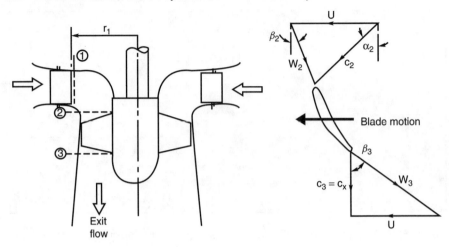

FIG. 9.17. Section of a Kaplan turbine and velocity diagrams at inlet to and exit from the runner.

EXAMPLE 9.4. A small-scale Kaplan turbine has a power output of 8 MW, an available head at turbine entry of 13.4 m and a rotational speed of 200 rev/min. The inlet guide vanes have a length of 1.6 m and the diameter at the trailing edge surface is 3.1 m. The runner diameter is 2.9 m and the hub to tip ratio is 0.4.

Assuming the hydraulic efficiency is 92 per cent and the runner design is "free-vortex", determine:

(1) the radial and tangential components of velocity at exit from the guide vanes;
(2) the component of axial velocity at the runner;
(3) the absolute and relative flow angles upstream and downstream of the runner at the hub, mid-radius and tip.

Solution. As $P = \eta_H \rho g Q H_E$, then the volume flow rate is

$$Q = P/(\eta_H \rho g H_E) = 8 \times 10^6/(0.92 \times 9810 \times 13.4) = 66.15 \text{ m/s}^2$$

$$\therefore c_{r1} = Q/(2\pi r_1 L) = 66.15/(2\pi \times 1.55 \times 1.6) = 4.245 \text{ m/s}^2$$

$$c_{x2} = \frac{4Q}{\pi D_{2t}^2(1 - v^2)} = 4 \times 66.15/(\pi \times 2.9^2 \times 0.84) = 11.922 \text{ m/s}^2.$$

As the specific work done is $\Delta W = U_2 c_{\theta 2}$ and $\eta_H = \Delta W/(gH_E)$, then at the tip

$$c_{\theta 2} = \frac{\eta_H g H_E}{U_2} = \frac{0.92 \times 9.81 \times 13.4}{30.37} = 3.892 \text{ m/s},$$

where the blade tip speed is, $U_2 = \Omega D_2/2 = (200 \times \pi/30) \times 2.9/2 = 30.37 \text{ m/s}$

$$c_{\theta 1} = c_{\theta 2} r_2/r_1 = 3.892 \times 1.45/1.55 = 3.725 \text{ m/s}^2$$

$$\alpha_1 = \tan^{-1}\left(\frac{c_{\theta 1}}{c_{r1}}\right) = \tan^{-1}\left(\frac{3.725}{4.245}\right) = 41.26 \text{ deg}.$$

TABLE 9.4. Calculated values of flow angles for Example 9.4.

Parameter	Ratio r/r_t		
	0.4	0.7	1.0
$c_{\theta 2}$ m/s	9.955	5.687	3.982
$\tan \alpha_2$	0.835	0.4772	0.334
α_2 (deg)	39.86	25.51	18.47
U/c_{x2}	1.019	1.7832	2.547
β_2 (deg)	10.43	52.56	65.69
β_3 (deg)	45.54	60.72	68.57

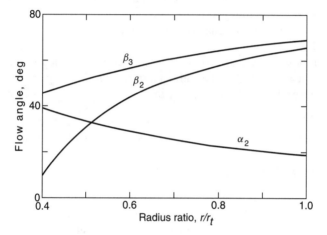

FIG. 9.18. Calculated variation of flow angles for Kaplan turbine of Example 9.4.

Values α_2, β_2 and β_3 shown in Table 9.4 have been derived from the following relations:

$$\alpha_2 = \tan^{-1}\left(\frac{c_{\theta 2}}{c_{x2}}\right) = \tan^{-1}\left(\frac{c_{\theta 2t}}{c_{x2}}\frac{r_t}{r}\right)$$

$$\beta_2 = \tan^{-1}\left(\frac{\Omega r}{c_{x2}} - \tan \alpha_2\right) = \tan^{-1}\left(\frac{U_{2t}}{c_{x2}}\frac{r}{r_t} - \tan \alpha_2\right)$$

$$\beta_3 = \tan^{-1}\left(\frac{U}{c_{x2}}\right) = \tan^{-1}\left(\frac{U_{2t}}{c_{x2}}\frac{r}{r_t}\right).$$

Finally, Figure 9.18 illustrates the variation of the flow angles, from which the large amount of blade twist mentioned earlier can be inferred.

Effect of size on turbomachine efficiency

Despite careful attention to detail at the design stage and during manufacture it is a fact that small turbomachines always have lower efficiencies than larger geometrically similar machines. The primary reason for this is that it is not possible to establish perfect dynamical similarity between turbomachines of different size. In order to obtain this condition, each of the the dimensionless terms in eqn. (1.2) would need to be the same for all sizes of a machine.

To illustrate this consider a family of turbomachines where the loading term, $\psi = gH/N^2D^2$ is the same and the Reynolds number, $Re = ND^2/\nu$ is the same for every size of machine, then

$$\psi Re^2 = \frac{gH}{N^2D^2} \cdot \frac{N^2D^4}{\nu^2} = \frac{gHD^2}{\nu^2}$$

must be the same for the whole family. Thus, for a given fluid (ν is a constant), a reduction in size D must be followed by an increase in the head H. A turbine model of $\frac{1}{8}$ the size of a prototype would need to be tested with a head 64 times that required by the prototype! Fortunately, the effect on the model efficiency caused by changing the Reynolds number is not large. In practice, models are normally tested at conveniently low heads and an empirical correction is applied to the efficiency.

With model testing other factors effect the results. Exact geometric similarity cannot be achieved for the following reasons:

(a) the blades in the model will probably be relatively thicker than in the prototype;
(b) the relative surface roughness for the model blades will be greater;
(c) leakage losses around the blade tips of the model will be relatively greater as a result of increased relative tip clearances.

Various simple corrections have been devised (see Addison 1964) to allow for the effects of size (or scale) on the efficiency. One of the simplest and best known is that due to Moody and Zowski (1969), also reported by Addison (1964) and Massey (1979), which as applied to the efficiency of reaction turbines is

$$\frac{1-\eta_p}{1-\eta_m} = \left(\frac{D_m}{D_p}\right)^n \tag{9.23}$$

where the subscripts p, m refer to prototype and model, and the index n is in the range 0.2 to 0.25. From comparison of field tests of large units with model tests, Moody and Zowski concluded that the best value for n was approximately 0.2 rather than 0.25 and for general application this is the value used. However, Addison (1964) reported tests done on a full-scale Francis turbine and a model made to a scale of 1 to 4.54 which gave measured values of the maximum efficiencies of 0.85 and 0.90 for the model and full-scale turbines, respectively, which agreed very well with the ratio computed with $n = 0.25$ in the Moody formula!

EXAMPLE 9.5. A model of a Francis turbine is built to a scale of 1/5 of full size and when tested it developed a power output of 3 kW under a head of 1.8 m of water, at a rotational speed of 360 rev/min and a flow rate of 0.215 m³/s. Estimate the speed, flow rate and power of the full-scale turbine when working under dynamically similar conditions with a head of 60 m of water.

By making a suitable correction for scale effects, determine the efficiency and the power of the full-size turbine. Use Moody's formula and assume $n = 0.25$.

Solution. From the group $\psi = gH/(ND)^2$ we get:

$$N_p = N_m(D_m/D_p)(H_p/H_m)^{0.5} = (360/5)(60/1.8)^{0.5} = 415.7 \,\text{rev/min}.$$

From the group $\phi = Q/(ND^3)$ we get:

$$Q_p = Q_m(N_p/N_m)(D_p/D_m)^3 = 0.215 \times (360/415.7) \times 5^3 = 23.27 \,\text{m}^3/\text{s}.$$

Lastly, from the group $\hat{P} = P/(\rho N^3 D^5)$ we get:

$$P_p = P_m(N_p/N_m)^3(D_p/D_m)^5 = 3 \times (415.7)^3 \times 5^5 = 14\,430 \,\text{kW} = 14.43 \,\text{MW}.$$

This result has still to be corrected to allow for scale effects. First we must calculate the efficiency of the model turbine. The efficiency is found from

$$\eta_m = P/(\rho QgH) = 3 \times 10^3/(10^3 \times 0.215 \times 9.81 \times 1.8) = 0.79.$$

Using Moody's formula the efficiency of the prototype is determined:

$$(1 - \eta_p) = (1 - \eta_m) \times 0.2^{0.25} = 0.21 \times 0.6687$$

hence

$$\eta_p = 0.8596.$$

The corresponding power is found by an adjustment of the original power obtained under dynamically similar conditions, i.e.

Corrected $P_p = 14.43 \times 0.8596/0.79 = 15.7 \,\text{MW}.$

Cavitation

A description of the phenomenon of cavitation, mainly with regard to pumps, was given in Chapter 1. In hydraulic turbines, where reliability, long life and efficiency are all so very important, the effects of cavitation must be considered. Two types of cavitation may be in evidence,

(a) on the suction surfaces of the runner blades at outlet which can cause severe blade erosion; and
(b) a twisting "rope-type" cavity that appears in the draft tube at off-design operating conditions.

Cavitation in hydraulic turbines can occur on the suction surfaces of the runner blades where the dynamic action of the blades acting on the fluid creates low pressure zones in a region where the static pressure is already low. Cavitation will commence when the local static pressure is less than the vapour pressure of the water, i.e. where the head is low, the velocity is high and the elevation, z, of the turbine is set too high above the tailrace. For a turbine with a horizontal shaft the lowest pressure will be located in the upper part of the runner, which could be of major significance in large machines. Fortunately, the runners of large machines are, in general, made so that their shafts are orientated vertically, lessening the problem of cavitation occurrence.

The cavitation performance of hydraulic turbines can be correlated with the Thoma coefficient, σ, defined as

$$\sigma = \frac{H_S}{H_E} = \frac{(p_a - p_v)/(\rho g) - z}{H_E}, \tag{9.24}$$

where H_S is the net positive suction head (NPSH), the amount of head needed to avoid cavitation, the difference in elevation, z, is defined in Figure 9.15 and p_v is the vapour pressure of the water. The Thoma coefficient was, strictly, originally defined in connection with cavitation in turbines and its use in pumps is not appropriate (see Yedidiah 1981). It is to be shown that σ represents the fraction of the available head H_E which is unavailable for the production of work. A large value of σ means that a smaller part of the available head can be utilised. In a pump, incidentally, there is no direct connection between the developed head and its suction capabilities, provided that cavitation does not occur, which is why the use of the Thoma coefficient is not appropriate for pumps.

From the energy equation, eqn. (9.20), this can be rewritten as

$$\frac{p_a - p_3}{\rho g} - z = \frac{1}{2g}(c_3^2 - c_4^2) - \Delta H_{DT}, \tag{9.25}$$

so that when $p_3 = p_v$, then H_S is equal to the rhs of eqn. (9.24).

Figure 9.19 shows a widely used correlation of the Thoma coefficient plotted against specific speed for Francis and Kaplan turbines, approximately defining the boundary between no cavitation and severe cavitation. In fact, there exists a wide range of critical values of σ for each value of specific speed and type of turbine due to the individual cavitation characteristics of the various runner designs. The curves drawn are meant to assist preliminary selection procedures. An alternative method for avoiding cavitation is to perform tests on a model of a particular turbine in which the value of p_3 is reduced until cavitation occurs or, a marked decrease in efficiency becomes apparent. This performance reduction would correspond to the production of large-scale cavitation bubbles. The pressure at which cavitation erosion occurs will actually be at some higher value than that at which the performance reduction starts.

For the centre-line cavitation that appears downstream of the runner at off-design operating conditions, oscillations of the cavity can cause severe vibration of the draft tube. Young reported some results of a "corkscrew" cavity rotating at 4 Hz. Air injected into the flow both stabilizes the flow and cushions the vibration.

EXAMPLE 9.6. Using the data in Example 9.3 and given that the atmospheric pressure is 1.013 bar and the water is at 25°C, determine the NPSH for the turbine. Hence, using Thoma's coefficient and the data shown in Figure 9.19, determine whether cavitation is likely to occur. Also using the data of Wislicenus verify the result.

Solution. From tables of fluid properties, e.g. Rogers and Mayhew (1995), or using the data of Figure 9.20, the vapour pressure for water corresponding to a

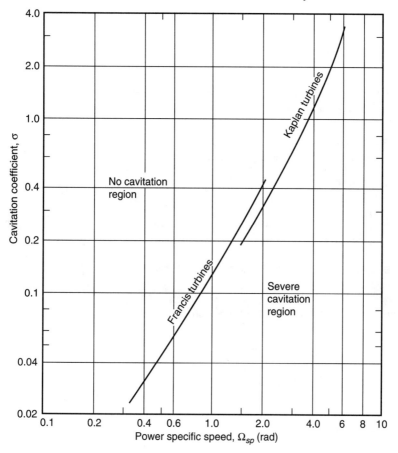

FIG. 9.19. Variation of critical cavitation coefficient with non-dimensional specific speed for Francis and Kaplan turbines (adapted from Moody and Zowski 1969).

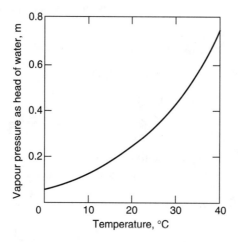

FIG. 9.20. Vapour pressure of water as head (m) versus temperature.

temperature of 25°C is 0.03166 bar. From the definition of NPSH, eqn. (9.24), we obtain:

$$H_s = \frac{p_a - p_v}{\rho g} - z = (1.013 - 0.03166) \times 10^5/(9810) - 2 = 8.003 \, \text{m}.$$

Thus, Thoma's coefficient is, $\sigma = H_S/H_E = 8.003/150 = 0.05336$.

At the value of $\Omega_{SP} = 0.8$ given as data, the value of the critical Thoma coefficient σ_c corresponding to this is 0.09 from Figure 9.19. From the fact that $\sigma < \sigma_c$, then the turbine *will* cavitate.

From the definition of the suction specific speed

$$\Omega_{SS} = \frac{\Omega Q^{1/2}}{(gH_S)^{3/4}} = \frac{44.9 \times 20^{1/2}}{(9.81 \times 8.003)^{3/4}} = 200.8/26.375 = 7.613.$$

According to eqn. (1.12b), when Ω_{SS} exceeds 4.0 (rad) then cavitation can occur, giving further confirmation of the above conclusion.

Connection between Thoma's coefficient, suction specific speed and specific speed

The definitions of suction specific speed and specific speed are

$$\Omega_{SS} = \frac{\Omega Q^{1/2}}{(gH_S)^{3/4}} \quad \text{and} \quad \Omega_S = \frac{\Omega Q^{1/2}}{(gH_E)^{3/4}}$$

Combining and using eqn. (9.24), we get:

$$\frac{\Omega_S}{\Omega_{SS}} = \left(\frac{gH_S}{gH_E}\right)^{3/4} = \sigma^{3/4}$$

$$\therefore \sigma = \left(\frac{\Omega_S}{\Omega_{SS}}\right)^{4/3}. \tag{9.26}$$

Exercise. Verify the value of Thoma's coefficient in the earlier example using the values of power specific speed, efficiency and suction specific speed given or derived.

We use as data $\Omega_{SS} = 7.613$, $\Omega_{SP} = 0.8$ and $\eta_H = 0.896$ so that, from eqn. (1.9c),

$$\Omega_S = \Omega_{SP}/\sqrt{\eta_H} = 0.8/\sqrt{0.896} = 0.8452$$

$$\therefore \sigma = (0.8452/7.613)^{4/3} = 0.05336.$$

Avoiding cavitation

By rearranging eqn. (9.24) and putting $\sigma = \sigma_c$, a critical value of z can be derived on the boundary curve between cavitation and no cavitation. Thus,

$$z = z_c = \frac{p_a - p_v}{\rho g} - \sigma_c H_E = (101.3 - 3.17)/9.81 - 0.09 \times 150 = -3.5 \, \text{m}.$$

This means that the turbine would need to be submerged at a depth of 3.5 m or more below the surface of the tailwater and, for a Francis turbine, would lead to problems with regard to construction and maintenance. Equation (9.24) shows that the greater the available head H_E at which a turbine operates, the lower it must be located relative to the surface of the tailrace.

Sonoluminescence

The collapse of vapour cavities generates both noise and flashes of light (called *sonoluminescence*). Young (1989) has given an extended and interesting review of experiments on sonoluminescence from hydrodynamic cavitation and its causes. The phenomenon has also been reported by Pearsall (1974) who considered that the collapse of the cavity was so rapid that very high pressures and temperatures were created. Temperatures as high as 10 000 K have been suggested. Shock waves with pressure differences of 4000 atm have been demonstrated in the liquid following the collapse of a cavity. The effect of the thermal and pressure shocks on any material in close proximity causes mechanical failure, i.e. erosion damage.

Light has been reported in large energy distributions in field installations. An example again quoted by Young is that of the easily visible light observed at night in the tailrace at Boulder Dam, USA. This occurs when sudden changes of load necessitate the release of large quantities of high-pressure water into an energy-dissipating structure. Under these conditions the water cavitates severely. In a further example, Young mentions the light (observed at night) from the tailrace of the hydroelectric power station at Erochty, Scotland. The luminescence appeared for up to ten seconds shortly after the relief valve was opened and was seen as a blue shimmering light stretching over an area of the water surface for several square metres.

References

Addison, H. (1964). *A Treatise on Applied Hydraulics* (5th edn), Chapman and Hall.

Cotillon, J. (1978). L'hydroélectricité dans le monde. *Houille Blanche*, **33**, no. 1/2, 71–86.

Danel, P. (1959). The hydraulic turbine in evolution. *Proc. Instn. Mech. Engrs.*, **173**, 36–44.

Douglas, J. F., Gasiorek, J. M. and Swaffield, J. A. (1995). *Fluid Mechanics* (3rd edn), Longman.

Durrer, H. (1986). Cavitation erosion and fluid mechanics. *Sulzer Technical Review*, Pt. 3, 55–61.

Massey, B. S. (1979). *Mechanics of Fluids* (4th edn), van Nostrand.

McDonald, A. T., Fox, R. W. and van Dewoestine, R. V. (1971). Effects of swirling inlet flow on pressure recovery in conical diffusers. *AIAA Journal*, **9**, No. 10, 2014–18.

Moody, L. F. and Zowski, T. (1969). Hydraulic machinery. In Section 26, *Handbook of Applied Hydraulics* (3rd edn), (C. V. Davis and K. E. Sorensen, eds). McGraw-Hill.

Nerz, K. P. (1986). Early detection and surveillance of material damaging processes by digital acoustic emission analysis. *Sulzer Technical Review*, Pt. 3. 62–4.

Pearsall, I. S. (1974). Cavitation. CME. *Instn. Mech. Engrs.*, July, 79–85.

Raabe, J. (1985). *Hydro Power. The Design, Use, and Function of Hydromechanical, Hydraulic, and Electrical Equipment*. VDI Verlag.

Rogers, G. F. C. and Mayhew, Y. R. (1995). *Thermodynamic and Transport Properties of Fluids (SI Units)* (5th edn). Blackwell.

Shames, I. H. (1992). *Mechanics of Fluids* (3rd edn), McGraw-Hill.

Yedidiah, S. (1981). The meaning and application-limits of Thoma's cavitation number. In *Cavitation and Polyphase Flow Forum – 1981* (J. W. Hoyt, ed.) pp. 45–6, Am. Soc. Mech. Engrs.
Young, F.R. (1989). *Cavitation*. McGraw-Hill.

Problems

1. A generator is driven by a small, single-jet Pelton turbine designed to have a power specific speed $\Omega_{SP} = 0.20$. The effective head at nozzle inlet is 120 m and the nozzle velocity coefficient is 0.985. The runner rotates at 880 rev/min, the turbine overall efficiency is 88 per cent and the mechanical efficiency is 96 per cent.

If the blade speed to jet speed ratio, $\nu = 0.47$, determine:

(1) the shaft power output of the turbine;
(2) the volume flow rate;
(3) the ratio of the wheel diameter to jet diameter.

2. (a) Water is to be supplied to the Pelton wheel of a hydroelectric power plant by a pipe of uniform diameter, 400 m long, from a reservoir whose surface is 200 m vertically above the nozzles. The required volume flow of water to the Pelton wheel is 30 m³/s. If the pipe skin friction loss is not to exceed 10% of the available head and $f = 0.0075$, determine the minimum pipe diameter.

(b) You are required to select a suitable pipe diameter from the available range of stock sizes to satisfy the criteria given. The range of diameters (m) available are: 1.6, 1.8, 2.0, 2.2, 2.4, 2.6, 2.8. For the diameter you have selected, determine:

(1) the friction head loss in the pipe;
(2) the nozzle exit velocity assuming no friction losses occur in the nozzle and the water leaves the nozzle at atmospheric pressure;
(3) the total power developed by the turbine assuming that its efficiency is 75% based upon the energy available at turbine inlet.

3. A multi-jet Pelton turbine with a wheel 1.47 m diameter, operates under an effective head of 200 m at nozzle inlet and uses 4 m³/s of water. Tests have proved that the wheel efficiency is 88 per cent and the velocity coefficient of each nozzle is 0.99.

Assuming that the turbine operates at a blade speed to jet speed ratio of 0.47, determine:

(1) the wheel rotational speed;
(2) the power output and the power specific speed;
(3) the bucket friction coefficient given that the relative flow is deflected 165°;
(4) the required number of nozzles if the ratio of the jet diameter to mean diameter of the wheel is limited to a maximum value of 0.113.

4. A four-jet Pelton turbine is supplied by a reservoir whose surface is at an elevation of 500 m above the nozzles of the turbine. The water flows through a single pipe 600 m long, 0.75 m diameter, with a friction coefficient $f = 0.0075$. Each nozzle provides a jet 75 mm diameter and the nozzle velocity coefficient $K_N = 0.98$. The jets impinge on the buckets of the wheel at a radius of 0.65 m and are deflected (relative to the wheel) through an angle of 160 deg. Fluid friction within the buckets reduces the relative velocity by 15 per cent. The blade speed to jet speed ratio, $\nu = 0.48$ and the mechanical efficiency of the turbine is 98 per cent.

Calculate, using an iterative process, the loss of head in the pipeline and hence, determine for the turbine:

(1) the speed of rotation;
(2) the overall efficiency (based on the effective head);
(3) the power output;
(4) the percentage of the energy available at turbine inlet which is lost as kinetic energy at turbine exit.

5. A Francis turbine operates at its maximum efficiency point at $\eta_0 = 0.94$, corresponding to a power specific speed of 0.9 rad. The effective head across the turbine is 160 m and the speed required for electrical generation is 750 rev/min. The runner tip speed is 0.7 times the spouting velocity, the absolute flow angle at runner entry is 72 deg from the radial direction and the absolute flow at runner exit is without swirl.

Assuming there are no losses in the guide vanes and the mechanical efficiency is 100 per cent, determine:

(1) the turbine power and the volume flow rate;
(2) the runner diameter;
(3) the magnitude of the tangential component of the absolute velocity at runner inlet;
(4) the axial length of the runner vanes at inlet.

6. The power specific speed of a 4 MW Francis turbine is 0.8, and the hydraulic efficiency can be assumed to be 90 per cent. The head of water supplied to the turbine is 100 m. The runner vanes are radial at inlet and their internal diameter is three-quarters of the external diameter. The meridional velocities at runner inlet and outlet are equal to 25 and 30 per cent, respectively, of the spouting velocity.

Determine:

(1) the rotational speed and diameter of the runner;
(2) the flow angles at outlet from the guide vanes and at runner exit;
(3) the widths of the runner at inlet and at exit.

Blade thickness effects can be neglected.

7. **(a)** Review, briefly, the phenomenon of cavitation in hydraulic turbines and indicate the places where it is likely to occur. Describe the possible effects it can have upon turbine operation and the turbine's structural integrity. What strategies can be adopted to alleviate the onset of cavitation?

(b) A Francis turbine is to be designed to produce 27 MW at a shaft speed of 94 rev/min under an effective head of 27.8 m. Assuming that the optimum hydraulic efficiency is 92 per cent and the runner tip speed to jet speed ratio is 0.69, determine:

(1) the power specific speed;
(2) the volume flow rate;
(3) the impeller diameter and blade tip speed.

(c) A 1/10 scale model is to be constructed in order to verify the performance targets of the prototype turbine and to determine its cavitation limits. The head of water available for the model tests is 5.0 m. When tested under dynamically similar conditions as the prototype, the net positive suction head H_S of the model is 1.35 m.

Determine for the model:

(1) the speed and the volume flow rate;

(2) the power output, corrected using Moody's equation to allow for scale effects (assume a value for $n = 0.2$);

(3) the suction specific speed Ω_{ss}.

(d) The prototype turbine operates in water at 30°C when the barometric pressure is 95 kPa. Determine the necessary depth of submergence of that part of the turbine mostly likely to be prone to cavitation.

8. The preliminary design of a turbine for a new hydro-electric power scheme has under consideration a vertical-shaft Francis turbine with a hydraulic power output of 200 MW under an effective head of 110 m. For this particular design a specific speed, $\Omega_s = 0.9$ (rad), is selected for optimum efficiency. At runner inlet the ratio of the absolute velocity to the spouting velocity is 0.77, the absolute flow angle is 68 deg and the ratio of the blade speed to the spouting velocity is 0.6583. At runner outlet the absolute flow is to be without swirl.

Determine:

(1) the hydraulic efficiency of the rotor;
(2) the rotational speed and diameter of the rotor;
(3) the volume flow rate of water;
(4) the axial length of the vanes at inlet.

9. A Kaplan turbine designed with a *shape factor* (power specific speed) of 3.0 (rad), a runner tip diameter of 4.4 m and a hub diameter of 2.0 m, operates with a net head of 20 m and a shaft speed of 150 rev/min. The absolute flow at runner exit is axial. Assuming that the hydraulic efficiency is 90% and the mechanical efficiency is 99%, determine:

(1) the volume flow rate and shaft power output;
(2) the relative flow angles at the runner inlet and outlet at the hub, the mean radius and at the tip.

Bibliography

Gostelow, J. P. (1984). *Cascade Aerodynamics*. Pergamon.

Greitzer, E. M. (1981). The stability of pumping systems – the 1980 Freeman Scholar Lecture. *J. Fluids Engineering, Trans Am. Soc. Mech. Engrs.*, **103**, 193–242.

Hawthorne, Sir William. (1978). Aircraft propulsion from the back room. *Aero. J.*, 93–108.

Horlock, J.H. (1973). *Axial Flow Turbines*. Kruger.

Horlock, J. H. and Marsh, H. (1982). Fluid mechanics of turbomachines: a review. *Int. J. Heat Fluid Flow*, **103**, No. 1, 3–11.

Johnson, I. A. and Bullock, R. O. (eds). (1965). Aerodynamic design of axial flow compressors (revised). NASA Report SP-36.

Jones, R. V. (1986). Genius in engineering. *Proc. Instn Mech Engrs.*, **200**, No. B4, 271–6.

Kerrebrock, J. L. (1981). Flow in transonic compressors – the Dryden Lecture. *AIAA J.*, **19**, 4–19.

Kerrebrock, J. L. (1984). *Aircraft Engines and Gas Turbines*. The MIT Press.

Kline, S. J. (1959). On the nature of stall. *Trans Am. Soc. Mech. Engrs., Series D*, **81**, 305–20.

Taylor, E. S. (1971). Boundary layers, wakes and losses in turbomachines. Gas Turbine Laboratory Rep. 105, MIT.

Ücer, A. S., Stow, P. and Hirsch, Ch. (eds) (1985). *Thermodynamics and Fluid Mechanics of Turbomachinery*, Vols 1 & 2. NATO Advanced Science Inst. Series, Martinus Nijhoff.

Ward-Smith, A. J. (1980). *Internal Fluid Flow: the Fluid Dynamics of Flow in Pipes and Ducts*. Clarendon.

Whitfield, A. and Baines, N. C. (1990). *Design of Radial Flow Turbomachines*. Longman.

Wilde, G. L. (1977). The design and performance of high temperature turbines in turbofan engines. *1977 Tokyo Joint Gas Turbine Congress*, co-sponsored by Gas Turbine Soc. of Japan, the Japan Soc. of Mech. Engrs and the Am. Soc. of Mech. Engrs., pp. 194–205.

Wilson, D. G. (1984). *The Design of High-Efficiency Turbomachinery and Gas Turbines*. The MIT Press.

Young, F. R. (1989). *Cavitation*. McGraw-Hill.

APPENDIX 1

Conversion of British and US Units to SI Units

Length
1 inch	$= 0.0254\,\text{m}$
1 foot	$= 0.3048\,\text{m}$

Area
1 in^2	$= 6.452 \times 10^{-4}\,\text{m}^2$
1 ft^2	$= 0.09290\,\text{m}^2$

Volume
1 in^3	$= 16.39\,\text{cm}^3$
1 ft^3	$= 28.32\,\text{dm}^3$
	$= 0.02832\,\text{m}^3$
1 gall (UK)	$= 4.546\,\text{dm}^3$
1 gall (US)	$= 3.785\,\text{dm}^3$

Velocity
1 ft/s	$= 0.3048\,\text{m/s}$
1 mile/h	$= 0.447\,\text{m/s}$

Mass
1 lb	$= 0.4536\,\text{kg}$
1 ton (UK)	$= 1016\,\text{kg}$
1 ton (US)	$= 907.2\,\text{kg}$

Density
1 lb/ft^3	$= 16.02\,\text{kg/m}^3$
1 slug/ft^3	$= 515.4\,\text{kg/m}^3$

Force
1 lbf	$= 4.448\,\text{N}$
1 ton f (UK)	$= 9.964\,\text{kN}$

Pressure
1 lbf/in^2	$= 6.895\,\text{kPa}$
1 ft H$_2$O	$= 2.989\,\text{kPa}$
1 in Hg	$= 3.386\,\text{kPa}$
1 bar	$= 100.0\,\text{kPa}$
1 atm	$= 101.3\,\text{kPa}$

Energy
1 ft lbf	$= 1.356\,\text{J}$
1 Btu	$= 1.055\,\text{kJ}$

Specific energy
1 ft lbf/lb	$= 2.989\,\text{J/kg}$
1 Btu/lb	$= 2.326\,\text{kJ/kg}$

Specific heat capacity
1 ft lbf/(lb°F)	$= 5.38\,\text{J/(kg°C)}$
1 ft lbf/(slug °F)	$= 0.167\,\text{J/(kg°C)}$
1 Btu/(lb °F)	$= 4.188\,\text{kJ/(kg°C)}$

Power
1 hp	$= 0.7457\,\text{kW}$

Some other units in use
1 tonne	$= 1000\,\text{kg}$
1 TWh	$= 10^6\,\text{MWh}$
	$= 3.6 \times 10^9\,\text{MJ}$

APPENDIX 2

Answers to Problems

Chapter 1

1. $6.277 \, \text{m}^3/\text{s}$.
2. $9.15 \, \text{m/s}$; 5.33 atmospheres.
3. $551 \, \text{rev/min}$, 1:10.8; $0.8865 \, \text{m}^3/\text{s}$; $17.85 \, \text{MN}$.
4. $4,035 \, \text{rev/min}$; $31.22 \, \text{kg/s}$.

Chapter 2

1. 88.1 per cent.
2. (1) $703.1 \, \text{K}$; (2) $751.9 \, \text{K}$; (3) $669 \, \text{K}$.
3. (1) $500 \, \text{K}$, $0.313 \, \text{m}^3/\text{kg}$; (2) 1.045.
4. $49.2 \, \text{kg/s}$; $24 \, \text{mm}$.
5. (1) $620 \, \text{kPa}$, $274°\text{C}$; $240 \, \text{kPa}$, $201°\text{C}$; $85 \, \text{kPa}$, $126'\text{C}$; $26 \, \text{kPa}$, $q = 0.988$; $7 \, \text{kPa}$, $q = 0 - 95$; (2) 0.619, 0.655, 0.699, 0.721, 0.750; (3) 0.739, 0.724; (4) 1.075.

Chapter 3

1. 49.8 deg.
2. 0.767; $C_D = 0.048$, $C_L = 2.245$.
3. -1.17 deg., 9.5 deg., 1.11.
4. (1) 53 deg and 29.5 deg; (2) 0.962; (3) $2.079 \, \text{kN/m}^2$.
5. (a) $s/l = 1.0$, $\alpha_2' = 24.8$ deg; (b) $C_L = 0.82$.
6. (b) 57.8 deg; (c) (1) $3.579 \, \text{kPa}$; (2) 0.96; (3) 0.0218, 1.075.
7. (a) $\alpha_1 = 73.2°$, $\alpha_2 = 68.1°$, (b) (1) $C_L = 0.696$, (2) $\eta_D = 0.8824$

Chapter 4

2. (a) 88 per cent; (b) 86.17 per cent, (c) $1170.6 \, \text{K}$
3. $\alpha_2 = 70$ deg., $\beta_2 = 7.02$ deg., $\alpha_3 = 18.4$ deg., $\beta_3 = 50.37$ deg.
4. $22.62 \, \text{kJ/kg}$; $420 \, \text{kPa}$, $177°\text{C}$.
5. 90.2 per cent.
6. (1) 1.50; (2) 39.9 deg, 59 deg; (3) 0.25; (4) 90.5 and 81.6 per cent.
7. (1) $488 \, \text{m/s}$; (2) $266.1 \, \text{m/s}$; (3) 0.83; (4) 0.128.
8. (1) $213.9 \, \text{m/s}$; (2) 0.10, 2.664; (3) 0.872; (4) $269°\text{C}$, $0.90 \, \text{Mpa}$.
9. (a) (1) $601.9 \, \text{m/s}$; (2) $282.8 \, \text{m/s}$; (3) 79.8 per cent. (b) 89.23 per cent.
10. (b) (1) $130.9 \, \text{kJ/kg}$; (2) $301.6 \, \text{m/s}$; (3) $707.6 \, \text{K}$ (c) (1) $10,200 \, \text{rev/min}$; (2) $0.565 \, \text{m}$ (3) 0.845.
11. (2) $0.2166 \, \text{m}^2$; (3) $8,740 \, \text{rev/min}$. (4) $450.7 \, \text{m/s}$, 0.846.

Chapter 5

1. 14 stages.
2. 30.35°C.
3. 132.1 m/s, 56.1 kg/s; 10.0 MW.
4. 86.5 per cent; 9.27 MW.
5. 0.59, 0.415.
6. 33.5 deg, 8.5 deg, 52.9 deg; 0.827; 34.5 deg, 0.997.
7. 56.9 deg, 41 deg; 21.8 deg.
8. (1) 229.3 m/s; (2) 23.47 kg/s, 15,796 rev/min; (3) 33.614 kJ/kg; (4) 84.7 per cent; (5) 5.856 stages, 0.789 MW; (6) With six stages and the same loading, the pressure ratio is then 6.209. However, to maintain a pressure ratio of 6.0, the specific work must be decreased to 32.81 kJ/kg which requires an absolute flow angle α_m to change from 30° to 30.26°. With five stages and a pressure ratio of six the weight and cost would be lower but the stage loading would increase to 39.37 kJ/kg which would require α_m to be changed to 28.08 deg.
9. (a) 16.22 deg., 22.08 deg., 33.79 deg. (b) 467.2 Pa, 7.42 m/s.
10. (1) $\beta_1 = 70.79°$, $\beta_2 = 68.24$ deg.; (2) 83.96 per cent; (3) 399.3 Pa; (4) 7.144 cm.
11. (1) 141.1 Pa, 0.588; (2) 60.48 Pa; (3) 70.14 per cent.

Chapter 6

1. 55 and 47 deg. 2. 0.602, 1.38, −0.08 (i.e. implies large losses near hub)
4. 70.7 m/s. 5. Work done is constant at all radii;

$$c_{x1}^2 = k_1 - 2a^2[r^2 - 2(b/a)\ln r]$$
$$c_{x2}^2 = k_2 - 2a^2[r^2 + 2(b/a)\ln r]$$
$$\beta_1 = 43.2 \text{ deg}, \beta_2 = 10.4 \text{ deg}.$$

6. (1) 469.3 m/s; (2) 0.798; (3) 0.079; (4) 3.244 MW; (5) 911.6 K, 897 K.
7. (1) 62 deg; (2) 55.3 and 1.5 deg; (3) 45.2 and 66 deg, (4) −0.175, 0.477.
8. See "*Solutions Manual*".

Chapter 7

1. (1) 27.9 m/s; (2) 882 rev/min, 0.604 m; (3) 182 W; (4) 0.333 (rad)
2. 579 kW; 169 mm; 5.273.
3. 0.8778; 5.62 kg/s.
4. 24,430 rev/min; 0.203 m, 0.5844.
5. 0.7324, 90.84 per cent.
6. (1) 542.5 kW; (2) 536 and 519 kPa; (3) 608.2 and 244 kPa, 1.22, 193.2 m/s; (4) 0.899; (5) 0.22; (6) 31,770 rev/min.
7. (1) 29.4 dm³/s; (2) 0.781; (3) 77.7 deg; (4) 7.8 kW.
8. (1) 14.11 cm; (2) 2.635 m; (3) 0.7664; (4) 17.73 m; (5) 13.8 kW; $\sigma_S = 0.722$, $\sigma_B = 0.752$.
9. (a) See text. (b) (1) 32,214 rev/min; (2) 5.246 kg/s; (c) (1) 1.254 MW; (2) 6.997.

Chapter 8

1. 587.4 m/s, 73.88 deg.
2. (1) 203.8 kPa, 977 K; (2) 0.40 m, 28,046 rev/min; (3) 1 MW.
3. (1) 90.54 per cent; (2) 0.2694 m; (3) 0.05316, 0.2009.
4. 1594 K.
5. (1) 2.159 m^3/s, 500 kW; (2) 0.0814 m and 0.1826 m; (3) 77 deg, 0.0995.
6. (a) 1.089 kg/s, 23,356 rev/min; (b) 9.063 × 10^5, 1.879 × 10^6.
7. (1) 81.82 per cent; (2) 890 K, 184.3 kPa; (3) 1.206 cm; (4) 196.3 m/s; (5) 0.492;
   (6) $r_{s3} = 6.59$ cm, $r_{h3} = 2.636$ cm.
8. (1) 308.24 m/s; (2) 56.42 kPa, 915.4 K; (3) 113.2 m/s, 0.2765 kg/s; (4) 5.472 cm;
   (5) 28.34 deg; (6) 0.7385 rad.

Chapter 9

1. (1) 224 kW; (2) 0.2162 m^3/s; (3) 6.423
2. (a) 2.138 m; (2) For $d = 2.2$ m, (1) 17.32 m; (2) 59.87 m/s, 40.3 MW
3. (1) 378.7 rev/min; (2) 6.906 MW, 0.252 (rad); (3) 0.793; (4) 3
4. Head loss in pipeline is 17.8 m. (1) 672.2 rev/min; (2) 84.5 per cent; (3) 6.735 MW;
   (4) 2.59 per cent.
5. (1) 12.82 MW, 8.69 m^3/s; (2) 1.0 m; (3) 37.6 m/s; (4) 0.226 m
6. (1) 663.2 rev/min; (2) 69.55 deg., 59.2 deg; (3) 0.152 m and 0,169 m.
7. (b) (1) 1.459 rad; (2) 107.6 m^3/s; (3) 3.153 m, 15.52 m/s; (c) (1) 398.7 rev/min,
   0.456 m^3/s; (2) 20.6 kW (uncorrected), 19.55 kW (corrected); (3) 4.06 (rad); (d)
   $z = 1.748$ m
8. (a) (1) 0.94; (2) 115.2 rev/min, 5.068 m; (3) 197.2 m^3/s; (4) 0.924 m.
9. (1) 11.4 m^3/s, 19.47 MW; (2) At hub, mean and tip radii the flow angles (deg)
   are as follows: Inlet 25.81, 62.99, 72.59; Outlet 59.55, 69.83, 75.04.

Index

Other Books from Butterworth-Heinemann

Analytical Instrumentation for the Water Industry
by T.R. Crompton
1991   272pp   hc   0-7506-1139-1

Finite Element Method in Engineering, Third Edition
by S.S. Rao
December 1998   550pp   pb   0-7506-7072-X

Flow, Level and Pressure Measurement in the Water Industry
by Graham Fowles
1994   224pp   hc   0-7506-1047-6

Mechanical Technology, Third Edition
by D.H. Bacon
June 1998   500pp   pb   0-7506-3886-9

Pumping Station Design, Second Edition
Edited by Robert Sanks
June 1998   1016pp   hc   0-7506-9483-1

. . . .

Detailed information on these and all other BH Engineering titles may be found in the BH Engineering catalog (Item #725). To request a copy, call 1-800-366-2665. You can also visit our web site at: http://www.bh.com

These books are available from all good bookstores or in case of difficulty call: 1-800-366-2665 in the U.S. or +44-1865-310366 in Europe.

E-Mail Mailing List
An e-mail mailing list giving information on latest releases, special promotions, offers and other news relating to BH Engineering titles is available. To subscribe, send an e-mail message to majordomo@world.std.com Include in message body (not in subject line): subscribe bh-engineering